An Introduction to Scientific Computing

Ionut Danaila · Pascal Joly ·
Sidi Mahmoud Kaber · Marie Postel

An Introduction to Scientific Computing

Fifteen Computational Projects Solved with MATLAB

Second Edition

 Springer

Ionut Danaila
Laboratoire de mathématiques
Raphaël Salem
Université de Rouen Normandie, CNRS
Rouen, France

Sidi Mahmoud Kaber
Laboratoire Jacques-Louis Lions
Sorbonne Université, CNRS
Université Paris Cité
Paris, France

Pascal Joly
Laboratoire Jacques-Louis Lions
Paris, France

Marie Postel
Laboratoire Jacques-Louis Lions
Sorbonne Université, CNRS
Université Paris Cité
Paris, France

ISBN 978-3-031-35031-3   ISBN 978-3-031-35032-0   (eBook)
https://doi.org/10.1007/978-3-031-35032-0

Mathematics Subject Classification: 65Mxx, 65Nxx, 65Kxx

1st edition: © Springer Science+Business Media, 2007
2nd edition: © The Editor(s) (if applicable) and The Author(s), under exclusive license to
Springer Nature Switzerland AG 2023

Cover image: Vortex interaction in a Navier-Stokes fluid (see Chap. 15)

This Springer imprint is published by the registered company Springer Nature Switzerland AG
The registered company address is: Gewerbestrasse 11, 6330 Cham, Switzerland

To   *Alice, Raphaël, Luminita*
        Lise, Baptiste
        Sarah, Thomas
        Camille, Paul

Preface

The purpose of this second edition remains the same: show the many facets (phenomenological, mathematical, and computational) of scientific computing by solving various computational projects of different degrees of difficulty. We always subscribed to the philosophy that the first experience in scientific computing must be based on realistic examples presented with sound theoretical background. This offers good motivation to students and triggers their curiosity to make the transition from the basis offered by this book to advanced books, monographs, and research literature.

During the 17 years since the first edition appeared, scientific computing has continued to grow and evolve, covering almost all domains of scientific or application/industrial fields. Becoming an expert in this field is nowadays very demanding in terms of mathematical and physical knowledge, as well as programming skills. The experience offered by solving the projects of this book was thus intended to be formative and challenging.

The difficulty in writing an introductory textbook based on projects is to find an even balance among physics/phenomenology, mathematics, programming, and interpretation of results. Our main concern was to make the entire content of a project bearable to students with different backgrounds. It is nevertheless true that a student who had calculus and differential equations will be more at ease in solving the exercises of some projects of level 2 or 3.

Based on these observations, and being aware of the fast evolution of numerical techniques and the need for more theoretical content, we reinforced in this second edition several mathematical aspects of numerical methods and added three new projects on timely topics.

As a complement of the project *Polynomial approximation*, the new project on *Fourier approximation* introduces the mathematical basis of Fourier spectral methods. With this new project and the one on *Legendre spectral methods*, we offer in this edition a more wide cover of spectral methods, which are interesting alternatives to the very popular finite difference or finite element methods already largely discussed in the first edition.

The second new project concerns *High-order finite difference methods*. The message it conveys is that thinking *high order* instead of customary *second order* is part of the modern design of a numerical method for a practical problem. We present implicit and compact finite difference schemes, that are not covered in many textbooks, despite the fact they offer high (spectral-like) accuracy with a negligible increase of the requested computational time. Ready-to-use schemes are provided for different settings and the efficiency of such schemes in solving differential equations is illustrated.

The third novel project entitled *Optimization applied to model fitting* gives the necessary foundations to understand numerical optimization challenges. Most of those difficulties are usually hidden deep within available packages. Optimization is however crucial in all domains of application, for instance, machine learning and, more generally, artificial intelligence. A thorough study of the most basic algorithms is necessary to understand the ins and outs of state-of-the-art methods. We apply optimization concepts to fit the parameters of an epidemiologic model, which is a timely application.

Concerning the 12 projects already presented in the first edition, some have been revised and updated. We acknowledge with gratitude the many readers of the first edition who sent corrections, comments, and suggestions. The order and numbering of projects have changed to welcome the three new projects. Even though it goes without saying that a book based on projects is not meant to be read from front to back, we would like to emphasize the importance of Project 1 on *Numerical Approximation of Model Partial Differential Equations*. We strongly advise the reader, wishing to acquire the basics of numerical simulation, or planning to tackle projects of medium (level 2) or advanced (level 3) difficulty, to first read the text and scripts of this first chapter.

Finally, the GIT repository https://plmlab.math.cnrs.fr/ai2sc/ai2sc contains the MATLAB scripts. In the future, it will also host the inevitable list of typos and possible new evolutions of the published projects. We are aware that the coverage of the book is rather broad and may interest instructors coming from different disciplines with different goals. This website will thus offer the possibility to post new exercises scoped to reinforce particular aspects (mathematical and computational) of some projects.

Paris, October 2022

Ionut Danaila   (ionut.danaila@univ-rouen.fr)
Pascal Joly   (pascal.joly4@wanadoo.fr)
Sidi Mahmoud Kaber   (sidi-mahmoud.kaber@sorbonne-universite.fr)
Marie Postel   (marie.postel@sorbonne-universite.fr)

Preface to the First Edition

Teaching or learning numerical methods in applied mathematics cannot be conceived nowadays without numerical experimentation on computers. There is a vast literature devoted either to theoretical numerical methods or numerical programming of basic algorithms, but there are few texts offering a complete discussion of numerical issues involved in the solution of concrete and relatively complex problems. This book is an attempt to fill this need. It is our belief that the advantages and disadvantages of a numerical method cannot be accounted for without experiencing all the steps of scientific computing, from physical and mathematical description of the problem to numerical formulation and programming, and finally, to critical discussion of numerical results.

The book provides 12 *computational projects* aimed at numerically solving problems selected to cover a broad spectrum of applications, from fluid mechanics, chemistry, elasticity, thermal science, computer-aided design, signal and image processing, etc. Even though the main volume of this text concerns the numerical analysis of computational methods and their implementation, we have tried to start, when possible, from realistic problems of practical interest for researchers and engineers.

For each project, an introductory record card summarizes the mathematical and numerical topics explained and the fields of application of the approach. A level of difficulty, scaling from 1 to 3, is assigned to each project. Most of the projects are of level 1 or 2 and can be easily tackled; the reader will no doubt realize that projects of level 3 require a solid background in both numerical analysis and computational techniques.

Except projects 1 and 3, which are more theoretical, all projects follow the typical steps of scientific computing: physical and mathematical modeling of the problem, numerical discretization, construction of a numerical algorithm, and finally, programming. We have placed considerable emphasis on practical issues of computational methods that are not usually available in basic textbooks. Numerical checking of accuracy or stability, the choice of boundary conditions, the effective solving of linear systems, and comparison to exact solutions when available are only a few examples of problems encountered in the application of numerical methods. The last section of each project contains solutions for all proposed exercises and guides the reader in using the MATLAB scripts that can be accessed via the publisher's website www.springer.com. Programming techniques such as vectorial programming and memory storage optimization are also addressed. We finally discuss the physical meaning of the obtained results. The complementary references given at the end of each chapter form a guide for further, more specialized, reading.

The text offers two levels of interest. The mathematical framework provides a basic grounding in the subject of numerical analysis of partial differential equations and main discretization techniques (finite differences, finite elements, spectral methods, and wavelets). Meanwhile, we hope that the information contained herein and the wide range of topics covered by the book will allow the reader to select the appropriate numerical method to solve his or her particular problem.

The book is based on material offered by the authors in courses at *Université Pierre et Marie Curie (Paris, France)* and different engineering schools. It is primarily intended as a graduate-level text in applied mathematics, but it may also be used by students in engineering or physical sciences. It will also be a useful reference for researchers and practicing engineers. Since different possible developments of the projects are suggested, the text can be used to propose assignments at different graduate levels.

Despite our efforts to avoid typing, spelling, or other errors, the reader will no doubt find some remaining. We shall appreciate all feedback notifying us of any mistakes, as well as comments and suggestions that will help us to improve the text. Please use the e-mail addresses given below for this purpose.

We conclude by saying a few words about the programs provided in this book. They are written in MATLAB, a widely used software environment for scientific computing produced by The MathWorks Inc. We consider that an interpreted language (such as MATLAB, SCILAB, and OCTAVE) is the ideal framework to start a scientific programming activity. Debugging is very simple and the wide variety of available numerical tools (for solving linear systems, integrating ordinary differential equations, etc.) allows one to concentrate on the main features of the resolution algorithm. The highly versatile graphical interface is also very important for easy visualization of the obtained results.

Our programs are written with a general concern for simplicity and efficiency on ordinary personal computers; program lines are commented in what we hope is sufficient detail for the reader to follow mathematical developments. Programming tricks are discussed in the text when they seem to be of general interest. Projects 11 and 12 are also provided with more elaborate versions of the programs, using interactive graphical user interfaces. The reader should try to modify these programs to test different suggested run cases or extensions of the projects. We believe that experience with these simple programs will be valuable in writing numerical codes using compiled languages (such as Fortran, C, or C++) to solve real industrial problems on mainframe computers.

Paris, October 2005

Ionut Danaila (danaila@ann.jussieu.fr)
Pascal Joly (joly@ann.jussieu.fr)
Sidi Mahmoud Kaber (kaber@ann.jussieu.fr)
Marie Postel (postel@ann.jussieu.fr)

Laboratoire Jacques-Louis Lions
Université Pierre et Marie Curie (Paris 6) and
Centre National de la Recherche Scientifique (CNRS)

Contents

Chapter 1
Numerical Approximation of Model Partial Differential Equations

Project Summary

Level of difficulty:	1
Keywords:	Linear differential equations; numerical integration methods; finite difference schemes: Euler schemes, Runge–Kutta schemes
Application fields:	Transport phenomena, diffusion, wave propagation

This first chapter is intended as a quick introduction to basic discretization techniques of time-dependent partial differential equations (PDEs). We consider it important that the reader, before tackling the complex problems of the next chapters, have some understanding of the mathematical and physical properties of the following model PDEs: the convection equation, the wave equation, and the heat equation. This chapter is therefore organized as a collection of several short exercises in which model PDEs are theoretically analyzed and numerically solved using the simplest discretization methods. The essential features of numerical methods are presented, with emphasis on fundamental ideas of accuracy, stability, convergence, and numerical dissipation. Particular care is devoted to the validation of numerical procedures by comparing to exact solutions available for these simple cases.

1.1 Discrete Integration Methods for Ordinary Differential Equations

We generally define a partial differential equation (PDE) as a relation between a function of several variables and its partial derivatives. In this section, we consider the simplest case of ordinary differential equations (ODE), with a solution depending on a single independent variable (time variable here).

© The Author(s), under exclusive license to Springer Nature Switzerland AG 2023
I. Danaila et al., *An Introduction to Scientific Computing*,
https://doi.org/10.1007/978-3-031-35032-0_1

We present discrete methods for the numerical integration of ODEs. These methods (or numerical schemes) will prove useful in the following sections when we discuss PDEs depending both on time and space variables.

Let us consider the following problem: find a differentiable function u : $[0, T] \mapsto \mathbb{R}^m$ that is a solution to the ODE

$$u'(t) = f(t, u(t)), \tag{1.1}$$

where T is a nonnegative real number and $f : [0, T] \times \mathbb{R}^m \mapsto \mathbb{R}^m$ a continuous function. This problem is not completely specified by its equation: for its integration, we need to know the initial value (at $t = 0$) of the unknown function.

Definition 1.1 A Cauchy (or initial value) problem is the coupling of the ODE (1.1) with an initial condition

$$u(0) = u_{\text{init}}, \tag{1.2}$$

where u_{init} is a given vector in $\mathbb{R}^m$.

Theoretical results on the existence and uniqueness of the solution to the problem (1.1)–(1.2) go back to Cauchy in 1824. The reader interested in a more mathematical approach to the problem will want to refer to many existing books on ODEs (see, for instance, the references at the end of this chapter). We adopt here a more practical point of view, and we start directly by presenting simple numerical methods to compute approximations of the solution in the scalar case, or 1D case, $m = 1$.

Since the computer can deal only with a finite number of discrete values, the numerical algorithm to solve the Cauchy problem (1.1)–(1.2) starts by setting the discrete points (or grid points) at which the solution will be computed. The equidistant or *regular* distribution of the grid points is the simplest and will be used in this chapter (see Fig. 1.1). Since the first index of arrays in MATLAB is 1, we set points $t_n = (n - 1)h, n = 1, \ldots, N + 1$, with $h = T/N$ the constant discretization step (or time step if t is regarded as a time variable). The interval $I = [0, T]$ is thus split in N subintervals $I_n = [t_n, t_{n+1}], n = 1, \ldots, N$ (notice that $t_1 = 0$ and $t_{N+1} = T$).

The numerical approximation of the Cauchy problem consists in building a sequence of numbers (depending on the grid size h) $u_1^{(h)}, \ldots, u_{N+1}^{(h)}$ that approximates the values $u(t_1), \ldots, u(t_{N+1})$ of the exact solution $u(t)$ at the same computation points. We always start with $u_1^{(h)} = u_{\text{init}}$ in order to satisfy the initial value condition $u(0) = u_{\text{init}}$. In order to simplify notation, we will refer, when possible, to $u_n^{(h)}$ by u_n.

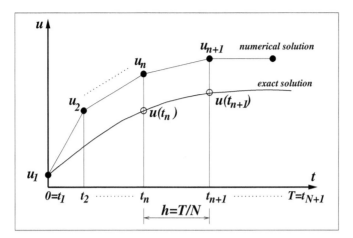

Fig. 1.1 Regular grid and numerical approximation of an ODE.

1.1.1 Construction of Numerical Integration Schemes

Having discretized the definition interval I, we must find a formula to compute values u_n, for $n = 1, \ldots, N + 1$. Such a formula, which is usually called a *numerical scheme*, is obtained by *discretizing* the differential operator in the ODE. We present here two types of methods that can be used to build numerical schemes for the ODE (1.1). Remember that any integration scheme will start from the value u_{init} imposed by the initial condition.

Methods Based on Finite Differences

This type of method consists in writing the equation (1.1) at time $t = t_n$ and replacing $u'(t_n)$ by a finite difference approximation. For this purpose, we use a Taylor series expansion to approximate the values of the unknown u for t close to t_n. We consider, for instance, the example of the first derivative.

Definition 1.2 The discretization step h being fixed, we define the following finite difference operators:

- *forward or progressive*

$$D^+ u(t) = \frac{u(t + h) - u(t)}{h}, \tag{1.3}$$

- *backward or regressive*

$$D^- u(t) = \frac{u(t) - u(t - h)}{h}, \tag{1.4}$$

- *central*

$$D^0 u(t) = \frac{u(t+h) - u(t-h)}{2h}. \tag{1.5}$$

Let us assume that the function u is twice continuously differentiable. Then there exists $\theta_n^+ \in]0, h[$ such that

$$u(t_{n+1}) = u(t_n) + h u'(t_n) + \frac{h^2}{2} u''(t_n + \theta_n^+). \tag{1.6}$$

We can derive from this expansion an approximation of $u'(t_n)$:

$$u'(t_n) = \frac{u(t_{n+1}) - u(t_n)}{h} - \frac{h}{2} u''(t_n + \theta_n^+) \approx D^+ u(t_n), \tag{1.7}$$

and define the local ($e_n = e_n(h)$) and global ($\varepsilon = \varepsilon(h)$) approximation (or truncation) errors

$$e_n = u'(t_n) - D^+ u(t_n), \quad \varepsilon = \max_n |e_n| \leq \frac{h}{2} \max_{t \in I_n} |u''(t)|. \tag{1.8}$$

Assuming that $|u''|$ is bounded, we infer that the truncation error ε decays to 0 with h. We use in the following Landau's notation $e_n = \mathcal{O}(h)$, meaning that $|e_n|$ is bounded by $C \cdot h$, with C a constant. In other terms, when $h \to 0$, the error converges to zero as h to the *power one*. We can finally write $u'(t_n) = D^+ u(t_n) + \mathcal{O}(h)$.

Definition 1.3 We say that $D^+ u(t_n)$ is a first-order approximation of $u'(t_n)$. We generally define the order of accuracy of the difference approximation as the power of h with which the approximation error ε tends to zero.

It is easy to see that $D^- u(t_n) = [u(t_n) - u(t_{n-1})]/h$ is also a first-order approximation of $u'(t_n)$, while $D^0 u(t_n) = [u(t_{n+1}) - u(t_{n-1})]/(2h)$ is a second-order approximation of $u'(t_n)$ (i.e., the order of accuracy is two).

More generally, it is possible to use linear combinations of several finite difference operators to find approximations of $u'(t_n)$. For instance, we can approximate

$$u'(t_n) \approx \alpha D^- u(t_n) + \beta D^0 u(t_n) + \gamma D^+ u(t_n), \tag{1.9}$$

with parameters α, β, and γ chosen such that the approximation has the highest possible order of accuracy. We present in Chap. 7 different methods to obtain high-order finite difference (explicit or implicit) schemes.

Taylor series expansion remains the basic tool for building approximations of higher order derivatives. For the second derivative, for instance, the simplest recipe is the following: continue the expansion (1.6) to the fourth-order, write a similar expansion for $u(t_{n-1})$, and sum the two relations (see also Chap. 7). A centered second-order approximation of the second derivative is thus obtained:

$$u''(t_n) \approx D^- D^+ u(t_n) = \frac{u(t_{n+1}) - 2u(t_n) + u(t_{n-1})}{h^2}. \tag{1.10}$$

We address now the problem of building numerical schemes for the ODE (1.1) using the previous finite difference approximations. Considering the ODE at time t_n and replacing $u'(t_n)$ by $D^+ u(t_n)$, we obtain the scheme

$$u_{n+1} = u_n + h f(t_n, u_n). \tag{1.11}$$

We recall that u_{n+1} and u_n are numerical approximations of $u(t_{n+1})$ and $u(t_n)$, respectively.

The scheme (1.11) is called the *explicit* Euler scheme, or simply the Euler scheme. This method is said to be explicit because u_{n+1} depends explicitly on t_n and on the *old* value u_n. More generally, a numerical scheme is explicit if u_{n+1} can be calculated explicitly from quantities that are already known (i.e., values of the solution at previous time instants).

Consider now the ODE (1.1) at time t_{n+1} and replace $u'(t_{n+1})$ by $D^- u(t_{n+1})$; we obtain the *implicit* Euler scheme

$$u_{n+1} = u_n + h f(t_{n+1}, u_{n+1}). \tag{1.12}$$

This time, u_{n+1}, is computed as the solution (if it exists!) of an implicit equation. This requires more work, in particular when the function $u \mapsto f(t, u)$ is nonlinear with respect to u.

The approximation $u'(t_n) \approx D^0 u(t_n)$ in (1.1) leads to the scheme

$$u_{n+1} = u_{n-1} + 2h f(t_n, u_n), \tag{1.13}$$

called the leapfrog (or midpoint) scheme.

Methods Based on Quadrature Formulas

Another way to build a numerical scheme is based on quadrature formulas (numerical integration is also called quadrature). Integrating the ODE (1.1) on the interval $I_n = [t_n, t_{n+1}]$, we obtain

$$u(t_{n+1}) - u(t_n) = \int_{t_n}^{t_{n+1}} f(s, u(s)) ds = \mathcal{I}_n. \tag{1.14}$$

We can hence compute $u(t_{n+1})$ starting from the old value $u(t_n)$ if we are able to approximate the integral $\mathcal{I}_n$. We go back to a quadrature problem.

Several quadrature rules can be used to estimate the integral in (1.14):

- the *left endpoint rule*

$$\mathcal{I}_n \approx h f(t_n, u_n), \tag{1.15}$$

leading to the explicit Euler scheme (1.11);
- the *right endpoint rule*

$$\mathcal{I}_n \approx hf(t_{n+1}, u_{n+1}), \tag{1.16}$$

defining the implicit Euler scheme (1.12);

- *the midpoint (or rectangle) rule*

$$\mathcal{I}_n \approx hf(t_n + h/2, u(t_n + h/2)), \tag{1.17}$$

leading, using the approximation

$$u(t_n + h/2) \approx u(t_n) + \frac{h}{2}u'(t_n) = u(t_n) + \frac{h}{2}f(t_n, u(t_n)), \tag{1.18}$$

to the modified explicit Euler scheme:

$$u_{n+1} - u_n = hf\left(t_n + \frac{h}{2}, u_n + \frac{h}{2}f(t_n, u_n)\right); \tag{1.19}$$

- *the trapezoid rule*

$$\mathcal{I}_n \approx \frac{h}{2}\left[f(t_n, u_n) + f(t_{n+1}, u_{n+1})\right], \tag{1.20}$$

yielding the *semi-implicit* Crank–Nicolson scheme.

1.1.2 General Form of Numerical Schemes

The general form of a numerical scheme for the ODE (1.1) is

$$u_{n+1} = F(h; t_{n+1}, u_{n+1}; t_n, u_n; \dots). \tag{1.21}$$

If F depends on q previous values u_{n-j+1}, $j = 1, \dots, q$, the scheme is said to be a q-step scheme. For instance, the leapfrog scheme is a two-step scheme. If F does not depend on the solution at time level t_{n+1}, the scheme is said to be explicit. Otherwise, the scheme is implicit.

Remark 1.1 To start a one-step scheme, a single value is needed; this is u_1, which is always set by the initial condition u_{init}. It goes differently in the case of a $(q > 1)$-step scheme; this scheme can be used to compute values u_n for $n \geq q$, once the first q values $u_1, \dots, u_q$ are known. Since only the initial condition u_{init} is provided, the missing intermediate values can be computed using lower step schemes. For example, a one-step scheme can be used to compute u_2, a two-step scheme to compute u_3, and a $(q - 1)$-step scheme to compute u_q.

Definition 1.4 For the numerical scheme (1.21), we define the formal local truncation error as

$$e_n = u(t_{n+1}) - F\left(h; t_{n+1}, u(t_{n+1}); t_n, u(t_n); \dots \right), \qquad (1.22)$$

where $u(t)$ is the solution to the ODE (1.1). The scheme has order of accuracy p if $e_n = \mathcal{O}(h^{p+1})$. The scheme is said to be consistent if $p \geq 1$.

The idea behind this definition is that, when applying the numerical scheme to the exact solution function $u(t)$, one should recover (by Taylor series expansions) the original ODE plus a reminder representing the truncation error. Let us illustrate this by the example of the leapfrog scheme (1.13), for which $F = u_{n-1} + 2hf(t_n, u_n)$. Using Taylor series expansions about $t = t_n$, we obtain for the truncation error (1.22) the expression

$$e_n = 2h\left[u'(t_n) - f(t_n, u(t_n))\right] + \frac{h^3}{3}u''(t_n) + \cdots . \qquad (1.23)$$

Since $u'(t_n) = f(t_n, u(t_n))$, we conclude that the leapfrog scheme is a second-order scheme, i.e., the order of accuracy is two. Some numerical schemes commonly used in practice are summarized in Table 1.1.

1.1.3 Application to the Absorption Equation

The model equation that describes an absorption (or production) phenomenon is the following: find a function $u : \mathbb{R}^+ \to \mathbb{R}$ that is the solution to the Cauchy problem

$$\begin{cases} u'(t) + \alpha u(t) = f(t), & \forall t > 0, \\ u(0) = u_{\text{init}}, \end{cases} \qquad (1.24)$$

where $\alpha \in \mathbb{R}$ is a given physical constant and the source term f takes into account the production in time of the quantity u.

Example 1.1 The intensity of the radiation emitted by a radioactive body is estimated by measuring the concentration $u(t)$ of an unstable isotope. This concentration decays by a factor of two during a time interval T (called the half-life) according to the law

$$u'(t) = \alpha u(t), \quad \text{with} \quad \alpha = -\frac{\ln 2}{T}.$$

Exercise 1.1 Consider the Cauchy problem (1.24).

1. We set $u(t) = e^{-\alpha t}v(t)$. Write the ordinary differential equation satisfied by v. Solve analytically this equation and verify that

Explicit Euler (first order)	$u_{n+1} = u_n + hf(t_n, u_n)$
Implicit Euler (first order)	$u_{n+1} = u_n + hf(t_{n+1}, u_{n+1})$
leapfrog (second order)	$u_{n+1} = u_{n-1} + 2hf(t_n, u_n)$
Modified Euler (second order)	$u_{n+1} = u_n + hf(t_n + \frac{h}{2}, u_n + \frac{h}{2}f(t_n, u_n))$
Crank–Nicolson (second order)	$u_{n+1} = u_n + \frac{h}{2}[f(t_n, u_n) + f(t_{n+1}, u_{n+1})]$
Adams–Bashforth (second order)	$u_{n+1} = u_n + h[\frac{3}{2}f(t_n, u_n) - \frac{1}{2}f(t_{n-1}, u_{n-1})]$
Adams–Bashforth (third order)	$u_{n+1} = u_n + h[\frac{23}{12}f(t_n, u_n) - \frac{16}{12}f(t_{n-1}, u_{n-1}) + \frac{5}{12}f(t_{n-2}, u_{n-2})]$
Adams–Moulton (third order)	$u_{n+1} = u_n + h[\frac{5}{12}f(t_{n+1}, u_{n+1}) + \frac{8}{12}f(t_n, u_n) - \frac{1}{12}f(t_{n-1}, u_{n-1})]$
Runge–Kutta (Heun) (second order)	$\begin{cases} k_1 = hf(t_n, u_n), \\ k_2 = hf(t_n + h, u_n + k_1), \\ u_{n+1} = u_n + \frac{1}{2}(k_1 + k_2) \end{cases}$
Runge–Kutta (fourth order)	$\begin{cases} k_1 = hf(t_n, u_n), \\ k_2 = hf(t_n + h/2, u_n + k_1/2), \\ k_3 = hf(t_n + h/2, u_n + k_2/2), \\ k_4 = hf(t_n + h, u_n + k_3), \\ u_{n+1} = u_n + \frac{1}{6}(k_1 + 2k_2 + 2k_3 + k_4) \end{cases}$

Table 1.1 Numerical schemes for the ODE $u'(t) = f(t, u)$.

$$u(t) = e^{-\alpha t}\left(u_{\text{init}} + \int_0^t e^{\alpha s} f(s)ds \right). \qquad (1.25)$$

2. Derive the exact solution in the case of α depending on t.

3. Assuming that α and f are constants, derive an expression for u and calculate $\lim_{t \to +\infty} u(t)$.

4. Consider $f = 0$, and a complex coefficient $\alpha \in \mathbb{C}$, with real part $\alpha_r > 0$. Show that $\lim_{t \to +\infty} u(t) = 0$.

5. Write a MATLAB function to implement the explicit Euler scheme (1.11). The definition header of the function will be as follows:

```
function u=PDE_F_EulerExp(fun,uinit,Ti,Tf,n)
% Input arguments:
% fun     the name of the right hand side EDO function
% Ti      the initial time
% uinit   the initial condition at  ti
% Tf      the final time
% N       the number of time subintervals between ti and tf
% Output arguments :
%   u     the dimension N+1 vector containing the numerical
%     solution at time instants  Ti+(i-1)*h, with h=(Tf-Ti)/N
```

Hint: use the MATLAB built-in function `feval` to evaluate the parameter function `fun` within the PDE_F_EulerExp function.

In a MATLAB program (or script), call the PDE_F_EulerExp function to solve the ODE $u'(t) + 4u(t) = 0$ with the initial condition $u(T_i) = 1$. Set $T_i = 0, T_f = 3$, and $N = 24$ ($h = 1/8$). Plot the results, both exact and numerical solutions superimposed on a single graph, and comment on the difference between solutions. Perform the same computation for $N = 6$ ($h = 1/2$). Comment on the results.

Use instead of the explicit Euler scheme the second-order and the fourth-order Runge–Kutta schemes (using as model PDE_F_EulerExp, write two other functions PDE_F_RKutta2 and PDE_F_RKutta4 using the formulas in Table 1.1). Comment on the results obtained for $h = 1/2$.

6. Check the order of convergence of each of the three schemes. For this purpose, it is necessary to solve the same ODE for different values of h and plot the global (or convergence) error ε as a function of h. Since the exact solution u_{ex} is known in this case, the global error ε will measure here the difference between the exact and numerical solutions at the final time T_f. If $\varepsilon \sim Ch^p$, the plot $\varepsilon(h)$ using logarithmic axes will display as a line of slope p, since $\log(\varepsilon) \sim \log(C) + p \log(h)$. Practically, start with a given h_0 and divide this value by two for each new run. If M is the number of runs, this means that the time step is varied as $h_m = h_0/2^{m-1}$, $m = 1, \ldots, M$. The corresponding number of discretization subintervals is then $N_m = (T_f - T_i)/h_m$. For each run m, compute the global error $\varepsilon_m = |u_{N_m+1} - u_{ex}(T_f)|$. Set $T_i = 0, T_f = 1$, and $M = 6$. Plot points (h_m, ε_m), $m = 1, \ldots, M$ using logarithmic axes (use the MATLAB function `loglog` instead of `plot`).

A solution of this exercise is proposed in Sect. 1.3 at page 21.

1.1.4 Stability of a Numerical Scheme

We consider here the absorption equation for $f = 0$ and $\alpha \in \mathbb{R}^+$. The exact solution is then $u(t) = e^{-\alpha t}u_{\text{init}}$ with the property $\lim_{t \to +\infty} u(t) = 0$. Assume that we want to compute this solution using the explicit Euler scheme (1.11). We obtain a sequence of values $u_n = (1 - \alpha h)^n u_{\text{init}}$. It is easy to see that

- if $h > 2/\alpha$, then $1 - \alpha h \leq -1$ and the sequence (u_n) diverges;
- if $0 < h < 2/\alpha$, then $|1 - \alpha h| < 1$ and the sequence (u_n) decays to 0 as $t \to \infty$, reproducing the behavior of the exact solution.

Let us assume at this point that the reader, pushed by curiosity, has already answered question 5 of Exercise 1.1. The above analysis explains the *strange* behavior of the numerical solution obtained for a discretization step $h = 1/2$ and $\alpha = 4$ (the solution takes alternatively the values $+1$ and -1; see Fig. 1.4). The real question behind this observation is how to be sure that the numerical scheme gives the correct solution. Part of the answer is related to the accuracy of the scheme: if the scheme is consistent (see Definition 1.4), we know that the discrete scheme commits local (at a given time) errors that vanish when $h \to 0$. Unfortunately, as can be seen from our example, consistency is not sufficient to achieve convergence to the exact solution. The stability of the numerical scheme is also required for a successful numerical computation. Intuitively, we can say that a numerical scheme will be stable if it does not magnify the errors appearing during the computation.

The fundamental concept of stability can be mathematically addressed in several ways (see, for example, Richtmyer and Morton, 1967; LeVeque, 1992; Hirsch, 1988; Trefethen, 1996). The widely used definition of the stability (also known as *zero-stability*, or Lax–Richtmyer stability for PDEs) requires that the computed values remain bounded when $h \to 0$ for a fixed integration interval $[0, T]$. This is an important concept since, as stated from the well-known equivalence theorem (due to Dahlquist for ODEs and to Lax and Richtmyer for PDEs; see Trefethen (1996)), the zero-stability is a necessary and sufficient condition for a consistent scheme to be convergent.

In some practical applications, it is not always possible to take h small enough for the zero-stability to apply. This is the case of *stiff* ODEs, i.e., involving different varying time scales (see Chap. 2). For this type of ODEs, we generally use the concept of *absolute stability* which considers the behavior of the numerical scheme when the time step h is held fixed and $t \to \infty$.

We illustrate in the following the concept of absolute stability by the example of the absorption equation $(u'(t) = -\alpha u)$.[1] We consider the simplest case of one-step schemes in Table 1.1. We can recast these numerical schemes into the general form

---

[1] The linear ODE $u'(t) = au(t)$ for some constant $a \in \mathbb{C}$ is generally used as model equation to investigate the absolute stability of a numerical scheme. For nonlinear systems of ODEs, a similar analysis can be applied after linearization and diagonalization (see Chap. 2)—this type of stability is often referred to as the *eigenvalue stability*.

$$u_{n+1} = G(-\alpha h)u_n = \ldots = [G(-\alpha h)]^{n+1} u_{\text{init}}. \qquad (1.26)$$

G is called the amplification function[2] and is supposed to reflect the behavior of the exact solution, since this satisfies the relation

$$u(t_{n+1}) = e^{-\alpha h} u(t_n). \qquad (1.27)$$

The reader is invited to derive the following expressions for the amplification function (we denote $z = -\alpha h$):

Explicit Euler:	$G(z) = 1 + z,$
Implicit Euler:	$G(z) = 1/(1 - z),$
Modified Euler:	$G(z) = (2 + z)/(2 - z),$
Runge–Kutta (second order):	$G(z) = 1 + z + z^2/2,$
Runge–Kutta (fourth order):	$G(z) = 1 + z + z^2/2 + z^3/6 + z^4/24.$

A sufficient condition for stability is now $|G(-\alpha h)| < 1$. This stability condition ensures that the numerical solution has the same behavior as the exact solution when $t \to \infty$, since

$$0 \le |u_n| \le |u_{init}| \, |G(-\alpha h)|^n \implies \lim_{n \to +\infty} |u_n| = \lim_{n \to +\infty} |u_{init}| \, |G(-\alpha h)|^n = 0.$$

Definition 1.5 The locus $\mathcal{S}$ of points $z \in \mathbb{C}$ for which $|G(z)| < 1$ is called the (absolute) stability region of the scheme.

For example, the stability region $\mathcal{S}$ of the explicit Euler 3cheme is the open disk of radius 1, centered at the point $(-1, 0)$. The scheme will hence be absolutely stable if the discretization step h is chosen such that $|1 - \alpha h| < 1$.

Remark 1.2 According to Definition 1.5, the absolute stability region $\mathcal{S}$ contains the points for which $|u_n| \to 0$ as $t \to \infty$. It is interesting to note that in some textbooks (e.g., Trefethen (1996)) the absolute stability region is defined as the locus $\bar{\mathcal{S}}$ of points $z \in \mathbb{C}$ for which $|G(z)| \le 1$, i.e., we ask that the numerical solution u_n be bounded as $t \to \infty$. In general, if $\mathcal{S}$ is not empty, $\bar{\mathcal{S}}$ is the closure of $\mathcal{S}$. But there are some special cases, as the leapfrog scheme for which $\mathcal{S}$ is empty and $\bar{\mathcal{S}} = [-i, i]$, where i is the imaginary unit ($i^2 = -1$).

   This second definition of the absolute stability is important since it makes the link with the zero-stability: a numerical scheme is zero-stable if and only if the origin $z = 0$ belongs to $\bar{\mathcal{S}}$ (for more details, see Trefethen 1996).

Exercise 1.2 Plot in the same figure the bounds of the stability regions of the following schemes: explicit Euler, second-order Runge–Kutta, and fourth-order Runge–Kutta. Hint: define a complex variable $z = (x, y)$ covering the rectangle $[-4, 1] \times [-4, 4]$ (use the MATLAB built-in function `meshgrid`) and plot the contour line corresponding to $|G(z)| = 1$ (function `contour`). A solution of this exercise is proposed in Sect. 1.3 at page 21.

---

[2] For multistep schemes, G becomes a matrix; for the analysis of the absolute stability of multistep schemes, see, for instance, Trefethen (1996).

1.2 Model Partial Differential Equations

The PDEs presented in this chapter model elementary physical phenomena: convection, wave propagation, and diffusion. For each of these problems, we present one or several model equations, derive an exact solution in particular cases, and compute approximate solutions using appropriate numerical schemes. We consider in this section, the following PDEs:

- the convection equation: $\partial_t u(x,t) + c\partial_x u(x,t) = f(x,t)$,
- the wave equation: $\partial_{tt}^2 u(x,t) - c^2\partial_{xx}^2 u(x,t) = 0$,
- the heat equation: $\partial_t u(x,t) - \alpha\partial_{xx}^2 u(x,t) = f(x,t)$.

1.2.1 The Convection Equation

The PDE describing the convection (or transport) of a quantity $u(x,t)$ at velocity c (assumed constant in the following) is

$$\partial_t u(x,t) + c\,\partial_x u(x,t) = f(x,t), \quad \forall x \in \mathbb{R}, \quad \forall t > 0, \qquad (1.28)$$

with the initial condition (at $t = 0$):

$$u(x,0) = u_{\text{init}}(x), \quad \forall x \in \mathbb{R}. \qquad (1.29)$$

The source term $f(x,t)$ generally models the production in time of u.

Example 1.2 The transport of a pollutant in the atmosphere is modeled by the PDE $\partial_t u + \partial_x(cu) = 0$, where $u(x,t)$ is the concentration of the pollutant and c the wind velocity. If c is assumed constant, we retrieve the form (1.28) of the PDE. Note that $f = 0$ implies that there is no further production of pollutant at time instants $t > 0$.

Exercise 1.3 We consider the convection equation (1.28) with the initial condition (1.29) in the case $f(x,t) = 0$ (no sources).

1. To compute the exact solution, we introduce the change of variables

$$X = \alpha x + \beta t, \quad T = \gamma x + \mu t, \qquad \alpha, \beta, \gamma, \mu \in \mathbb{R}, \qquad (1.30)$$

and define the function U by $U(X,T) = u(x,t)$. Is this change of variables a bijection? Write the PDE satisfied by U. What happens to this equation if $\beta = -c\alpha$? Solve the last equation analytically. Deduce that the solution is constant along the lines (C_ξ) defined in the (x,t) plane by

$$x = \xi + ct, \quad \xi \in \mathbb{R}. \qquad (1.31)$$

Definition 1.6 The lines (1.31) are called characteristic curves of the convection equation (1.28).

2. We now want to find the exact solution of (1.28)–(1.29) on a finite real interval $[a, b]$. We assume that $c > 0$ and proceed geometrically. After drawing in the (x, t) plane, the characteristic curves C_ξ for $\xi \in [a, b]$ show that the solution $u(x, t)$ for all $x \in [a, b]$ and all $t > 0$ is completely determined by the initial condition u_{init} and an additional boundary condition

$$u(a, t) = \varphi(t), \qquad \forall t > 0. \qquad (1.32)$$

With the help of Fig. 1.2, show that for a given T the exact solution is

$$u(x, T) = \begin{cases} u_{\text{init}}(x - cT), & \text{if } x \geq a + cT, \\ \varphi\left(T - \dfrac{x - a}{c}\right), & \text{if } x < a + cT. \end{cases} \qquad (1.33)$$

Note that in the case $u_{\text{init}}(a) \neq \varphi(0)$, the solution is discontinuous along the characteristic curve $x - cT = a$. Find the value of T after which the initial condition u_{init} has completely left the interval $[a, b]$, i.e., $u(x, t)$ can be written as a function of $\varphi(t)$ only. Determine the boundary condition necessary to calculate u in the case of $c < 0$.

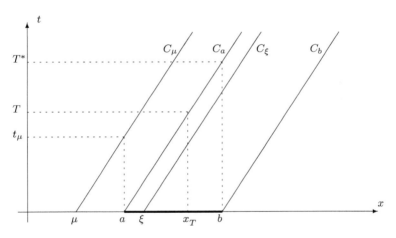

Fig. 1.2 Characteristics used to calculate the solution of the convection equation (1.28).

3. For the numerical solution of the convection equation (1.28), we define a regular space discretization

$$x_j = a + (j - 1)\delta x, \quad \delta x = \frac{b - a}{J}, \quad j = 1, 2, \ldots, J + 1, \qquad (1.34)$$

and a time discretization $(T > 0$ is fixed)

$$t_n = n\delta t, \quad \delta t = \frac{T}{N}, \quad n = 0, 1, \ldots, N. \tag{1.35}$$

Write a MATLAB function to compute the exact solution (1.33) following the model:

```
function uex=PDE_F_conv_exact(a,b,c,x,T,fun_in,fun_bc)
% Input arguments:
%       a,b   the interval [a,b]
%       c>0   the convection speed
%       x     the vector x(j)=a+(j-1)*delta x, j=1,2,...,J+1
%       T     the time at which the solution is computed
%       fun_in(x)   the initial condition for t=0
%       fun_bc(x) the  boundary condition for x=a
% Output argument:
%       uex   the vector of length J+1 containing the
%       exact solution
```

4. We assume that $c > 0$ and denote by u_j^n an approximation of $u(x_j, t_n)$. The following numerical scheme is proposed to compute u_j^n:

   - For $n = 0$ the initial condition is imposed: $u_j^0 = u_{\text{init}}(x_j), j = 1, 2, \ldots, J+1$.
   - For $n = 1, \ldots, N$ (loop in time):
     - For $j = 2, \ldots, J+1$ (compute u^{n+1} in the interior of the domain)

$$u_j^{n+1} = u_j^n - \frac{c\delta t}{\delta x}(u_j^n - u_{j-1}^n). \tag{1.36}$$

   Hint: in programs, we use a single vector solution u (of dimension $J+1$) that is updated after each time step. In other words, the time index (n) does not appear in programs. As a consequence, formula (1.36) has to be implemented using a back loop $(j = J+1, \ldots, 2)$ to ensure that u_{j-1} is taken at previous time instant n.
   - Set boundary value: $u_1^{n+1} = \varphi(t_{n+1})$.

(a) Justify geometrically (draw the characteristic starting from point u_j^{n+1}) that the previous algorithm is well defined if

$$\sigma = \frac{c\delta t}{\delta x} \leq 1. \tag{1.37}$$

Definition 1.7 The inequality (1.37) gives a sufficient condition for the stability of the upwind scheme (1.36) for the convection equation and is called the CFL (Courant–Friedrichs–Lewy) condition.

(b) Write a program using this algorithm to solve the convection equation for the following data:

$$a = 0, \quad b = 1, \quad c = 4, \quad f = 0,$$

$$u_{\text{init}}(x) = x, \quad \varphi(t) = \sin(10\pi t).$$

Choose $J = 40$ and compute δt from (1.37) with $\sigma = 0.8$.
Plot the solutions obtained after $n = 10, 20, 30, 40, 50$ time steps. Compare them to the exact solution.
What happens if $\sigma = 1$, or $\sigma = 1.1$? Comment on the influence of the value of σ on the stability of the scheme.

A solution of this exercise is proposed in Sect. 1.3 at page 23.

1.2.2 The Wave Equation

Acoustic (or elastic, or seismic) wave propagation is modeled by the following second-order PDE:

$$\partial_{tt}^2 u(x,t) - c^2 \partial_{xx}^2 u(x,t) = 0, \quad t > 0, \tag{1.38}$$

where $c > 0$ is the wave propagation velocity. The corresponding Cauchy problem requires two initial conditions:

$$u(x,0) = u_{\text{init}}(x), \quad \partial_t u(x,0) = v_{\text{init}}(x). \tag{1.39}$$

Example 1.3 The oscillations of an elastic string are described by the equation (1.38), where $u(x,t)$ represents the displacement of the string in the vertical plane. The propagation speed depends on the tension τ in the string and on its linear density ρ according to the law $c = \sqrt{\tau/\rho}$. Relations (1.39) provide the initial position and velocity of the string.

If the string is considered infinite, the equation is defined on the whole set $\mathbb{R}$. For a string of finite length ℓ, boundary conditions must be imposed in addition. For instance, if the string is fixed at both ends, the corresponding boundary conditions will be

$$u(0,t) = u(\ell,t) = 0, \quad \forall t > 0. \tag{1.40}$$

Definition 1.8 Boundary conditions (1.40) are called Dirichlet conditions (the values of the solution are imposed at the boundary of the computational domain). When the imposed values are null, the boundary conditions are said to be homogeneous.

Infinite string. We first consider the case of an infinite vibrating string.

Exercise 1.4 *Exact solution for the infinite string.* Using the change of variables (1.30), we define the function $U(X, T) = u(x, t)$ and attempt to derive the exact solution:

- Write $\partial_{tt}^2 u$ and $\partial_{xx}^2 u$ as functions of the derivatives of U. Derive the PDE satisfied by U.
- Write this PDE for $\mu = c\gamma$ and $\beta = -c\alpha$. Show that there exist two functions $F(X)$ and $G(T)$ such that $U(X, T) = F(X) + G(T)$.
- Conclude that the general solution of the wave equation can be written as

$$u(x, t) = f(x - ct) + g(x + ct). \tag{1.41}$$

- Using the initial conditions (1.39), show that

$$u(x, t) = \frac{u_{\text{init}}(x - ct) + u_{\text{init}}(x + ct)}{2} + \frac{1}{2c} \int_{x - ct}^{x + ct} v_{\text{init}}(s) ds. \tag{1.42}$$

A solution of this exercise is proposed in Sect. 1.3 at page 27.

Domain of dependence, CFL condition. The expression (1.42) shows that the value $u(x, t)$ depends only on initial conditions u_{init} and v_{init} restricted to the interval $[x - ct, x + ct]$ (see Fig. 1.3).

Definition 1.9 The lines of equations $x - ct = \xi$ and $x + ct = \xi$, with $\xi \in \mathbb{R}$ a given constant, are the characteristic curves of the wave equation (1.38).

We now intend to use a numerical scheme to solve the wave equation. At time $t_{n+1} = (n + 1)\delta t$, the value of the solution u_j^{n+1} at point $x_j = (j - 1)\delta x$ depends on the information transported from the level t_n along the two characteristics starting from the point (x_j, t_{n+1}) (see Fig. 1.3). The region located between the two characteristics is called the domain of dependence of the wave equation.

Exercise 1.5 Justify the following numerical scheme for the wave equation:

$$\frac{u_j^{n+1} - 2u_j^n + u_j^{n-1}}{\delta t^2} = c^2 \frac{u_{j+1}^n - 2u_j^n + u_{j-1}^n}{\delta x^2}. \tag{1.43}$$

Show that this scheme is second-order accurate in time and space (use (1.10)) and that the stability (CFL) condition is the same as that found for the convection equation:

$$\sigma = c\frac{\delta t}{\delta x} \leq 1. \tag{1.44}$$

A solution of this exercise is proposed in Sect. 1.3 at page 27.

Exercise 1.6 *Periodic initial conditions.* Let us now assume that the initial conditions $u_{\text{init}}(x)$ and $v_{\text{init}}(x)$ are periodic (with the same period τ). Show that the solution $u(x, t)$ of the wave equation is periodic in space (with period τ) and in time (with period τ/c).

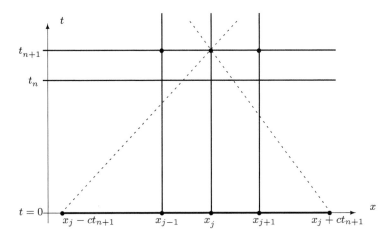

Fig. 1.3 Domain of dependence for the wave equation.

1. Justify the following algorithm:

   - for given initial conditions u_{init} and v_{init}, compute u_j^0 and u_j^1 as

$$u_j^0 = u_{\text{init}}(x_j), \quad u_j^1 = u_j^0 + \delta t\, v_{\text{init}}(x_j), \tag{1.45}$$

   - for $n \geq 1$, compute

$$u_j^{n+1} = 2(1 - \sigma^2)u_j^n + \sigma^2(u_{j-1}^n + u_{j+1}^n) - u_j^{n-1}. \tag{1.46}$$

2. Write a program to implement this algorithm. Hint: as for the convection equation, the time index (n) will not be used in programs. To implement the numerical scheme (1.46), only three vectors are necessary: u0 corresponding to u^{n-1}, u1 to u^n, and u2 to u^{n+1}. After each time step, the new value of u2 is calculated and the other two vectors are updated in this order: u0 ← u1, u1 ← u2.

3. Test the program for a string of length $\ell = 1$ and wave velocity $c = 2$. The initial data are $u_{\text{init}}(x) = \sin(2\pi x) + \sin(10\pi x)/4$ and $v_{\text{init}}(x) = 0$, corresponding to a string initially at rest. What is the time period of the solution?

4. Using $J = 50$ subintervals for the space discretization and $N = 50$ subintervals to discretize one period of time, superimpose on a single graph the exact and numerical solutions corresponding to one and two time periods. Verify that the numerical scheme preserves the periodicity of the solution. The same question goes for $J = 51$. Comment on the results.

A solution of this exercise is proposed in Sect. 1.3 at page 28.

Finite-length vibrating string. Consider the wave equation (1.38) with initial conditions (1.39) and boundary conditions (1.40). We seek a solution of the following form (also called the Fourier or elementary waves expansion):

$$u(x,t) = \sum_{k \in \mathbb{N}^*} \hat{u}_k(t)\phi_k(x), \quad \phi_k(x) = \sin\left(\frac{k\pi}{\ell}x\right). \tag{1.47}$$

For each wave ϕ_k, k is the wave number and $\hat{u}_k$ the wave amplitude.

Exercise 1.7

1. Derive and solve the PDE satisfied by a function $\hat{u}_k$.
2. Show that the exact solution for the finite-length vibrating string is

$$u(x,t) = \sum_{k \in \mathbb{N}^*} \left[A_k \cos\left(\frac{k\pi}{\ell}ct\right) + B_k \sin\left(\frac{k\pi}{\ell}ct\right)\right]\phi_k(x), \tag{1.48}$$

   with

$$A_k = \frac{2}{\ell}\int_0^\ell u_{\text{init}}(x)\phi_k(x)dx, \quad B_k = \frac{2}{k\pi c}\int_0^\ell v_{\text{init}}(x)\phi_k(x)dx. \tag{1.49}$$

   Find the time and space periods of the solution.
3. Write a program to solve the finite-length vibrating string problem, using the centered scheme (1.43). Hint: start from the program previously implemented and modify the boundary conditions.
   Find the exact solution corresponding to the following initial conditions:

$$u_{\text{init}}(x) = \sin\left(\frac{\pi}{\ell}x\right) + \frac{1}{4}\sin\left(10\frac{\pi}{\ell}x\right), \quad v_{\text{init}}(x) = 0. \tag{1.50}$$

   Plot the exact and numerical solutions at several times over one spatial period. Use the following numerical values: $c = 2$, $\ell = 1$, $J = 50$, and $N = 125$.

A solution of this exercise is proposed in Sect. 1.3 at page 29.

1.2.3 The Heat Equation

Diffusion phenomena such as molecular and heat diffusion can be described mathematically using the heat equation model

$$\partial_t u - \alpha \partial_{xx}^2 u = f(x,t), \quad \forall t > 0, \tag{1.51}$$

with the initial condition

$$u(x,0) = u_{\text{init}}(x). \tag{1.52}$$

Example 1.4 The temperature θ of a heated body is a solution of the equation

$$\partial_t \theta - \partial_x(\alpha \partial_x \theta) = f(x,t), \tag{1.53}$$

where α is the thermal diffusivity of the material and the function f models the heat source. In a homogeneous medium, α does not depend on the space position x, and we retrieve the model equation (1.51).

Consider the problem of a wall of thickness ℓ, initially at uniform temperature θ_0 (the room temperature). At time $t = 0$, the outside temperature (at $x = 0$) suddenly rises to the value $\theta_s > \theta_0$, which is afterwards maintained constant. The temperature at $x = \ell$ is kept at its initial value θ_0. The heat propagation within the wall is described by the heat equation (1.51), with the unknown $u(x,t) = \theta(x,t) - \theta_0$, $f(x,t) = 0$, the initial condition $u_{\text{init}}(x) = 0$, and Dirichlet boundary conditions

$$u(0,t) = \theta_s - \theta_0 = u_s, \quad u(\ell,t) = 0, \quad \forall t > 0. \tag{1.54}$$

Infinite domain. For an infinitely thick wall ($\ell \to \infty$), we search for a solution of the form

$$u(x,t) = g(\eta), \quad \text{with} \quad \eta = \frac{x}{2\sqrt{\alpha t}}. \tag{1.55}$$

Exercise 1.8 Show that the function g defined above satisfies the following PDE:

$$\frac{d^2 g}{d\eta^2} + 2\eta \frac{dg}{d\eta} = 0. \tag{1.56}$$

We introduce the following function, called the *error function*:

$$\text{erf}(z) = \frac{2}{\sqrt{\pi}} \int_0^z e^{-\zeta^2} d\zeta, \tag{1.57}$$

which satisfies $\text{erf}(0) = 0$ and $\text{erf}(\infty) = 1$. Find that the solution of the heat equation for $\ell \to \infty$ is

$$u(x,t) = u_s \left[1 - \text{erf}\left(\frac{x}{2\sqrt{\alpha t}}\right)\right]. \tag{1.58}$$

A solution of this exercise is proposed in Sect. 1.3 at page 30.

Remark 1.3 A change in the value of the initial condition at point $x = 0$ has as consequence the modification of the solution everywhere in the domain. In other words, the perturbation introduced at $x = 0$ is instantaneously propagated in the computation domain ($u(x,t) > 0, \forall x$ in formula (1.58)). The propagation speed is said to be infinite. We can also prove that the solution at any point depends on all initial values $u_{\text{init}}(x)$. This implies that the domain of dependence for the heat equation is the whole domain of definition.

We recall, for comparison, that for the wave equation the domain of dependence is restricted to the area bounded by the characteristics and that the propagation speed of the solution is finite.

Finite domain. For a wall of finite thickness ℓ, the elementary waves expansion (1.47) is used.

Exercise 1.9 Write and solve the equation satisfied by the functions $\hat{u}_k$ in the case of the heat equation. Verify that the solution of the heat equation with boundary conditions (1.54) is

$$u(x,t) = \left(1 - \frac{x}{\ell}\right) u_s + \sum_{k \in \mathbb{N}^*} A_k \exp\left(-\left(\frac{k\pi}{\ell}\right)^2 \alpha t\right) \phi_k(x). \qquad (1.59)$$

Show that $A_k = -2u_s/(k\pi)$.
A solution of this exercise is proposed in Sect. 1.3 at page 31.

Remark 1.4 Let us compare the exact solution (1.59) to the exact solution of the wave equation (1.48). The wave equation describes the transport in time of the initial condition. The amplitude of each spatial wave $\phi_k(x)$ oscillates over one time period without damping. The diffusion phenomenon described by the heat equation is characterized by a fast decrease in time of the amplitude of each wave $\phi_k(x)$ due to the presence of the exponential factor in (1.59). This *smoothing* effect of the heat operator increases as the wave number k becomes larger.

Exercise 1.10 *Numerical solution.* Consider first the following explicit centered scheme for the heat equation:

$$\frac{u_j^{n+1} - u_j^n}{\delta t} - \alpha \frac{u_{j+1}^n - 2u_j^n + u_{j-1}^n}{\delta x^2} = 0. \qquad (1.60)$$

The stability condition for this scheme is

$$\alpha \frac{\delta t}{\delta x^2} \le \frac{1}{2}. \qquad (1.61)$$

1. Write a program to solve the problem of the heat propagation in a finite thickness wall. Set $\alpha = 1$, $\ell = 1$, $u_s = 1$, and take $J = 50$ subintervals for the space discretization. The time step δt is calculated using (1.61). Plot the numerical solution for different times and compare it to the exact solution (1.59). Compare it also to the solution obtained for an infinite domain (1.58). Comment on the results for small t and then for large t. Hint: the exact solution is computed to a fair degree of approximation by considering the first 20 wave numbers k in the expansion (1.59).
2. *Smoothing effect.* Run the previous program for $u_s = 0$ and

$$u_{\text{init}}(x) = u(x,0) = \sin\left(\frac{\pi}{\ell}x\right) + \frac{1}{4}\sin\left(10\frac{\pi}{\ell}x\right). \qquad (1.62)$$

Compare the numerical solution to the exact solution (1.59). Comment on the results by comparing with those obtained for the wave equation with the same initial condition. Describe the damping of the waves defining the initial condition.

A solution of this exercise is proposed in Sect. 1.3 at page 31.

1.3 Solutions and Programs

Solution of Exercises 1.1 and 1.2 (the Absorption Equation)

1. Using the substitution $u(t) = e^{-\alpha t} v(t)$ in (1.24), we obtain the differential equation $v'(t) = e^{\alpha t} f(t)$, with initial condition $v(0) = u(0) = u_{\text{init}}$. It is easy to integrate this ODE to obtain the expression for v and finally (1.25).

2. If $\alpha = \alpha(t)$, we obtain the following general solution for linear ODEs:

$$u(t) = e^{-\int_0^t \alpha(s)ds} \left[u_{\text{init}} + \int_0^t e^{\int_0^z \alpha(s)ds} f(z)dz \right]. \tag{1.63}$$

3. Let us assume that the function $f(t) = f$ is constant. Expression (1.25) becomes

$$u(t) = \frac{f}{\alpha} + e^{-\alpha t} \left(u_{\text{init}} - \frac{f}{\alpha} \right). \tag{1.64}$$

If $u_{\text{init}} = f/\alpha$, the solution is constant: $u(t) = u_{\text{init}}$, $\forall t$.
For $\alpha > 0$, $u(t) \to f/\alpha$ for $t \to \infty$.
For $\alpha < 0$, $u(t) \to +\infty \times \text{sign}(u_{\text{init}} - f/\alpha)$.
4. If $f = 0$ and $\alpha = \alpha_r + i\alpha_i$, we obtain

$$|u(t)| = |e^{-\alpha t} u_{\text{init}}| = |e^{-(\alpha_r + i\alpha_i)t} u_{\text{init}}| = |e^{-\alpha_r t} u_{\text{init}}| \to 0.$$

5. To integrate numerically the ODE $u'(t) = f(t, u)$, we provide MATLAB functions PDE_F_EulerExp.m, PDE_F_RKutta2.m, and PDE_F_RKutta4.m which implement the explicit Euler scheme, the second-order Runge–Kutta, and fourth-order Runge–Kutta methods, respectively (see Table 1.1). The right-hand side $f(t, u)$ is identified by the generic name **fun** inside these functions; the real name of this function is specified by the user when the functions are called. These three functions return a vector holding numerical values computed at discrete times uniformly distributed between T_i and T_f.

The MATLAB program PDE_M_absorption.m calls successively the three functions implementing numerical schemes, sending as input argument the name PDE_F_absorption_source, which is the function implementing the right-hand side of the absorption equation. The results are displayed in three separate figures (Fig. 1.4).

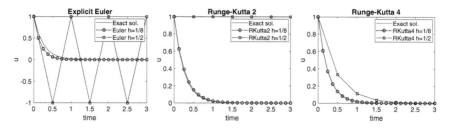

Fig. 1.4 Numerical solution of the absorption equation $u'(t) + 4u(t) = 0$, $u(0) = 1$, obtained using the explicit Euler scheme, the second- and fourth-order Runge–Kutta schemes. Solid line represents the exact solution.

Let us comment on the results displayed in Fig. 1.4. Everything goes well for $h = \frac{1}{8} < \frac{2}{\alpha} = \frac{1}{2}$: the numerical solution approaches the exact solution with a better approximation for Runge–Kutta schemes (for the fourth-order scheme, the exact and numerical solutions are not distinguishable in the graph). On the other hand, for $h = \frac{1}{2} = \frac{2}{\alpha}$ the stability limit of the explicit Euler scheme is reached. The numerical solution remains bounded (which is no longer true when $h > \frac{1}{2}$, a case to be tested) but does not converge. For the same value of $h = \frac{1}{2}$, the second-order Runge–Kutta scheme is not better and the numerical solution remains constant, of value $u = 1$ (this could be surprising, but can be easily checked *by hand* using the relations in Table 1.1).

The fourth-order Runge–Kutta (*RKutta4*) scheme has a wider stability region and remains stable for the two values of the discretization step h considered in this computation. The program PDE_M_absorption_stability.m computes the stability regions for the three numerical schemes, following the indications of Exercise 1.2. Figure 1.5 (left) shows that the stability region of the *RKutta4* scheme includes both regions of stability of the explicit Euler scheme and *RKutta2*.

Finally, we check using the program PDE_M_absorption_convergence.m the convergence order p of the considered three schemes (Exercise 1.1, question 6). The orders of convergence ($p = 1$ for explicit Euler, $p = 2$ for *RKutta2*, and $p = 4$ for *RKutta4*) are displayed in Fig. 1.5 (right). They are identical to theoretical orders of accuracy (see Definition 1.4). Note that we computed the approximation error ε with respect to the exact solution of the problem, which is different from the definition of the error of consistence (1.22). The error ε has an obvious practical importance for assessing the accuracy of the calculation, while the error of consistence has a theoretical interest for designing mathematically sound (consistent) numerical schemes. Each scheme was supposed to be consistent and stable. Consistence and stability imply convergence of the numerical scheme (see Sect. 1.1.4). The relation between convergence and consistency errors can be found in many textbooks (Trefethen, 1996; Delabrière and Postel, 2004; Demailly, 1996).

In light of these results, the *RKutta4* scheme seems to be the best choice, offering the best stability and accuracy. In fact, the choice of one scheme or

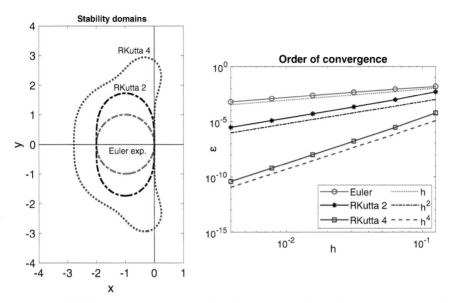

Fig. 1.5 Stability region (left) and order of convergence (right) for three numerical schemes: explicit Euler, second-order Runge–Kutta, and fourth-order Runge–Kutta.

another for a practical application is motivated by a compromise between its characteristics (accuracy, stability) and its computational costs (the *RKutta4* scheme is approximately four times as expensive as the explicit Euler scheme for the same discretization step).

Solution of Exercise 1.3 (the Convection Equation)

1. The change of variables can be written as

$$\begin{pmatrix} X \\ T \end{pmatrix} = \begin{pmatrix} \alpha & \beta \\ \gamma & \mu \end{pmatrix} \begin{pmatrix} x \\ t \end{pmatrix},$$

and it is one-to-one and onto (i.e., bijective) if $\alpha\mu \neq \beta\gamma$. Differentiation with respect to the new variables gives

$$\partial_t u = \beta \partial_X U + \mu \partial_T U, \quad \partial_x u = \alpha \partial_X U + \gamma \partial_T U, \tag{1.65}$$

and we find that U is the solution of the PDE:

$$(\beta + c\alpha)\, \partial_X U + (\mu + c\gamma)\, \partial_T U = 0.$$

Setting $\beta = -c\alpha$ results in $(\mu + c\gamma)\, \partial_T U = 0$, and, since $\mu \neq -c\gamma$ from above, we obtain that $\partial_T U = 0$. The last equation has $U(X,T) = F(X)$ as a solution, where F is an arbitrary smooth function. Therefore, we can present $u(x, t) = F(X) = F(\alpha x + \beta t) = F(\alpha x - \alpha c t) = G(x - ct)$, where G is again an arbitrary function. Finally, imposing the initial condition (1.29), we get $u(x, t) = u_{\text{init}}(x - ct)$. This implies, in particular, that the solution remains unchanged along the characteristic lines, that is,

$$\text{if} \quad (x, t) \in C_\xi \implies u(x, t) = u_{\text{init}}(\xi), \quad \xi = x - ct.$$

In the plane (x, t), the characteristic curves C_ξ are straight lines of positive slopes $1/c$ (see Fig. 1.2).

2. To derive the solution $u(x, T)$ for $x \in [a, b]$ and $T < T^* = (b - a)/c$, we draw the characteristics through the points (x, T) and use the fact that the solution $u(x, t)$ is constant along a characteristic line. Two cases are possible, as shown in Fig. 1.2:

- the characteristic (C_ξ in the figure) crosses the segment $[a, b]$; this is the case for points located at $x \geq x_T$, with $x_T = a + cT$. The solution will be therefore determined by the initial condition:

$$u(x, T) = u_{\text{init}}(\xi) = u_{\text{init}}(x - cT).$$

- the characteristic (C_μ in Fig. 1.2) does not cross the segment $[a, b]$; in this case, a boundary condition is needed. If we impose $u(a, t) = \varphi(t)$, since the information will be searched through the characteristics back to this boundary condition, the solution is calculated as

$$u(x, T) = \varphi(t_\mu) = \varphi\left(T - \frac{x - a}{c}\right).$$

Note that the initial condition u_{init} is completely "evacuated" from the domain $[a, b]$ after a time value $T^* = (b - a)/c$.

For $c < 0$, the boundary condition must be imposed on the right-hand side of the domain by setting $u(b, t) = \varphi(t)$.

3. The function PDE_F_conv_exact.m computes the exact solution for a given time T. Note in particular the use of the MATLAB command **find** to implement formula (1.33).

4. We note that (1.36) is a discretization of Eq. (1.28) where the time derivative $\partial_t u$ is approximated by $D^+ u(t_n)$ and the space derivative $\partial_x u$ by $D^- u(x_j)$. The choice of the upwind approximation for $\partial_x u$ is imposed by the fact that $c > 0$ and therefore the information comes from the left.

The characteristic going through the point u_j^{n+1} is drawn in Fig. 1.6. This line cuts the horizontal line $t = t_n$ at a point P located between x_{j-1} and x_j, at a distance $c\delta t$ from x_j and $\delta x - c\delta t$ from x_{j-1}. Since the solution is constant along a characteristic, necessarily $u_j^{n+1} = u_P^n$.

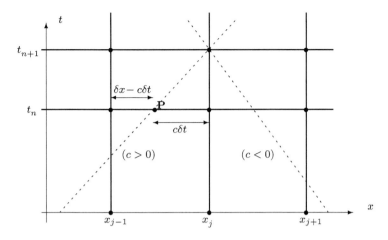

Fig. 1.6 Geometrical interpretation of the CFL condition (1.37) for the upwind scheme (1.36) used to solve the convection equation (1.28).

It is interesting to note from these geometrical considerations that the scheme (1.36) is nothing else but a linear interpolation between the values of the solution at points x_{j-1} and x_j:

$$u_j^{n+1} = \frac{\delta x - c\delta t}{\delta x} u_j^n + \frac{c\delta t}{\delta x} u_{j-1}^n.$$

The CFL condition $c\delta t \leq \delta x$ can be thus regarded as a criterion imposing the positivity of the interpolation coefficients; in other words, the point P must lie within the interval $[x_{j-1}, x_j]$.

Some computing tips may be useful at this point. First of all, we must be careful and make all array indices start from 1. Then, the solution at time t_{n+1} will be computed according to

$$u_j^{n+1} = (1 - \sigma)u_j^n + \sigma u_{j-1}^n, \quad \sigma = \frac{c\delta t}{\delta x}.$$

Since we intend to compute the solution at the final time, we can save storage memory by using a single array u for the calculation. If the solution is needed at intermediate times, it will be either written to a file or graphically displayed. With this programming trick, the previous relation becomes

$$u(j) = (1 - \sigma)u(j) + \sigma u(j - 1),$$

and the values at time t_n will be replaced by new values at time t_{n+1}. We also must be careful to use in the numerical scheme the values u_{j-1} before they are modified (i.e., at previous time t_n). This is achieved by using an inverse loop ($j = J + 1, J, \ldots, 2$); for $j = 1$ the boundary condition is imposed.

This algorithm is implemented in the program PDE_M_convection.m. This program calls the functions PDE_F_conv_bound_cond (defining boundary condition φ) and PDE_F_conv_uinit (defining initial condition u_{init}).

The solution for chosen intermediate times is represented in Fig. 1.7 and compared to the exact solution (computed by the function PDE_F_conv_exact). Note that starting from time $T^* = \frac{1}{4}$, the initial data leaves the computation domain, and the solution depends only on the boundary condition.

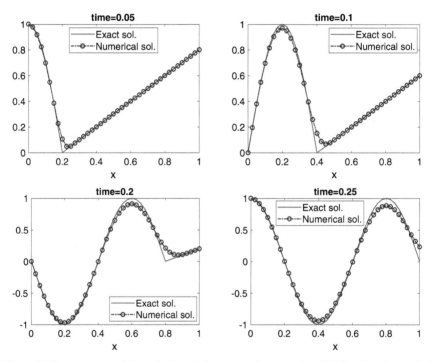

Fig. 1.7 Computation of the solution of the convection equation (1.28) using the upwind scheme (1.36) with CFL= 0.8. Solid line represents the exact solution.

An interesting phenomenon can be observed when one looks closer at the solution in the region where the derivative of the exact solution is discontinuous (i.e., where the part of the solution depending on the initial condition is connected to that depending on the boundary condition). There is a numerical *smoothing* of this sharp transition. What is interesting here is that this observed dissipation has no physical meaning and is exclusively due to the numerical scheme. The upwind scheme is therefore said to be *dissipative*. We shall discuss in detail dissipation effects, in a more physical context, when analyzing the heat equation.

A computation performed for $\sigma = 1$ (or CFL $= 1$) gives results that are perfectly superimposed on the exact solution. This is not surprising, since the upwind scheme becomes in this case an exact relation: $u_j^{n+1} = u_{j-1}^n$. In practice, the convection speed c and the space discretization step δx are generally not constant, making impossible a computation with $\sigma = 1$.

The computation for $\sigma = 1.1$ illustrates the loss of stability of the upwind scheme when $\sigma > 1$.

Solution of Exercise 1.4 (the Wave Equation)

Starting from (1.65), we obtain for the second derivatives

$$\begin{cases} \partial_{tt}^2 u = \beta^2 \partial_{XX}^2 U + \mu^2 \partial_{TT}^2 U + 2\beta\mu\partial_{TX}^2 U, \\ \partial_{xx}^2 u = \alpha^2 \partial_{XX}^2 U + \gamma^2 \partial_{TT}^2 U + 2\alpha\gamma\partial_{TX}^2 U, \end{cases}$$

and conclude that U is a solution of the PDE:

$$\left(\mu^2 - c^2\gamma^2\right)\partial_{TT}^2 U + \left(\beta^2 - c^2\alpha^2\right)\partial_{XX}^2 U + 2\left(\beta\mu - c^2\alpha\gamma\right)\partial_{XT}^2 U = 0.$$

For $\mu = c\gamma$ and $\beta = -c\alpha$, the equation becomes $-4c^2\alpha\gamma\partial_{XT}^2 U = 0$, which implies that $\partial_{XT}^2 U = 0$, or again $\partial_T(\partial_X U) = 0$. From the previous relation, we infer that there exist two functions $F(X)$ and $G(T)$ such that $U(X,T) = F(X) + G(T)$. Since $X = \alpha(x - ct)$ and $T = \gamma(x + ct)$, we can choose $\alpha = \gamma = 1$, and the solution becomes

$$u(x, t) = f(x - ct) + g(x + ct).$$

Imposing initial conditions for $u(x, t)$ and $\partial_t u(x, t) = -cf'(x - ct) + cg'(x + ct)$, we obtain the system of equations

$$\begin{cases} f'(x) + g'(x) = u'_{\text{init}}(x), \\ -f'(x) + g'(x) = (1/c)v_{\text{init}}(x), \end{cases}$$

giving expressions for f' and g' and finally the formula (1.42) for u.

Solution of Exercise 1.5

In the discretization (1.43) of the wave equation (1.39), the second partial derivatives $\partial_{xx}^2 u$ and $\partial_{tt}^2 u$ are approximated by centered finite differences (see Eq. (1.10)). One can easily show that the scheme (1.43) is of second order in space and also in time.

The stability condition (1.44) expresses that the domain of dependence (see Fig. 1.3) bounded by the two characteristics starting from the point (x_j, t_{n+1}) must lie inside the triangle $\{(x_j, t_{n+1}), (x_{j-1}, t_n), (x_{j+1}, t_n)\}$ defined by the three-point stencil used by the numerical scheme. In other words, if the time step is larger than the critical value $\delta x/c$, the information searched for by the characteristics will be found outside the interval $[x_{j-1}, x_{j+1}]$ used by the scheme. This is not in agreement with the physics described by the wave equation and therefore results in the instability of the numerical scheme.

From formula (1.42), it can be easily checked that $u(x + \tau, t) = u(x, t)$ and $u(x, t + \tau/c) = u(x, t)$. The solution $u(x, t)$ is hence periodic in time and space, with period τ in space and τ/c in time.

Solution of Exercise 1.6

In the proposed algorithm, the scheme (1.43) is used together with the computation of the solution for the first time step based on the approximation $v_{\text{init}}(x) = \partial_t u(x, 0) \approx (u(x, \delta t) - u(x, 0))/\delta t$.

This algorithm is implemented in the program PDE_M_wave_infstring.m and the initial conditions in files PDE_F_wave_infstring_uinit.m and, respectively, PDE_F_wave_infstring_vinit.m.

It is worth explaining some programming tricks used in this program. The periodicity condition (also satisfied by the initial condition $u_{\text{init}}(x) = u_{\text{init}}(x + \tau)$ with $\tau = 1$!) is translated in discrete form by $u_{J+1} = u_1$, since the spatial discretization is built such that $x_1 = 0$ and $x_{J+1} = 1$. In order to fully exploit the capabilities of MATLAB in terms of vector programming, we define the arrays jp and jm corresponding to indices $j + 1$, respectively $j - 1$, for all discretization points. The periodicity is expressed then by setting $jp(J) = 1$, $jm(1) = J$, and the numerical scheme (1.46) is written within a single line of code:

```
u2=-u0+2*(1-sigma2)*u1+sigma2*(u1(jm)+u1(jp)).
```

Here u2 corresponds to the array $(u^{n+1})_j$, u1 to $(u^n)_j$, and u0 to $(u^{n-1})_j$. The advantage of this compact programming is to avoid loops interrupted by specific treatments of the points on the boundaries. We shall use this simple programming tip in a more complicated project (Chap. 15).

The numerical results are displayed in Fig. 1.8. The period of the solution in time is $1/c = 0.5$. For $J = N = 50$, the CFL number is $\sigma = 1$. The numerical scheme propagates the initial condition correctly and preserves the periodicity in time. The solution after one time period coincides with the exact solution (which is in fact the initial condition u_{init}). Unlike the upwind scheme used for the convection equation, this centered scheme does not generate any numerical diffusion (even for smaller J corresponding to CFL < 1). For $J = 51$, the scheme becomes unstable because CFL > 1. It is also interesting to note that the instability of the scheme appears only after some time (here after one time period).

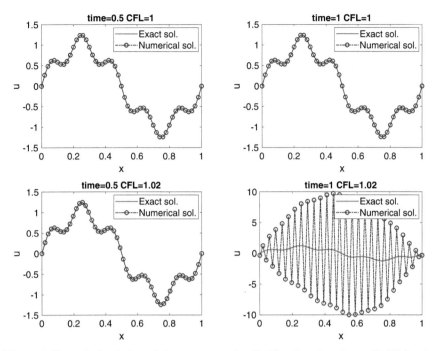

Fig. 1.8 Numerical solution of the wave equation (1.39) using the scheme (1.43) for the infinite vibrating string (periodicity conditions). Initial condition: $u(x,0) = \sin(2\pi x) + \sin(10\pi x)/4$ and $\partial_t u(x,0) = 0$. Comparison with the exact solution for CFL = 1 (top) and CFL > 1 (bottom) after one period in time (left) and two periods in time (right).

Solution of Exercise 1.7

The amplitude $\hat{u}_k$ satisfies the PDE

$$\frac{d^2}{dt^2}\hat{u}_k + c^2 \left(\frac{k\pi}{\ell}\right)^2 \hat{u}_k = 0, \tag{1.66}$$

which is often encountered in physics. It models, for example, the oscillations of a pendulum. The general solution of this PDE being

$$\hat{u}_k(t) = A_k \cos\left(\frac{k\pi}{\ell}ct\right) + B_k \sin\left(\frac{k\pi}{\ell}ct\right), \tag{1.67}$$

the expression (1.48) is straightforward to obtain. The coefficients A_k, B_k are computed using the orthogonality of the trigonometric functions ϕ_k on $[0, \ell]$, with $\int_0^\ell \phi_k(x)\phi_j(x)dx = (\ell/2)\delta_{kj}$, with δ the Kronecker symbol.

We observe that the solution is periodic, of period 2ℓ in space and $2\ell/c$ in time. The initial condition (1.50) corresponds to a decomposition in elementary waves with

$$k = \{1, 10\}, \quad A_k = \{1, 1/4\}, \quad B_k = \{0, 0\}.$$

The exact solution is given by Eq. (1.48) with these values.

The program PDE_M_wave_fstring.m computes the solution for the finite-length string. Since $v_{\text{init}} = 0$, only the initial condition u_{init} is computed in PDE_F_wave_fstring_uinit.m. The computed solution is compared to the exact solution, implemented in PDE_F_wave_fstring_exact.m.

Note that, once again, we use vector notation (avoiding **for** loops) for the centered scheme. Computation points corresponding to the boundaries are not modified during the loop in time, and their initial values are preserved (which respect the imposed boundary conditions). Figure 1.9 displays a comparison between the exact solution and the numerical solution for two different time instants.

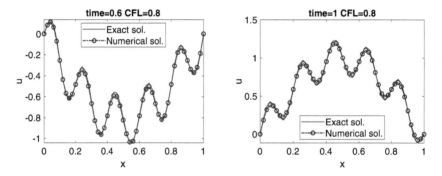

Fig. 1.9 Numerical solution of the wave equation (1.39) using the scheme (1.43) for the finite-length string. Initial condition: $u(x, 0) = \sin(\pi x) + \sin(10\pi x)/4$ and $\partial_t u(x, 0) = 0$. Comparison with the exact solution (one time period corresponds to $t = 1$).

Solution of Exercise 1.8

The partial derivatives can be written in terms of f as

$$\frac{\partial u}{\partial t} = -\frac{\eta}{2t} \frac{dg}{d\eta}, \quad \frac{\partial^2 u}{\partial x^2} = \frac{1}{4\alpha t} \frac{d^2 g}{d\eta^2},$$

hence the PDE (1.56) is satisfied by g. After integration, we obtain

$$\frac{dg}{d\eta} = Ae^{-\eta^2} \implies u(x, t) = g(\eta) = B + A\,\mathrm{erf}(\eta).$$

Taking into account the properties of the function erf when imposing boundary conditions, we can easily obtain the formula (1.58).

Solution of Exercise 1.9

The amplitude $\hat{u}_k$ is a solution of the ODE:

$$\frac{d\hat{u}_k}{dt} + \alpha \left(\frac{k\pi}{\ell}\right)^2 \hat{u}_k = 0,$$

and has the analytical form

$$\hat{u}_k(t) = A_k \exp\left(-\left(\frac{k\pi}{\ell}\right)^2 \alpha t\right).$$

We can easily check that any elementary wave $\hat{u}_k(t)\phi_k(x)$ is a solution of the heat equation, but it does not satisfy boundary conditions (1.54). This is the reason why a linear function of x (which is also a solution of the heat equation) has been added to obtain the final form of the exact solution (1.59). We note that this is possible since the differential operators in the heat equation are linear.

Finally, the coefficients A_k are calculated using the orthogonality of ϕ_k functions:

$$A_k = -\frac{2u_s}{\ell} \int_0^l \left(1 - \frac{x}{\ell}\right) \sin\left(\frac{k\pi}{\ell}x\right) dx = -\frac{2u_s}{k\pi}.$$

Solution of Exercise 1.10

The MATLAB program PDE_M_heat.m answers questions 1 and 2. The initial condition is computed in PDE_F_heat_uinit.m and the exact solution in PDE_F_heat_exact.m (the erf function is already available in the standard MATLAB package). Numerical results (see Fig. 1.10) confirm the fact that the erf-solution (1.58) obtained for an infinite domain is a good approximation for small times t (this is the main reason why it is often used in practice by engineers). For longer times t, the exact solution (and hopefully the numerical one as well) converges to the *steady-state* solution (i.e., independent of time) $u(x) = (1 - \frac{x}{\ell})u_s$.

The diffusion phenomenon described by the heat equation is characterized by a time scale $t_0 = \ell^2/\alpha$ (see the expression of η in (1.55)). Consequently, the effective speed of propagation $c_0 = \ell/t_0 = \alpha/\ell$ of a thermal perturbation decreases with the distance to the source. This accounts for the poor efficiency of diffusion systems to propagate heat for large distances or time!

Let us imagine a domestic heating system based on diffusion only. The thermal diffusivity of air being $\alpha \approx 20$ [mm^2/s], the heating effect will be felt at a distance of 1 cm after 5 seconds and at 1 meter after $5 \cdot 10^4$ s ≈ 14 hours!

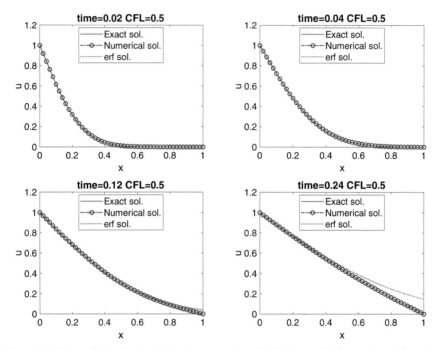

Fig. 1.10 Numerical solution of the heat equation (1.51) for $x \in [0, \ell]$, $\alpha = 1$, and boundary conditions $u(0, t) = 1, u(\ell, t) = 0$. Initial condition $u(0, 0) = 1, u(x, 0) = 0, x > 0$. Comparison with the exact (1.59) and the erf (1.58) solutions.

Fortunately, real heating systems are more efficient due to other phenomena (such as air convection and radiation).

For the next question, it is easy to return to the previous program (PDE_M_heat.m) to implement the new initial condition (1.62). The user can choose between the two cases using an integer flag. The results (see Fig. 1.11) clearly show that the wave of highest wave number, equivalent to highest frequency ($k = 10$ in our case), is first damped. The solution tends to the constant steady-state solution $u(x) = 0$. Recall, for comparison, the behavior of the wave equation, for which the same initial condition was transported without damping of the wave amplitudes (see Fig. 1.9).

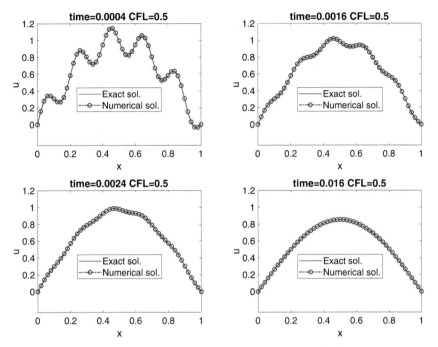

Fig. 1.11 Numerical solution of the heat equation (1.51) for $x \in [0, \ell]$, $\alpha = 1$, and boundary conditions $u(0, t) = 0, u(\ell, t) = 0$. Initial condition: $u(x, 0) = \sin(\pi x) + \sin(10\pi x)/4$. Comparison with the exact solution (1.59). Note the early damping of high-frequency waves.

Chapter References

Extensive analysis of numerical methods for solving ODEs or PDEs can be found in a large number of books, ranging from the classic text by Richtmyer and Morton (1967) to Lambert (1973), John (1978), Mitchell and Griffiths (1980), Butcher (1987), and more recently Trefethen (1996). Introductions to finite difference methods can be found in specialized books, such as Strikwerda (1989), LeVeque (2007), or several texts on computational fluid dynamics, such as Hirsch (1988) and LeVeque (1992).

The reader acquainted with French literature may refer to Crouzeix and Mignot (1989), Delabrière and Postel (2004), and Demailly (1996) for the numerical analysis of ODEs and Lucquin (2004), Mohammadi and Saïac (2003) for PDEs; the implementation of numerical schemes using object-oriented programming is discussed in Danaila, Hecht, and Pironneau (2003).

J.C. Butcher, *The Numerical Analysis of Ordinary Differential Equations* (Wiley, 1987)

M. Crouzeix, A. Mignot, *Analyse numérique des équations différentielles* (Masson, Paris, 1989)

I. Danaila, F. Hecht, O. Pironneau, *Simulation numérique en C++* (Dunod, Paris, 2003)

S. Delabrière, M. Postel, *Méthodes d'approximation, Equations différentielles, Applications Scilab* (Ellipses, Paris, 2004)

J.P. Demailly, *Analyse numérique et équations différentielles* (Presses Universitaires de Grenoble, 1996)

C. Hirsch, *Numerical Computation of Internal and External Flows* (Wiley, 1988)

F. John, *Partial Differential Equations* (Springer, 1978)

J.D. Lambert, *Computational Methods in Ordinary Differential Equations* (Wiley, 1973)

R. LeVeque, *Numerical Methods for Conservation Laws* (Birkhäuser, 1992)

R. LeVeque, *Finite Difference Methods for Ordinary and Partial Differential Equations* (SIAM, Philadelphia, 2007)

B. Lucquin, *Équations aux dérivées partielles et leurs approximations* (Ellipses, Paris, 2004)

A.R. Mitchell, D.F. Griffiths, *Computational Methods in Partial Differential Equations* (Wiley, 1980)

B. Mohammadi, J.-H. Saïac, *Pratique de la simulation numérique* (Dunod, 2003)

R.D. Richtmyer, K.W. Morton, *Difference Methods for Initial Value Problems*, 2nd ed. (Wiley Interscience, 1967)

J.C. Strikwerda, *Finite Difference Schemes and Partial Differential Equations* (Wadsworth and Brooks/Cole, 1989)

L.N. Trefethen, *Finite Difference and Spectral Methods for Ordinary and Partial Differential Equations*, unpublished text. http://people.maths.ox.ac.uk/trefethen/pdetext.html (1996)

Chapter 2
Nonlinear Differential Equations: Application to Chemical Kinetics

Project Summary

Level of difficulty:	1
Keywords:	Nonlinear system of differential equations, Stability, integration schemes, Euler explicit scheme, Runge–Kutta scheme, Delayed differential equation
Application fields:	Chemical kinetics, Biology

2.1 Physical Problem and Mathematical Modeling

The laws governing chemical kinetics can be written as systems of ordinary differential equations. In the case of complex reactions with several different participating molecules, these equations are nonlinear and present interesting mathematical properties (stability, periodicity, bifurcation, etc.). The numerical solution of this type of system is a domain of study in itself with flourishing literature. Very efficient numerical methods to solve systems of ODEs are implemented in MATLAB, as in most other software. The first model of reaction that we shall study in this chapter can be completely solved using standard packages. We will, therefore, use the `ode` solvers provided by MATLAB, assuming that the user masters the underlying theory and the basic concepts such as convergence, stability, and precision (see Chap. 1).

The other model includes a delay term. We choose here not to use the MATLAB delay equation solver `dde23` and describe a specific numerical method. Both are examples of models presented in Hairer, Norsett, and Wanner (1987).

We first study the so-called *Brusselator* model, which involves six reactants and products A, B, D, E, X, and Y. The chemical reactions can be modeled

I. Danaila et al., *An Introduction to Scientific Computing*,
https://doi.org/10.1007/978-3-031-35032-0_2

by

$$\begin{cases} A \xrightarrow{v_1} X, \\ B + X \xrightarrow{v_2} Y + D, & \text{bimolecular reaction,} \\ 2X + Y \xrightarrow{v_3} 3X, & \text{trimolecular autocatalytic reaction,} \\ X \xrightarrow{v_4} E, \end{cases} \qquad (2.1)$$

where v_i are the constant chemical reaction rates. The concentrations of the species as functions of time t are denoted by $A(t)$, $B(t)$, $D(t)$, $E(t)$, $X(t)$, and $Y(t)$. Mass conservation in the chemical reactions leads to the following ODEs:

$$\begin{cases} A' = -v_1 A, \\ B' = -v_2 BX, \\ D' = v_2 BX, \\ E' = v_4 X, \\ X' = v_1 A - v_2 BX + v_3 X^2 Y - v_4 X, \\ Y' = v_2 BX - v_3 X^2 Y. \end{cases}$$

We start by eliminating the two equations governing the production of species D and E, since the four others do not depend on them. The system can be furthermore simplified by assuming that A and B are kept constant and taking all reaction rates equal to 1. The resulting system of two equations with two unknowns can be written as the initial value problem

$$\begin{cases} U'(t) = F(U(t)), \\ U(0) = U_0 = (X_0, Y_0)^T, \end{cases} \qquad (2.2)$$

where $U(t) = (X(t), Y(t))^T$ is the vector modeling the variations of concentration of substances X and Y, and

$$F(U) = \begin{pmatrix} A - (B+1)X + X^2 Y \\ BX - X^2 Y \end{pmatrix}.$$

2.2 Stability of the System

The stability of the system is its propensity to evolve toward a *constant* or *steady* solution. This steady solution $U(t) = U_c$, if it exists, satisfies $U'(t) = 0$, and can, therefore, be calculated by solving $F(U_c) = 0$. The solution U_c is called a *critical* or *stationary point*. In the above example, it is easy to compute: $U_c = (A, B/A)^T$. The stability of the system can also be regarded as its ability to relax to the steady state when the initial condition consists of a small perturbation of the critical state, $U_0 \approx U_c$. In order to study the influence of these perturbations, the right-hand side of the system is linearized around the critical point using a Taylor expansion

$$U'(t) = F(U) = F(U_c) + \nabla F_{U=U_c}(U - U_c) + \mathcal{O}(\|U - U_c\|^2),$$

where ∇F is the Jacobian matrix

$$\nabla F = \begin{pmatrix} \dfrac{\partial F_1}{\partial X} & \dfrac{\partial F_1}{\partial Y} \\[2mm] \dfrac{\partial F_2}{\partial X} & \dfrac{\partial F_2}{\partial Y} \end{pmatrix} = \begin{pmatrix} 2XY - (B+1) & X^2 \\ B - 2XY & -X^2 \end{pmatrix}.$$

Assuming that $U(t) = U_c + \Delta(t)$, with $\Delta(t)$ a small variation, the term $\mathcal{O}(\|U - U_c\|^2)$ can be neglected, leading to the linear differential system

$$\begin{cases} \Delta'(t) = J\Delta(t), \\ \Delta(0) = \Delta_0, \end{cases} \tag{2.3}$$

where J is the Jacobian matrix

$$J = \nabla F_{U=U_c} = \begin{pmatrix} B - 1 & A^2 \\ -B & -A^2 \end{pmatrix}.$$

When J is diagonalizable, it can be decomposed as $J = MDM^{-1}$, where D is the diagonal matrix containing the eigenvalues of J, $D_{ij} = \lambda_i \delta_{ij}$. Its integer powers are $J^n = MD^n M^{-1}$ for $n > 0$. We recall the definition of the exponential of a matrix J (see for instance Allaire and Kaber (2006)):

$$e^J = \sum_{n=0}^{\infty} \frac{1}{n!} J^n = \sum_{n=0}^{\infty} \frac{1}{n!} MD^n M^{-1} = M \left(\sum_{n=0}^{\infty} \frac{1}{n!} D^n \right) M^{-1} = Me^D M^{-1},$$

where e^D is the diagonal matrix $(e^D)_{ij} = \delta_{ij} e^{\lambda_i}$, formed out of the exponential of the eigenvalues of the matrix J. The differential system (2.3) can be directly integrated

$$\Delta(t) = e^{tJ} \Delta_0. \tag{2.4}$$

The long-time behavior of $\Delta(t)$ is obtained by making $t \longrightarrow +\infty$ in (2.4). If all eigenvalues λ of J have negative real part, then $e^{\lambda t} \to 0$ as $t \to +\infty$. Therefore, the matrix $e^{Jt} = Me^{Dt} M^{-1} \to 0$ as $t \to +\infty$ and the solution $\Delta(t)$ goes to 0. The Taylor expansion around the critical point is valid in this case, and the solution of the nonlinear system (2.2) tends toward the critical point.

In this very simple example, the eigenvalues of the matrix J can be explicitly calculated as the roots of the characteristic polynomial. The reader can easily verify that they are

$$\lambda_\pm = \frac{B - A^2 - 1 \pm \sqrt{\Delta}}{2}, \quad \text{with } \Delta = (A^2 - B + 1)^2 - 4A^2,$$

and that their real part is negative if $B < A^2 + 1$.

The numerical method described in the following exercise can also be used to provide a stability criterion. This can be useful in a more general case when the eigenvalues cannot be calculated explicitly.

Exercise 2.1 Write a program to display, as a function of B, the maximum of the real part of the eigenvalues of the matrix J. The parameter A is kept constant. Mark out the value of the stability criterion, which is the abscissa at which the curve crosses the horizontal axis.

A solution of this exercise is proposed in Sect. 2.5 at page 44.

In order to solve the system of differential equations (2.2), we could implement one of the numerical integration schemes proposed in Table 1.1, Chap. 1. Another possibility is to use already available programs, for instance, one of the ODE solvers proposed in MATLAB.

Exercise 2.2 Compute the approximated solutions for different choices of parameter B corresponding to stability and instability. In each case, display graphically the solutions X and Y as a function of time, and in another figure, Y as a function of X, that is, the parametric curve $(X(t), Y(t))_t$.

A solution of this exercise is proposed in Sect. 2.5 at page 44.

2.3 Model for the Maintained Reaction

Consider now the system (2.1) with the hypothesis that component B is injected in the mixture at rate v. The concentration of B as a function of time is denoted by $Z(t)$. The system of chemical reactions reduces to a new system of three equations:

$$\begin{cases} X' = A - (Z + 1)X + X^2Y, \\ Y' = XZ - X^2Y, \\ Z' = -XZ + v. \end{cases} \tag{2.5}$$

2.3.1 Existence of a Critical Point and Stability

The problem (2.5) now admits a steady solution corresponding to the critical point $U_c = (A, v/A^2, v/A)^T$. The Jacobian matrix of the right-hand-side function of the system (2.5) is

$$\nabla F = \begin{pmatrix} -(Z+1)+2XY & Z-2XY & -Z \\ X^2 & -X^2 & 0 \\ -X & X & -X \end{pmatrix}.$$

In order to study the stability of the system, this matrix is evaluated at the critical point

$$J = \begin{pmatrix} v/A-1 & 1 & -1 \\ -v & -1 & 1 \\ -v & 0 & -1 \end{pmatrix}.$$

Exercise 2.3 Find numerical values of v corresponding to the stable or unstable behavior of the system. Hint: the numerical method proposed in Exercise (2.1) can be used again.
A solution of this exercise is proposed in Sect. 2.5 at page 45.

2.3.2 Numerical Solution

Exercise 2.4 Solve the system (2.5) numerically for the following values of v: 0.9, 1.3, and 1.52. For each case, display in separate figures the three concentrations versus time and concentrations Y and Z versus X.
A solution of this exercise is proposed in Sect. 2.5 at page 46.

2.4 Model of Reaction with a Delay Term

An example of a more complicated chemical reaction is proposed in Hairer, Norsett, and Wanner (1987). An additional component I is introduced at a constant rate into the system, initiating a chain reaction.

$$I \longrightarrow Y_1 \xrightarrow{z} Y_2 \xrightarrow{k_2} Y_3 \xrightarrow{k_3} Y_4 \xrightarrow{k_4}$$

The quantity of the final product Y_4 slows down the first step of the reaction $Y_1 \rightarrow Y_2$. Fine modeling of this process, taking into account the transport time and diffusion properties of molecules, leads to a *delayed* ODE system:

$$\begin{cases} y_1'(t) = I - z(t)y_1(t), \\ y_2'(t) = z(t)y_1(t) - y_2(t), \\ y_3'(t) = y_2(t) - y_3(t), \\ y_4'(t) = y_3(t) - 0.5y_4(t), \\ z(t) \;\; = \dfrac{1}{1 + \alpha y_4(t - t_d)^3}, \end{cases} \tag{2.6}$$

where t_d is the time delay parameter. This system has a critical point Y_c, which is, once again, determined by solving $F(Y_c) = 0$:

$$Y_c = \begin{pmatrix} I(1 + 8\alpha I^3) \\ I \\ I \\ 2I \end{pmatrix}. \tag{2.7}$$

As in the previous section, the system can be linearized around this point. The stability of the resulting system can then be studied by introducing a fifth variable $y_5(t) = y_4(t - t_d)$. The Jacobian of the right-hand-side function

$$F(y) = \begin{pmatrix} I - \dfrac{y_1}{1 + \alpha y_5^3} \\ \dfrac{y_1}{1 + \alpha y_5^3} - y_2 \\ y_2 - y_3 \\ y_3 - 0.5y_4 \end{pmatrix}$$

can then be easily calculated at the critical point

$$\nabla F(Y_c) = \begin{pmatrix} -\bar{z} & 0 & 0 & 0 & 12\alpha I^3 \bar{z} \\ \bar{z} & -1 & 0 & 0 & -12\alpha I^3 \bar{z} \\ 0 & 1 & -1 & 0 & 0 \\ 0 & 0 & 1 & -0.5 & 0 \end{pmatrix},$$

where

$$\bar{z} = \frac{1}{1 + 8\alpha I^3}.$$

The small variations $\Delta(t)$ around the critical point Y_c satisfy to a first-order approximation the linear system of ODEs

$$\begin{aligned} \Delta_1'(t) &= -\bar{z}\Delta_1(t) + 12\alpha I^3 \bar{z}\Delta_4(t - t_d), \\ \Delta_2'(t) &= \bar{z}\Delta_1(t) - \Delta_2(t) - 12\alpha I^3 \bar{z}\Delta_4(t - t_d), \\ \Delta_3'(t) &= \Delta_2(t) - \Delta_3(t), \\ \Delta_4'(t) &= \Delta_3(t) - 0.5\Delta_4(t). \end{aligned} \tag{2.8}$$

An expression for $\Delta(t)$ is sought of the form $\Delta(t) = Ve^{xt}$, where V is a constant vector of $\mathbb{R}^4$. Plugging this ansatz into (2.8) leads to the characteristic equation

$$(x+1)^2(x+0.5)(x+\bar{z}) + 12\alpha\bar{z}I^3 xe^{-xt_d} = 0. \tag{2.9}$$

The corresponding system is stable if all roots have a negative real part.

Exercise 2.5 Set $\alpha = 0.0005$, $t_d = 4$, and solve numerically the characteristic equation for $x \in \mathbb{C}$ for different values of I between 0 and 20. Estimate a minimum value of the parameter I beyond which unstable equilibrium solutions can be obtained.
A solution of this exercise is proposed in Sect. 2.5 at page 48.

In order to illustrate numerically the instability phenomenon, the full system (2.6) has to be integrated. Standard solvers for ODE cannot be used, since they assume the generic form

$$\begin{cases} y'(t) = F(t, y(t)), \\ y(0) = u_0, \end{cases} \tag{2.10}$$

whereas in our case the right-hand side depends on the solution at a previous time,

$$\begin{cases} y'(t) = G(t, y(t), y(t - t_d)), \\ y(t) = u_0, \quad \text{for} \quad t \leq 0. \end{cases} \tag{2.11}$$

In this example, the function G is the vector function

$$G : \mathbb{R} \times \mathbb{R}^4 \times \mathbb{R}^4 \longrightarrow \mathbb{R}^4,$$

$$(t, u, v) \longmapsto G(t, u, v) = \begin{pmatrix} I - \dfrac{u_1}{1 + \alpha v_4^3} \\ \dfrac{u_1}{1 + \alpha v_4^3} - u_2 \\ u_2 - u_3 \\ u_3 - 0.5u_4 \end{pmatrix}. \tag{2.12}$$

Numerical schemes well adapted to systems of standard type (2.10) have to be modified to handle the time delay. We start with the simplest case of the explicit Euler scheme

$$\begin{Vmatrix} \text{initialization}: & y_0 = u_0 \\ \text{for} \quad i = 0, 1, \ldots, n - 1 \quad \text{do} \\ \qquad y_{i+1} = y_i + hF(t_i, y_i) \\ \text{end} \end{Vmatrix}$$

This scheme provides a first-order approximation $y_n \approx y(t_n)$, with $t_n = nh$ for h sufficiently small (see Butcher (1987)). It can be easily adapted to the system with delay term (2.11) if the time delay t_d (here $t_d = 4$) is an integer multiple of the time step h, i.e., $t_d = dh$ with $d \in \mathbb{N}$,

$$\left\|\begin{array}{l} \text{initialization}: \quad y_0 = u_0 \\[4pt] \text{for} \quad i = 0, 1, \ldots, n-1 \quad \text{do} \\[6pt] \qquad \gamma_i = \begin{cases} u_0 & \text{if} \quad i < d \\ y_{i-d} & \text{elsewhere} \end{cases} \\[10pt] \qquad y_{i+1} = y_i + h G(t_i, y_i, \gamma_i) \\[4pt] \text{end} \end{array}\right. \qquad (2.13)$$

Exercise 2.6 1. The solution is supposed to be constant and equal to y_0 for all $t \le 0$. Write a function ODE_F_DelayEnzyme(t,Y,h,y0) returning the value $G(t, y(t), y(t - t_d))$ given by (2.12). Hints: the values of y are discretized with the time step $h = t_{max}/n$ and stored in an array Y. The delay parameter t_d is a global variable set equal to 4 in the calling script.
2. Write a function ODE_F_EulerDelay(fdelay,tmax,nmax,y0) implementing the algorithm (2.13) to compute an approximation of the solution at time tmax in nmax time steps. The name of the right-hand-side function ODE_F_DelayEnzyme is passed as input argument fdelay.
3. Write a main script to integrate the system using the algorithm (2.13) up to $t_{max} = 160$. The value of α is fixed at 0.0005 and I is chosen in the range corresponding to instability. The initial condition u_0 should be chosen close to the equilibrium solution Y_c. Display graphically the four components of the solution versus time in one figure and the components y_i, $i = 2, \ldots, 4$, versus the component y_1 in another figure.
A solution of this exercise is proposed in Sect. 2.5 at page 49.

In the case of a Runge–Kutta-type scheme, intermediate values needed for the computation of y_{i+1} must be stored. The standard fourth-order Runge–Kutta scheme presented in Chap. 1 will be adapted to our problem. We start by rewriting the scheme such as to compute explicitly the intermediate parameters of the right-hand-side function instead of the values of the function itself

$$
\begin{aligned}
\text{initialization:}\quad & y_0 = u_0 \\
\text{for}\quad & i = 0, 1, \ldots, n-1 \quad \text{do} \\
& g^1 = y_i, \\
& g^2 = y_i + \frac{h}{2} F(t_i, g^1), \\
& g^3 = y_i + \frac{h}{2} F(t_i, g^2), \\
& g^4 = y_i + h F(t_i, g^3), \\
& y_{i+1} = y_i + \frac{h}{6} \left(F(t_i, g^1) + 2F\left(t_i + \frac{h}{2}, g^2\right) \right. \\
& \qquad\qquad \left. + 2F\left(t_i + \frac{h}{2}, g^3\right) + F\left(t_i + h, g^4\right) \right).
\end{aligned}
$$

$$\text{end}$$

To adapt this algorithm to the case (2.11), the values g^k, $k = 1, \ldots, 4$, should be stored as functions of time. They are needed to compute the intermediate values for the third input argument of the system function G, which holds the values of the solution at the delayed time $t - t_d$. This leads to the following algorithm:

$$
\begin{aligned}
\text{initialization:}\quad & y_0 = u_0 \\
\text{for}\quad & i = 0, 1, \ldots, n-1 \quad \text{do} \\
& g_i^1 = y_i, \\
& g_i^2 = y_i + \frac{h}{2} G(t_i, g_i^1, \gamma_i^1), \\
& g_i^3 = y_i + \frac{h}{2} G(t_i, g_i^2, \gamma_i^2), \\
& g_i^4 = y_i + h G(t_i, g_i^3, \gamma_i^3), \\
& y_{i+1} = y_i + \frac{h}{6} \left(G(t_i, g_i^1, \gamma_i^1) + 2G\left(t_i + \frac{h}{2}, g_i^2, \gamma_i^2\right) \right. \\
& \qquad\qquad \left. + 2G\left(t_i + \frac{h}{2}, g_i^3, \gamma_i^3\right) + G(t_i + h, g_i^4, \gamma_i^4) \right), \\
& \text{where}\quad \gamma_i^k = \begin{cases} u_0 & \text{if } i + c_k \leq d, \\ g_{i-d}^k & \text{otherwise,} \end{cases} \\
& \text{with}\quad c = \begin{pmatrix} 0 & 0.5 & 0.5 & 1 \end{pmatrix}^T.
\end{aligned}
$$

$$\text{end}$$

(2.14)

Exercise 2.7 1. Write a function ODE_F_DelayRungeKutta with input arguments fdelay,tmax,nmax, and y0 implementing the algorithm (2.14) to compute an approximation of the solution at time tmax in nmax time steps.

2. Compare graphically the solutions obtained using respectively the Euler and Runge–Kutta schemes. For $t_{max} = 16$, plot the two solutions obtained using $n_{max} = 100$ and $n_{max} = 1000$. Compute the solution for $n_{max} = 5000$ and store it as a reference solution. Compute the error in L^∞ norm, as a function of h, by performing several calculations for different values of n_{max} varying between 100 and 2000.

3. Study the influence of the initial condition: plot the trajectories y_i, $i = 2,\dots,4$, as functions of y_1 with different colors for different initial conditions.

A solution of this exercise is proposed in Sect. 2.5 at page 49.

2.5 Solutions and Programs

Solution of Exercise 2.1

To illustrate graphically the stability criterion, the script ODE_F_stab2comp.m computes the eigenvalues of the Jacobian matrix of F at the critical point, using the MATLAB built-in function eigen. The maximum value of the real part of these eigenvalues is computed and stored for different values of the parameter B, the parameter A remaining fixed. An approximation of the stability criterion is the abscissa where the maximum value of the real part changes sign.

Solution of Exercise 2.2

The script ODE_F_Chemistry2.m uses the ODE solver ode45 available in MATLAB main distribution to integrate numerically the system of ODEs (2.2):

```
global A
global B
fun='ODE_F_2comp' ;
A=1;
B=0.9;
%
U0=[2;1]; % Initial condition
t0=0;      % initial time
t1=10;    % final time
[timeS1,solS1]=ode45(fun,[t0,t1],U0);
```

The above example corresponds to a stable case. The MATLAB function ode45 requires as input the following parameters:

- the right-hand-side vector function of the differential system (written in ODE_F_2comp.m),
- the time interval [t0,t1] over which the system is integrated,
- the solution U0 at initial time t0.

It returns as output the array timeS1 of discrete intermediate times at which the solver has computed the corresponding solution solS1.

The A and B parameters of the differential system are declared as global in the main script and in the right-hand-side function ODE_F_2comp. Therefore, they do not need to be included in the list of input parameters of ODE_F_2comp when calling ode45. We first run the script with parameters corresponding to stability ($A = 1$ and $B = 0.9$), then with parameters corresponding to instability ($A = 1$ and $B = 3.2$). In the first run case, the concentrations tend, for large times, to a constant value, which is the critical point. This behavior is illustrated in Fig. 2.1. In the left panel, the trend of concentrations versus time is represented, showing that they rapidly stabilize to their critical values. The right panel shows the behavior of component Y versus the component X for two different initial conditions. The two trajectories converge toward the same critical point of coordinates $(A, B/A) = (1, 0.9)$.

The second choice of parameters, corresponding to instability, is illustrated in Fig. 2.2. In the left panel, the concentrations are displayed as functions of time. They remain bounded but exhibit periodic behavior. If the simulation is run over a long enough time, the graph of Y versus X represents the limit cycle. The right panel of Fig. 2.2 numerically illustrates that this cycle does not depend on initial conditions, but only on the parameters A and B. As they get closer to the instability limit ($B = A^2 + 1$), the limit cycle becomes smaller and eventually collapses into the critical point. This phenomenon is called *Hopf bifurcation* (see Hairer, Norsett, and Wanner (1987) for details).

Solution of Exercise 2.3

The script ODE_M_stab2comp.m that was written for Exercise 2.1 is modified to find the values of the parameter v for which all eigenvalues of the Jacobian matrix J have a negative real part. From the figure displayed by the script ODE_F_stab3comp.m, we find that only the first value $v = 0.9$ proposed in Exercise 2.4 corresponds to a stable case. For the values $v = 1.3$ and 1.52, some of the eigenvalues have a positive real part.

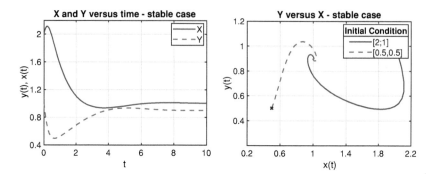

Fig. 2.1 Simplified Brusselator model, stable case $A = 1$, $B = 0.9$. Left: concentrations X and Y as a function of time. Right: parametric curves $(X, Y)_t$ for two different initial conditions (crosses), $(2, 1)^T$ and $(0.5, 0.5)^T$.

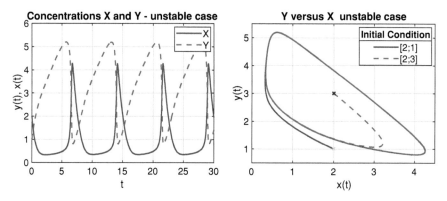

Fig. 2.2 Simplified Brusselator model, unstable case $A = 1$, $B = 3.2$. Left: concentrations X and Y as a function of time. Right: parametric curves $(X, Y)_t$ for two different initial conditions (crosses), $(2, 1)^T$ and $(2, 3)^T$.

Solution of Exercise 2.4

The integration of the full system with three equations is performed in the script ODE_F_Chemistry3.m. The right-hand-side function is defined in ODE_F_3comp.m. For $v = 0.9$, the system is stable; therefore, all three concentrations tend toward their equilibrium value ($U_c = (1, 0.9, 0.9)^T$) as shown in the left panel of Fig. 2.3. The right panel, which shows the variations of Y as a function of X, also points out the convergence to the critical point, starting from several different initial conditions.

For $v = 1.3$, the system is unstable, but the concentrations remain bounded. Their variation as a function of time tends toward periodic behavior, *i.e.*, a limit cycle, as displayed in Fig. 2.4. The solution can tend toward the limit cycle from inside or from outside, depending on the initial condition. Eventually, for $v > 1.5$, the system is unstable and divergent, and the

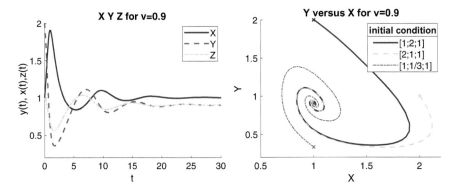

Fig. 2.3 Brusselator model, stable case $v = 0.9$. Left: concentrations X, Y, and Z as a function of time. Right: parametric curves $(X, Y)_t$ for three different initial conditions (crosses), $(1, 2, 1)^T$, $(2, 1, 1)^T$ and $(1; 1/3; 1)^T$.

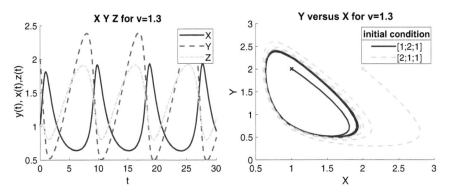

Fig. 2.4 Brusselator model, unstable periodic case $v = 1.3$. Left: concentrations X, Y, and Z as a function of time. Right: parametric curves $(X, Y)_t$ for two different initial conditions (crosses), $(1, 2, 1)^T$ and $(2, 2, 2)^T$.

values of the concentrations y and z are unbounded for large times while the concentration x goes to 0. The global behavior is completely different from the previous case. In particular, there is no limit cycle of y as a function of x or of z as a function of x (Fig. 2.5).

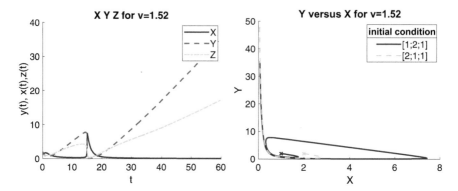

Fig. 2.5 Brusselator model, unstable divergent case $v = 1.52$. Left: concentrations X, Y, and Z as a function of time. Right: parametric curves $(X, Y)_t$ for two different initial conditions (crosses), $(1, 2, 1)^T$ and $(2, 2, 2)^T$.

Solution of Exercise 2.5

This nonlinear equation can be solved numerically using the MATLAB built-in function fsolve, as proposed in the script ODE_F_StabDelay.m displayed below

```
clear %important to reinitialize the Matlab  square root of (-1)
td=4;
alpha=0.0005;
for  I=0:20
 bz=1/(1+8*alpha*I^3);
 funtext='(x+1)^2*(x+0.5)*(x+bz)+12 *alpha*I^3*bz*x*exp(-td*x)';
 funequi=inline(funtext,'x','bz','I','alpha','td');
 guess=i/2;
 x0=fsolve(funequi,guess,optimset('Display','off'),bz,I,alpha,td);
 fprintf('I=%f x0=%f+i%f n',I,real(x0),imag(x0))
end
```

Running the script with a real value as initial guess for the fsolve function (guess=2 for instance) will provide a negative real solution. This corresponds to a stable equilibrium, since deviations from the critical point decay exponentially to zero. Conversely, if we run the script with a pure imaginary initial guess, the root found by the solver is complex, with a nonzero imaginary part. Choose guess=i/2 as in the above example, and let I vary to obtain a solution with a real part that will be positive for values of $I > 9$. The equilibrium for this parameter choice is unstable since the deviations increase exponentially. On the other hand, stability is not ensured for $I < 9$, since not all the roots of equation (2.9) necessarily have a negative real part.

Solution of Exercise 2.6

The main script ODE_M_Enzyme.m calls the function ODE_F_EulerDelay, which implements the Euler scheme (2.13) adapted to the delayed equation. The right-hand-side function $G(t, y, \gamma)$ is programmed in the file ODE_F_DelayEnzyme.m. The selected value I=10 corresponds to an instability. The initial condition is fixed by adding a small deviation to the unstable equilibrium solution (2.7). In the right panel of Fig. 2.6, the trajectories are superimposed, which indicates the periodic character of the solution. The length of the period can be graphically estimated in the left panel of Fig. 2.6 to a value close to 13, which roughly corresponds to one of the phases obtained in solving the characteristic equation (2.9).

In contrast, when setting I=5, we observe that the solutions tend to equilibrium.

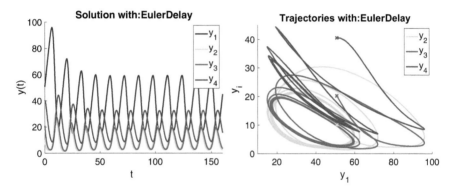

Fig. 2.6 Solutions of the system (2.6) obtained using the Euler scheme. Left: $y(t)$ versus t. Right: trajectories y_i, $i = 2, \ldots, 4$, versus y_1. Crosses indicate initial conditions.

Solution of Exercise 2.7

The delayed system of equations is now integrated with the fourth-order Runge–Kutta scheme programmed in the function ODE_F_RungeKuttaDelay. Calls to ODE_F_EulerDelay should be replaced by ODE_F_RungeKuttaDelay in the script ODE_M_Enzyme.m. This is done by changing the assignment of the variable scheme. In order to implement the Runge–Kutta scheme for delayed ODEs (2.14), we introduce a triple-index array g(:,n,k) for $k = 1, \ldots, 4$. It is used to store the four intermediate values $(g^k)_{k=1}^4$ as a function of time, so that it can be passed as input parameter to the right-hand-side function ODE_F_DelayEnzyme.

A more detailed study of the convergence order of the two schemes is proposed in the script ODE_M_ErrorEnzyme.m. Unlike the case studied in Exercise

1.1, we do not have access here to the exact solution to calculate the numerical error. Reference solutions for each scheme are computed with a very fine discretization. They are used as exact solutions to evaluate the error on the solution at the final time $t_{max} = 50$ when coarser discretizations are used. In Fig. 2.7, the variations of the error with the discretization parameter h are represented in logarithmic scale, along with the theoretical convergence orders $\mathcal{O}(h)$ and $\mathcal{O}(h^4)$ for comparison.

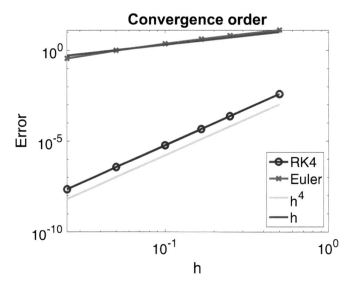

Fig. 2.7 Error in L^∞ norm as a function of the time step at time $t = 50$ for Euler and fourth-order Runge–Kutta schemes.

Finally, the influence of the initial condition is investigated by the script ODE_M_EnzymeCondIni. We display in the same figure the trajectories starting from different initial conditions, randomly chosen in the vicinity of the unstable equilibrium. Figure 2.8 shows that after an initial phase (of different lengths), they all converge to the same periodic trajectory. Indeed on the right panel, where only the last part of the trajectories are displayed, they are not distinguishable.

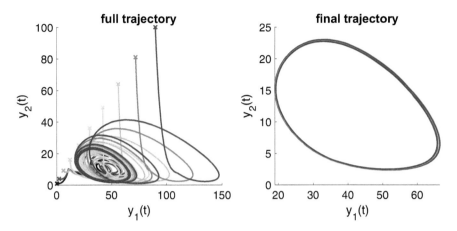

Fig. 2.8 Trajectories y_2 versus y_1 of solutions of system (2.6) obtained for four different initial conditions using the Runge–Kutta scheme. Left panel displays full trajectories (2000 time steps), with crosses at initial condition locations. Right panel displays only the last 300 time steps.

Chapter References

Numerous references for solving ODEs have already been cited in the previous chapter. The numerical examples treated in this project are directly selected from Hairer, Norsett, and Wanner (1987). We also recommand the book by Iserles (1996) for a comprehensive introduction to numerical methods for ODEs and the book by Bellen and Zennaro (2003) for advanced reading on delay equations.

G. Allaire, S. M. Kaber, *Numerical Linear Algebra*, forthcoming (Springer, New York, 2007)

A. Bellen, M. Zennaro, *Numerical Methods for Delay Differential Equations*, Numerical mathematics and scientific computation. (The Clarendon Press, Oxford University Press, New York, 2003)

J.C. Butcher, *The Numerical Analysis of Ordinary Differential Equations* (Wiley, 1987)

E. Hairer, S.P. Norsett, G. Wanner, *Solving Ordinary Differential Equations I, Nonstiff Problems*, Springer series in computational mathematics, vol. 8 (Springer, 1987)

A. Iserles, *A First Course in the Numerical Analysis of Differential Equations*, Cambridge texts in applied mathematics (Cambridge University Press, Cambridge, 1996)

Chapter 3
Fourier Approximation

Project Summary

Level of difficulty:	1
Keywords:	Polynomial approximation for periodic functions, interpolation, best approximations, Fourier series expansions
Application fields:	Approximation of functions

This project is devoted to the approximation by trigonometric polynomials of a given periodic real function f defined on an interval $I \subset \mathbb{R}$. Function f is assumed to be $2\pi-$periodic and I denotes either $[0, 2\pi]$ or $[-\pi, \pi]$. The set of trigonometric polynomials spanned by e^{ikx} for $k = -n, \cdots, n$ is denoted by $\mathbb{S}_n$. Definitions and results of this chapter are widely used in other projects of this book (see Chaps. 7 and 15). For further reading, we refer the reader to monographs on analysis and applications of Fourier approximation, for instance, Zygmund (2002), Trefethen (2000), Gottlieg and Orszag (1997), and many others.

3.1 Introduction

Approximation of a given function by a trigonometric polynomial is an efficient tool to solve many problems arising in applied mathematics: signal and image processing, partial differential equations, etc. *Approximation* may mean either *interpolation* or *best approximation*. We present in this section the main features of these approaches. Numerical algorithms are discussed in the following sections.

Interpolation

For integer $n \geq 0$, we define $2n+1$ interpolation or collocation points spanning $[0, 2\pi[$

$$x_j = jh, \quad h = \frac{2\pi}{2n+1}, \quad j = 0, \ldots, 2n. \tag{3.1}$$

The problem is to find a polynomial $\mathcal{I}_n f$ in $\mathbb{S}_n$ that coincides with the continuous function f at collocation points. Such a polynomial exists and is unique. Furthemore if $f \in \mathbb{S}_n$, then $\mathcal{I}_n f = f$. Associated to collocation points are the Lagrange (trigonometric) polynomials or fundamental polynomials defined by $\ell_j \in \mathbb{S}_n$ and $\ell_j(x_k) = \delta_{j,k}$ with δ the Kronecker symbol. In the periodic setting, all the Lagrange polynomials are just a translation of one of them

$$\ell_j(x) = \ell_0(x - x_j).$$

Indeed, we have $\ell_j(x) = \ell_0(0) = 1$, for $k > j$, $\ell_j(x_k) = \ell_0(x_k - x_j) = \ell_0(x_{k-j}) = 0$, and for $k < j$, $\ell_j(x_k) = \ell_0(x_k - x_j) = \ell_0(x_k - x_j + 2\pi) = \ell_0(x_m) = 0$, with $m = k - j + 2n + 1 \in [1, 2n]$. So, it is sufficient to define ℓ_0. We present here several equivalent definitions of this polynomial.

1. Geometric series. We have

$$\ell_0(x) = \frac{1}{2n+1} \sum_{k=-n}^{n} e^{ikx}. \tag{3.2}$$

Clearly, $\ell_0 \in \mathbb{S}_n$, $\ell_0(0) = 1$, and more generally $\ell_0(x) = 1$ if $e^{ix} = 1$. If $e^{ix} \neq 1$, we write the sum as a geometric series

$$(2n+1)\ell_0(x) = e^{-inx} \sum_{k=0}^{2n} (e^{ix})^k = e^{-inx} \frac{1 - e^{i(2n+1)x}}{1 - e^{ix}},$$

and infer that $\ell_0(x_k) = 0$ for $k = 1, \cdots, 2n$.
2. The Dirichlet kernel D_n is defined for $x \neq 0$ by

$$D_n(x) = \frac{\sin[(n + \frac{1}{2})x]}{\sin(\frac{1}{2}x)}, \tag{3.3}$$

and $D_n(0) = \lim_{x \to 0} D_n(x) = 2n + 1$. For $x \in]0, 2\pi[$, we have

$$\ell_0(x) = \frac{e^{-inx} - e^{i(n+1)x}}{(2n+1)(1 - e^{ix})} = \frac{e^{-i(n+1/2)x} - e^{i(n+1/2)x}}{(2n+1)(e^{-ix/2} - e^{ix/2})} = \frac{D_n(x)}{2n+1}. \tag{3.4}$$

3. In terms of the cardinal sine function $\mathrm{sinc}(x) = (\sin x)/x$, we have

$$\ell_0(x) = \frac{\mathrm{sinc}((n + 1/2)x)}{\mathrm{sinc}(x/2)}.$$

4. Explicit form using the collocation points. We have

$$\ell_0(x) = \prod_{k=1}^{2n} \frac{\sin(\frac{x-x_k}{2})}{\sin(\frac{x_0-x_k}{2})}.$$

Indeed, this function takes the value 1 for $x = 0$ and 0 at all the other collocation points. Furthermore, it belongs to $\mathbb{S}_n$. The reader is invited to check this symmetric form of ℓ_j

$$\ell_j(x) = \ell_0(x - x_j) = \prod_{k \neq 0} \frac{\sin(\frac{x-x_j-x_k}{2})}{\sin(\frac{x_0-x_k}{2})} = \prod_{k \neq j} \frac{\sin(\frac{x-x_k}{2})}{\sin(\frac{x_j-x_k}{2})}.$$

Expressions (3.2) and (3.4) of ℓ_0 are suitable for the analysis while the two last ones are more convenient for computations. The set $(\ell_j)_{j=0}^{2n}$ forms a basis of $\mathbb{S}_n$. Examples of trigonometric Lagrange polynomials are displayed in Fig. 3.1.

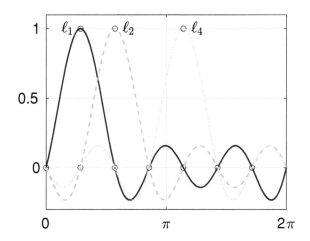

Fig. 3.1 Examples of Lagrange polynomials in the periodic case (7 interpolation points).

The Lagrange basis is mainly used to write in a very simple way the Lagrange polynomial interpolant as

$$\mathcal{I}_n f(x) = \sum_{i=0}^{2n} f(x_i) \ell_i(x). \tag{3.5}$$

We first address the stability of the Lagrange interpolation. The Lebesgue constant associated to points x_i is

$$\Lambda_n = \max_{x \in I} \sum_{i=0}^{2n} |\ell_i(x)|. \tag{3.6}$$

It is important to notice that Λ_n does not depend on the function being interpolated, but only on interpolation points. From (3.5) and (3.6), we obtain the bound $\|\mathcal{I}_n f\|_\infty \leq \Lambda_n \|f\|_\infty$, where $\|.\|_\infty$ denotes the supremum norm on I. Hence, for any continuous functions f and g, we have

$$|\mathcal{I}_n f(x) - \mathcal{I}_n g(x)| \leq \Lambda_n \max_{0 \leq i \leq 2n} |f(x_i) - g(x_i)|. \tag{3.7}$$

Suppose that function g is very close to f at collocation points $(g(x_i) = f(x_i) + \varepsilon_i)$, then

$$\max_{x \in I} |\mathcal{I}_n f(x) - \mathcal{I}_n g(x)| \leq \varepsilon \Lambda_n,$$

with $\varepsilon = \max_i |\varepsilon_i|$. Therefore, the Lebesgue constant is a measure of the amplification of the error in the Lagrange interpolation. It is the *stability constant* of the Lagrange interpolation procedure. The following result holds.

Proposition 3.1 *Independently of the interpolation points*

$$\lim_{n \to +\infty} \Lambda_n = +\infty. \tag{3.8}$$

Hence, small perturbations on the data (small ε) can lead to very large variations in the solution $\mathcal{I}_n f$. This is the typical case of an *ill-conditioned* problem.

Best approximation

In this section, we search for a polynomial that is closest to f for a prescribed norm $\|.\|_\mathcal{X}$, where $\mathcal{X}$ is a linear functional space that contains the polynomials. For $f \in \mathcal{X}$, the problem is to find a polynomial that is a best polynomial approximation of f in $\mathbb{S}_n$, i.e. $p_n \in \mathbb{S}_n$ such that

$$\|f - p_n\|_\mathcal{X} = \inf_{q \in \mathbb{S}_n} \|f - q\|_\mathcal{X}. \tag{3.9}$$

Two questions are of interest: what class of functions satisfy $\lim_{n \to +\infty} \|f - p_n\|_\mathcal{X} = 0$? For such functions, how fast this quantity decreases to 0 as n increases? The most commonly considered best approximation problems are the least square approximation and the uniform approximation.

- <u>Least square approximation</u> (or Hilbertian approximation, or L^2 approximation). In the periodic case, $\mathcal{X}$ is $L^2_\#(I)$ the space of square integrable 2π-periodic functions. The Hermitian inner product and associated norms are

$$\langle f, g \rangle = \int_0^{2\pi} f(x)\overline{g(x)}dx, \quad \|f\|_2 = \sqrt{\langle f, f \rangle}, \quad f, g \in L^2_\#(I), \quad (3.10)$$

with $\overline{z}$ the complex conjugate of $z \in \mathbb{C}$. The solution is a polynomial $\pi_n f \in \mathbb{S}_n$ such that $\pi_n f - f$ is orthogonal to $\mathbb{S}_n$. In other words, $\pi_n f$ is the orthogonal projection of f onto the polynomial space $\mathbb{S}_n$, see Fig. 3.2.

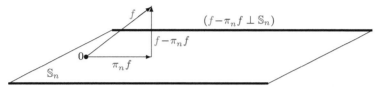

Fig. 3.2 The best Hilbertian approximation of f is its orthogonal projection on $\mathbb{S}_n$.

- Uniform approximation. In this case, $\mathcal{X} = \mathcal{C}_\#(I)$ the set of 2π−periodic continuous functions over I. The associated norm is the uniform norm:

$$\|f\|_\infty = \max_{x \in I} |f(x)|, \quad (f \in \mathcal{C}_\#(\mathcal{I})).$$

Uniform approximation is also called L^∞ or Chebyshev approximation. Let us recall the celebrated Weierstrass approximation theorem.

Theorem 3.1 (Weierstrass approximation theorem (Fourier)) *Every real-valued continuous periodic function is a uniform limit of trigonometric polynomials.*

Thus, for $f \in \mathcal{C}_\#(I)$, the best approximation error of f in $\mathbb{S}_n$, in the uniform norm

$$E_{n,\infty}(f) = \inf_{q \in \mathbb{S}_n} \|f - q\|_{\mathcal{X}} \tag{3.11}$$

decreases to 0 as $n \to +\infty$. For every $q \in \mathbb{S}_n$, we can write $f(x) - \mathcal{I}_n f(x) = f(x) - q(x) - (\mathcal{I}_n f - q)(x)$ and derive the following uniform bound of the error.

Proposition 3.2 *For any continuous function f defined on I,*

$$\|f - \mathcal{I}_n f\|_\infty \le (1 + \Lambda_n)E_{n,\infty}(f).$$

Remark 3.1 The global error $\|f - \mathcal{I}_n f\|_\infty$ is bounded by the product of two terms:

- Λ_n, that does not depend on the function f and always tends to $+\infty$ when $n \to +\infty$ (by Proposition 3.1),
- $E_{n,\infty}(f)$, that does not depend on interpolation points and decreases to 0 (by Theorem 3.1). The rate of convergence of $E_{n,\infty}(f)$ toward 0 increases with the regularity of function f.

In conclusion, the Lagrange interpolation process converges uniformly if the product $\Lambda_n E_{n,\infty}(f) \to 0$.

3.2 Fourier Series

We point out in this section some properties of the Fourier series. Let f be a 2π-periodic real function, $f : \mathbb{R} \to \mathbb{R}$, that admits a Fourier series expansion

$$f = \frac{1}{2}a_0 + \sum_{k=1}^{+\infty}(a_k \cos(kx) + b_k \sin(kx)), \tag{3.12}$$

with real Fourier coefficients

$$a_k = \frac{1}{\pi}\int_0^{2\pi} f(x)\cos(kx)dx, \quad b_k = \frac{1}{\pi}\int_0^{2\pi} f(x)\sin(kx)dx.$$

For practical reasons, the complex basis $(e^{ikx})_k$ is mostly used. The Fourier series expansion is then

$$f(x) = \sum_{k=-\infty}^{+\infty} \hat{f}_k e^{ikx}, \tag{3.13}$$

with complex Fourier coefficients

$$\hat{f}_k = \frac{1}{2\pi}\int_0^{2\pi} f(x)e^{-ikx}dx, \quad k = 0, \pm 1, \pm 2, \cdots \tag{3.14}$$

Expansions (3.12) and (3.13) are related by: $a_0 = 2\hat{f}_0$ and for $k \geq 1$ $a_k = \hat{f}_k + \hat{f}_{-k}$ and $b_k = i(\hat{f}_k - \hat{f}_{-k})$. It is useful to take advantage of the symmetries of the function f in order to save computations: for odd functions, $a_k = 0$, and for even functions, $b_k = 0$. In the following exercise, we show examples of Fourier series expansions.

Example 3.1 Consider the following two 2π-periodic functions with different regularity. A continuous, but not differentiable, function defined on $[0, 2\pi]$ by $f_1(x) = |x - \pi|$ and a differentiable, but not twice differentiable, function defined on $[-\pi, \pi]$ by $f_2(x) = 1 - x(\pi^2 - x^2)$. Indeed, the derivative of this function is $f_2'(x) = 3x^2 - \pi^2$, therefore, $f_2'(-\pi) = f_2'(\pi)$ and f_2' is continuous on $\mathbb{R}$, but the second derivative $f_2''(x) = 6x$ is not continuous at the ends of the interval $f_2''(-\pi) \neq f_2''(\pi)$.

Functions in Example 3.1, as well as their truncated Fourier series, are displayed in Fig. 3.3.

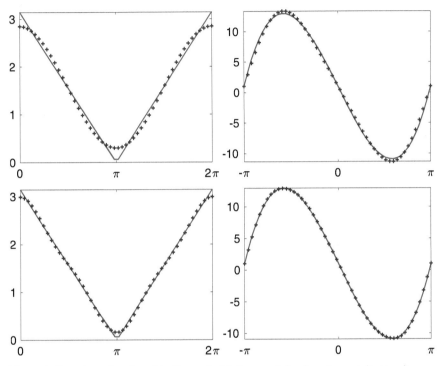

Fig. 3.3 Functions f and $\pi_n f$ in Example 3.1 for $n = 2$ (top) and $n = 4$ (bottom).

Remark 3.2 Differentiating (3.13) p times, we note that the Fourier coefficients of $f^{(p)}$, if they exist, are simply given by $\widehat{f^{(p)}}_k = (ik)^p \, \hat{f}_k$.

As the reader has noticed in previous examples, in both cases, we have $\lim_{|k| \to +\infty} \hat{f}_k = 0$. This is the Riemann–Lebesgue lemma for Fourier series. A natural question is: how fast the Fourier coefficients of a function tend to zero as $k \to +\infty$? The answer could be summarized as follows:

$$\text{smoother } f \iff \text{faster decay of } \hat{f}_k.$$

For example, if $f \in C^{p-1}$ (as a 2π−periodic function) and $f^{(p)}$ is piecewise differentiable then $\lim_{k \to +\infty} k^p \, \hat{f}_k = 0$, which means that $\hat{f}_k$ decreases faster than $1/|k|^p$ as k goes to $+\infty$. Since functions $(e^{ikx})_k$ are orthogonal on $[0, 2\pi]$ with respect to the Hermitian inner product (3.10), we have for $f \in L^2_\#$

$$\|f\|_2^2 = 2\pi \sum_{k \in \mathbb{Z}} |\hat{f}_k|^2.$$

This is Parseval's identity. The truncated Fourier series of order $n \in \mathbb{N}$ of f is

$$\pi_n f(x) = \sum_{k=-n}^{n} \hat{f}_k e^{ikx}. \tag{3.15}$$

Due to the orthogonality of the exponentials, we obtain that for all $g \in \mathbb{S}_n$

$$\langle f - \pi_n f, g \rangle = 0. \tag{3.16}$$

That is to say, $f - \pi_n f$ is orthogonal to $\mathbb{S}_n$ and $\pi_n f$ is the $L_\#^2$ orthogonal projection of f on $\in \mathbb{S}_n$ (see Fig. 3.2). Lastly, we recall Bessel's inequality:

$$\|\pi_n f\|_2 \leq \|f\|_2. \tag{3.17}$$

Exercise 3.1 (Fourier Series) What are the Fourier series expansions of functions f_1 and f_2 in Example 3.1? Write a program computing the Fourier coefficients of these functions and the corresponding truncated Fourier series $\pi_n f$. Obtain the plots displayed in Fig. 3.3.
A solution of this exercise is proposed in Sect. 3.8 at page 68.

3.3 Trigonometric Interpolation

The general expression (3.5) of $\mathcal{I}_n f$ is mostly used in theoretical analysis. For practical purposes, it is better to use another expression of $\mathcal{I}_n f$ that we introduce now. The discrete Fourier coefficients of a continuous function f associated to collocation points (3.1) are

$$\tilde{f}_k = \frac{1}{2n+1} \sum_{j=0}^{2n} f(x_j) e^{-ikx_j}, \quad k = -n, \dots, n. \tag{3.18}$$

These discrete Fourier coefficients $\tilde{f}_k$ are nothing but an approximation of the continuous Fourier coefficient $\hat{f}_k$ using the simplest quadrature formula, the rectangle rule. Indeed

$$\hat{f}_k = \sum_{j=0}^{2n} \frac{1}{2\pi} \int_{x_j}^{x_{j+1}} f(x) e^{-ikx} dx \simeq \frac{1}{2n+1} \sum_{j=0}^{2n} f(x_j) e^{-ikx_j} = \tilde{f}_k. \tag{3.19}$$

In practical applications, the Fast Fourier Transform (FFT) algorithm should be used to compute the Fourier coefficients, (see Sect. 3.6). Here, we found it more convenient to deal with an odd number of Fourier coefficients, just as in the mathematical analysis of Fourier series. It is easy to check that the Lagrange interpolant of f at collocation points x_j is

$$\mathcal{I}_n f(x) = \sum_{k=-n}^{n} \tilde{f}_k e^{ikx}. \tag{3.20}$$

Functions in Example 3.1, as well as their interpolants, are displayed in Fig. 3.4.

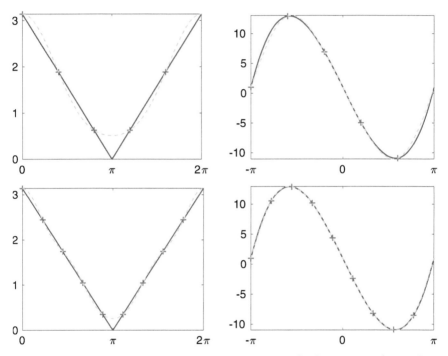

Fig. 3.4 Functions f and its interpolants $\mathcal{I}_n f$, for $n = 2$ (top) and $n = 4$ (bottom) for the functions used in Example 3.1. The interpolated values are marked with the symbol '+'.

Exercise 3.2 (Trigonometric Interpolation)

1. Discrete Fourier coefficients. The aim is to compare two ways to compute the discrete Fourier coefficients (3.18). The input consists of two arrays of length $N = 2n + 1$; the first one contains the collocation points x_j, and the second one, the values y_j of a function at these points.

   a. Method 1 (using loops). Write a function that computes the discrete
      Fourier coefficients using two loops `for`: the inner loop computes $\tilde{f}_k$ for
      $k = -n, \cdots, n$.
   b. Method 2 (using vectorization). Let $C \in \mathbb{R}^N$ be the vector with elements
      $\tilde{f}_k$ and $F \in \mathbb{R}^N$ the vector with elements y_j. What are the elements of
      the squared matrix A such that $C = AF$? Compute the discrete Fourier
      coefficients this way.
   c. Compare the two methods in terms of the required computing time. Use
      the MATLAB function `cputime`. *Application:* take $f(x) = x \exp(\sin(2x))$
      in $[0, 2\pi[$ and test several values of n. Conclude that vectorization is
      faster than using loops. It is good practice to use vectorization when it
      is possible.

2. Fourier Interpolation. We denote by $(z_m)_{m=1}^M$ the points where $\mathcal{I}_n f$ should
   be computed. $G \in \mathbb{R}^M$ defines the vector with elements $\mathcal{I}_n f(z_m)$. What
   are the elements of the matrix B such that $G = BC$? Write a program
   computing $\mathcal{I}_n f(x)$. The input consists in two arrays of the same length
   $N = 2n + 1$; one contains the discrete Fourier coefficients and the other
   contains the points z_m. Mark the points $(z_m, f(z_m))$ as in Fig. 3.4. *Application:* $f(x) = x \exp(\sin(2x))$, select $M = 6$ points in $[0, 2\pi[$ and several
   values for n.

A solution of this exercise is proposed in Sect. 3.8 at page 69.

Exercise 3.3 In 1846, Urbain Le Verrier discovered the planet Neptune, after
more than one year of tremendous computations explaining the discrepancies
observed in Uranus's orbit. *"Mr. Le Verrier saw the new star without needing
to cast a single glance toward the sky; he saw it at the tip of his pen"* said
François Arago. A major ingredient in the search for the new planet was the
computation of the inverse of the distance of two planets. It can be expressed
in a simplified way[1] as a function $f(\theta)$ depending on angles $\theta \in [0, 2\pi]$ and a
parameter ϱ related to the ratio α of distances to the Sun of the two planets.
More specifically

$$f(\theta) = \frac{1}{\sqrt{1 + \varrho\cos(\theta)}}, \qquad \varrho = \frac{2\alpha}{1 + \alpha^2}, \qquad \alpha = \frac{a}{a'} < 1.$$

We may fix $\alpha = 0.6$. Instead of computing $f(\theta)$ (that was hard to do accurately), Le Verrier used a truncation of the Fourier series expansion of f with
discrete Fourier coefficients computed by (3.18).

1. Fix $n = 7$ and $\alpha = 0.6$. Plot the function f and its truncated series $\pi_n f$
   on one graph and the absolute value of their difference on another graph.
2. For each n, the discrete Fourier coefficients $(\tilde{f}_k)_{k=-n}^n$ are now computed
   using a quadrature formula with $\bar{n}+1$ points: $\tilde{f}_k = \frac{1}{\bar{n}+1} \sum_{j=0}^{\bar{n}} f(x_j)e^{-ikx_j}$.

---

[1] J. Laskar, private communication.

Note that $\bar{n} = 2n$ corresponds to the previous case. The supremum norm of the difference $f - \pi_n f$ computed at 100 equidistant points in $[0, 2\pi]$ will be denoted $e_{n,\bar{n}}$. Fix again $n = 7$ and plot $e_{n,\bar{n}}$ as a function of $\bar{n}_0 = \bar{n}/2$ for $\bar{n}_0 = 5, 6, \cdots, 10$. Add to the plot a horizontal line $y = e_{n,n}$. Repeat computations for different values of $n = 3, 4, \cdots, 10$ and $\bar{n}_0 = n - 2, n - 1 \cdots, n + 3$. Comment on the results obtained.

A solution of this exercise is proposed in Sect. 3.8 at page 70.

3.4 L^2-Approximation

What is the best approximation, in the L^2-norm, of a given periodic function by trigonometric polynomials? From Eq. (3.16), we deduce that for all f and g in $\mathbb{S}_n$,

$$\|f - g\|_2^2 = \|f - \pi_n f\|_2^2 + \|\pi_n f - g\|_2^2.$$

Hence, the truncated series of order n of a function $f \in L_\#^2$ is its best least square approximation by polynomials in $\mathbb{S}_n$:

$$\|f - \pi_n f\|_2 \leq \|f - g\|_2, \qquad \forall g \in \mathbb{S}_n, \tag{3.21}$$

with strict inequality if $g \neq \pi_n f$. Let us now investigate how fast $\pi_n f$ converges to f in the L^2-norm. Consider the case of a function $u \in C^{p-1}$ such that $u^{(p)}$ admits a Fourier series expansion. By Parseval's identity and Remark 3.2, we obtain

$$\|f - \pi_n f\|_2^2 = 2\pi \sum_{|k|=n+1}^\infty |\hat{f}_k|^2 = \sum_{|k|=n+1}^\infty \frac{2\pi}{|k|^{2p}} |\widehat{f^{(p)}}_k|^2 \leq \frac{2\pi}{n^{2p}} \sum_{|k|=n+1}^\infty |\widehat{f^{(p)}}_k|^2.$$

It follows that

$$\|f - \pi_n f\|_2 \leq \frac{C}{n^p} \|f^{(p)}\|_2,$$

with a constant C. Thus, the error $\|f - \pi_n f\|_2$ decreases to 0 as a power of $1/n$. This power increases with the regularity of f. The regularity of f is measured here by the largest derivative of f that belongs to L^2. Figure 3.5 represents the error for three functions with different regularity: u_1 (continuous) and u_2 (differentiable) from Example 3.1, and f_3 (twice differentiable) defined on $[-\pi, \pi]$ by $f_3(x) = (x^2 - \pi^2)^2/8$. The choice of a log–log plot is motivated by the following argument: if the error $e_n = \|f - \pi_n f\|_2$ behaves like C/n^p, then $\log e_n$ should behave like a linear function of $\log(n)$. Hence, a log–log plot will produce a straight line with slope $(-p)$. To reproduce Fig. 3.5, it is important to clarify how the error is computed. First, the interval I is split into J subintervals $[\theta_j, \theta_{j+1}]$ of length $h = 1/J$; then each integral on subintervals is computed by a basic rectangle rule

$$\|\varphi\|_2^2 = \int_I \varphi(x)^2 dx = \sum_{j=1}^{J} \int_{I_j} \varphi(x)^2 dx \simeq h \sum_{j=1}^{J} \varphi(\theta_j)^2.$$

So, if $\varphi_J \in \mathbb{R}^J$ is the vector of elements $(\varphi(\theta_j))_{i=1}^{J}$ then $\|\varphi\|_2 \simeq \sqrt{h}\|\varphi_J\|_2$.

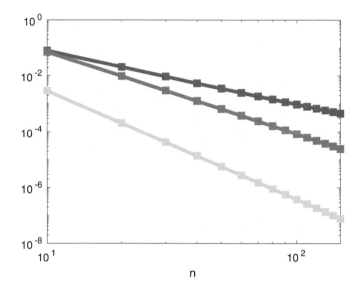

Fig. 3.5 Log–log plot of the error $\|f - \pi_n f\|_2$ versus n. Functions have different regularity: continuous function f_1 (top), C^1 function f_2 (middle), and C^2 function f_3 (bottom).

Exercise 3.4 Obtain a plot similar to Fig. 3.5. Take $n = 10, 20, \cdots, 150$. Compute the three slopes using the MATLAB function `polyfit`. Comment on the results.
A solution of this exercise is proposed in Sect. 3.8 at page 70.

Exercise 3.5 Same as Exercise 3.4 with the derivatives of functions f_2 and f_3. Use Remark 3.2 to compute the Fourier coefficients of the derivatives. Take $n = 10, 20 \cdots, 100$.
A solution of this exercise is proposed in Sect. 3.8 at page 71.

Exercise 3.6 Compute the Fourier coefficients of $f_4(x) = \frac{4\cos x - 2}{5 - 4\cos x}$. Add the corresponding error to the figure obtained for Exercise 3.4. Take $n = 10, 20, \cdots, 100$. Explain why the curve corresponding to f_4 is not a straight line. Plot $\log e_n$ as a function of n. A straight line should be obtained. What is its slope? Comment on this result.
A solution of this exercise is proposed in Sect. 3.8 at page 71.

3.5 The Dirichlet kernel

The Dirichlet kernel is a key tool in the analysis of Fourier approximation (see Eqs. (3.2)-(3.4))

$$D_n(x) = \sum_{k=-n}^{n} e^{ikx} = \frac{\sin[(n+\frac{1}{2})x]}{\sin(\frac{1}{2}x)}.$$

We have studied the convergence of $\pi_n f$ to f in the L^2-norm. What about pointwise convergence and the uniform convergence? Let us see how the use of this kernel allows us to handle this problem. Bessel inequality (3.17) tells us that $\pi_n f$ is smaller than f in the Hilbertian norm. Does the same hold with the maximum norm? To answer this question, we write $\pi_n f$ as a convolution with the Dirichlet kernel D_n. Recall that convolution of two 2π-functions φ and ψ is the function denoted by $\varphi * \psi$ and defined by $(\varphi * \psi)(x) = \frac{1}{2\pi} \int_0^{2\pi} \varphi(y)\psi(x-y)dy$.

Exercise 3.7

1. Show that *"convolution in the physical space x is equivalent to multiplication in the Fourier space k"*, that is to say $\widehat{(\varphi * \psi)}_k = \hat{\varphi}_k \hat{\psi}_k$. Show the identity $\pi_n f = D_n * f$.
2. The Lebesgue constant for the Fourier approximation is defined by $\Lambda_n = \frac{1}{\pi} \int_0^\pi |D_n(x)| dx$. Not to be mixed up with the Lebesgue constant for interpolation defined in (3.6). Prove the inequality $\|\pi_n f\|_\infty \le \Lambda_n \|f\|_\infty$. Compare with the Bessel inequality (3.17). Comment on.

A solution of this exercise is proposed in Sect. 3.8 at page 72.

Since $D_n = (2n+1)\ell_0$, we have $|\mathcal{I}_n f(x)| \le \Lambda_n \|f\|_\infty$. We plot in Fig. 3.6 several Dirichlet kernels D_n and observe that, as n increases, the kernel becomes more and more concentrated at the origin. Indeed, D_n tends to the Dirac function as $n \to +\infty$ and the integral of its absolute value tends to $+\infty$. That is to say, $\lim_{n \to +\infty} \Lambda_n \to +\infty$. More precisely, the Lebesgue constant grows logarithmically with n as we show in the next exercise.

Exercise 3.8 Lebesgue constant in Fourier approximation.

1. Use the MATLAB function `integral` to compute Dirichlet kernels. Plot the curve Λ_n as a function of $\ln(n)$. Take $n = 5, 10, \cdots, 100$. A straight line should be obtained. Compute its slope (it should be close to $4/\pi^2$).
2. Show (graphically) that the distance between Λ_n and $\frac{4}{\pi^2} \ln(2n+1)$ tends asymptotically to a constant. Find numerically an approximation of this constant by plotting the two curves.
A solution of this exercise is proposed in Sect. 3.8 at page 73.

What about pointwise convergence of $\pi_n f(x)$ to $f(x)$ for a fixed $x \in I$? For n large enough, we can neglect the contribution of D_n outside a small interval $[-\varepsilon, \varepsilon]$ (see Fig. 3.6) and approximate $\pi_n f(x)$ by

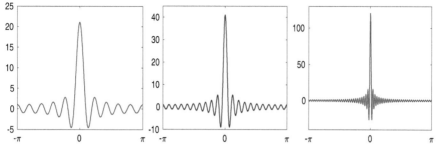

Fig. 3.6 Dirichlet kernels for $n = 10$, $n = 20$, and $n = 60$.

$$\frac{1}{2\pi} \int_{-\varepsilon}^{\varepsilon} D_n(y) f(x-y) dy = \frac{1}{\pi} \int_{-\varepsilon}^{\varepsilon} \frac{\sin[(n+\frac{1}{2})y]}{y} f(x-y) dy.$$

Using a change of variable, we obtain

$$\pi_n f(x) \simeq \frac{f(x^+)}{\pi} \int_{-(n+\frac{1}{2})\varepsilon}^{0} \frac{\sin y}{y} dy + \frac{f(x^-)}{\pi} \int_{0}^{(n+\frac{1}{2})\varepsilon} \frac{\sin y}{y} dy.$$

Hence, using the identity $\int_{-\infty}^{\infty} \frac{\sin y}{y} dy = \pi$, we get

$$\lim_{n \to \infty} \pi_n f(x) = \frac{f(x^-) + f(x^+)}{2}.$$

This is the pointwise convergence insured by the Dirichlet Theorem: if f is continuous 2π-periodic and has derivatives on both sides of the point x, then $\lim_{n \to +\infty} \pi_n f(x) = f(x)$. Consider now the uniform convergence of $\pi_n f$ to f. For all $v_n \in \mathbb{S}_n$, we can write

$$\|f - \pi_n f\|_\infty = \|f - v_n - \pi_n(f - v_n)\|_\infty \leq (1 + \Lambda_n)\|f - v_n\|_\infty.$$

Hence, we obtain the bound

$$\|f - \pi_n f\|_\infty \leq (1 + \Lambda_n) E_{n,\infty}(f), \tag{3.22}$$

with $E_{n,\infty}(f)$ the best uniform approximation error of f in $\mathbb{S}_n$. As already observed for interpolation (see Remark 3.1), there should be a balance between $E_{n,\infty}(u) \to 0$ and $\Lambda_n \to +\infty$. Hence, the uniform convergence of $\pi_n f$ to f is not guaranteed for a function that is only continuous. More regularity is required. For example, uniform convergence is guaranteed for Hölder continuous functions.

3.6 The Fast Fourier Transform

Let N be the total number of interpolation points. For the sake of simplicity, we will take into account only multiplications and divisions and neglect additions. Computing each of the N discrete Fourier coefficients $\hat{f}_k$ by (3.19) requires about $\mathcal{O}(N)$ operations (precisely $N + 1$). Computation of all the discrete spectrum requires $\mathcal{O}(N^2)$ operations. This is huge for large N. The Fast Fourier Transform (FFT) developed by J. Cooley and J. Tuckey (1965) reduces the computation of the Fourier coefficients to only $\mathcal{O}(N \ln N)$ operations, which is considerably lower than $\mathcal{O}(N^2)$. We will use two MAT-LAB functions, `fft` and `ifft`, computing the direct and inverse FFT, respectively.

- If X is a vector (an array) with components $F_1, \ldots F_N$, then the MAT-LAB instruction C=fft(F) produces a vector C whose N components are

$$C_k = \sum_{j=1}^{N} F_j e^{-i(k-1)\theta_j}, \qquad 1 \le k \le N, \tag{3.23}$$

with $\theta_j = \frac{2\pi}{N}(j-1)$ the interpolation points. These C_k are the discrete Fourier coefficients computed by the FFT.
- Conversely, F=ifft(C) produces a vector X with N components

$$F_j = \frac{1}{N} \sum_{k=1}^{N} C_k e^{i(k-1)\theta_j}. \tag{3.24}$$

As a reminder, note that the function `fft` goes from the physical space representation to the Fourier space (or spectral space) and the function `ifft` does the opposite transfer.

Exercise 3.9 Compare, in terms of computational time, the vectorization method used for Exercice 3.2 and the FFT. Take $f(x) = x \exp(\sin(2x))$ and vary n between 100 and 500 with increments of 50. Use the MATLAB function `cputime` to estimate the CPU time used for each case.
A solution of this exercise is proposed in Sect. 3.8 at page 73.

As you may have noticed from the previous exercise, the FFT is extremely fast. It is not simple to exhibit numerically the $\mathcal{O}(N \ln N)$ computational complexity of the FFT. We investigate in the next exercise the computational complexity of the vectorization method used for Exercice 3.2.

Exercise 3.10 Take $f(x) = x \exp(\sin(2x))$ and let n vary between 5 000 and 10 000 by steps of 1000. Use again the MATLAB function `cputime`. Plot the logarithm of the CPU time as a function of $\ln(n)$. Compute the slope of this straight line using `polyfit`. Comment on the results.
A solution of this exercise is proposed in Sect. 3.8 at page 73.

Let us emphasize the connection between the discrete Fourier coefficients $(\tilde{f}_k)_{k=-n}^n$ defined in (3.18) and those computed by the FFT (3.23) using the same number of points $N = 2n + 1$. For negative indices $-n \leq k \leq -1$, $\tilde{f}_k = \frac{1}{N}C_{N+k+1}$ and for $0 \leq k \leq n$, $\tilde{f}_k = \frac{1}{N}C_{k+1}$. Of course, the interpolated values computed by (3.20) or (3.24) are the same.

Remark 3.3 The FFT is more efficient when the number of parameters N is even, especially if it is a power of 2. That is why, in practical situations, the range of the indices of the Fourier coefficients is $-\frac{N}{2}+1, \cdots, \frac{N}{2}$, with $N = 2^{N_0}$ (or any even number). In that case, the array C computed by the function fft contains the coefficients in this order $0, \cdots, N/2, -N/2 + 1, \cdots, -1$.

3.7 Further Reading

For the general theory of Fourier approximation, consult the resource book Zygmund (2002). For numerical applications, mainly for Partial Differential Equations, see Trefethen (2000), Gottlieg and Orszag (1997), and many others. In Chap. 15, Fourier approximation is used in the algorithm to solve the Navier–Stokes equations. Wavelets (generalization of Fourier approximation) are used in Chap. 8 for image processing purposes. See Cohen (2003) for the numerical analysis of wavelets methods.

3.8 Solutions and Programs

Solution of Exercise 3.1.

The Fourier series of functions f_1 and f_2 are respectively

$$\frac{\pi}{2} + \frac{1}{\pi}\sum_{k \in \mathbb{Z}^*}\frac{e^{ikx}}{k^2} = \frac{\pi}{2} + \frac{2}{\pi}\sum_{k=1}^{+\infty}\frac{\cos(kx)}{k^2},$$

and

$$1 - 6i\sum_{k \in \mathbb{Z}^*}\frac{(-1)^k}{k^3}e^{ikx} = 1 + 12\sum_{k=1}^{+\infty}\frac{(-1)^k}{k^3}\sin(kx).$$

We answer the questions for f_2 and $n = 2$. The following script calls the function Pnf which calls the function fk. Both functions are defined in the same file as the script, in APP1_M_FourierSeries.m

Solution of Exercise 3.2.

1. Discrete Fourier coefficients. The following script is available in APP1_M_FourierInterp1.m.

   a. The first method to compute the Fourier coefficients uses the function APP1_F_FourierColoc defined in a separate file.

   ```
   f=inline('x.*exp(sin(2*x))');% or f= @(x) (x.*exp(sin(2*x));
   n=6;
   X0= APP1_F_FourierColoc(n); % computes collocation points xi
   Y0=f(X0);                   % f(xi)
   N=2*n+1;
   % Method 1 using loops
   Coef1=zeros(N,1);
   for k=-n:n
       s=0;
       for j=1:N
           s=s+Y0(j)*exp(-i*k*X0(j));
       end
       Coef1(k+n+1)=s/N;
   end;
   ```

   Function APP1_F_FourierColoc is defined in file APP1_F_FourierColoc.m

   ```
   function y=APP1_F_FourierColoc(n)
   % compute 2n+1 collocation points in [0,2 pi[
   y=2*pi*(0:2*n)'/(2*n+1);
   ```

   b. The second method uses the MATLAB vector operations. The $(2n + 1) \times (2n + 1)$ square matrix A is defined by $A_{k,j} = \frac{1}{2n+1}e^{-ikx_j}$ with $k = -n, \cdots , n$ and $j = 0, \cdots , 2n$. Warning: MATLAB indices start at 1, this must be given prime consideration! Here is the corresponding script

   ```
   % Method 2 using vectorization
   A=exp(-i*(-n:n)'*X0')/N;
   Coef2=A*Y0;
   % compare the two sets of coefficients
   if norm(Coef1-Coef2) > 1.e-12
       disp('Problem !!')
   end
   ```

   c. The end of the script compares graphically the computing times needed by both methods. It shows that for n increasing from 3 to 4500, the vectorized method is much faster.

2. If M is the number of points z_m, then B is a $M \times (2n + 1)$ matrix whose coefficients are $B_{m,k} = e^{ikx_m}$. This following script is available

in APP1_M_FourierInterp2.m. It calls Functions APP1_F_FourierDCoef.m (computation of the discrete coefficients) and APP1_F_FourierInterp.m (interpolation) defined in two separate files.

```
f=inline('x.*exp(sin(2*x))');
n=10;
X0=APP1_F_FourierColoc(n);
fX0=f(X0);
cf=APP1_F_FourierDCoef(X0,fX0);
M=6; Z=rand(M,1)*2*pi;        % M points randomly in (0,1)
IZ=APP1_F_FourierInterp(cf,Z);
X=linspace(0,2*pi);Y=f(X);
plot(X,Y,'b-',Z,IZ,'ro','Linewidth',2);grid on;
```

Solution of Exercise 3.3.

1. ```
   alf=6;
   rho=2*alf/(1+alf*alf);
   f = @(x)((1)./(sqrt(1+ rho*cos(x))));
   n0=7
   X0 = APP1_F_FourierColoc(n0); fX0=f(X0);
   cf=APP1_F_FourierDCoef(X0,fX0);
   X=linspace(0,2*pi,50)';   fX=f(X);
   IX=APP1_F_FourierInterp(cf,X);
   IX=real(IX);     % imaginary part shoud be almost 0
   figure(1);plot(X,fX,'b-',X,IX,'ro','Linewidth',2);grid on;
   figure(2);plot(X,abs(fX-IX),'k-','Linewidth',2);grid on;
   err0=norm(IX-f(X),Inf)
   ```
2. See the script in file APP1_M_FourierLeverrier.m. Taking more than $2n+1$ points in the quadrature formula does not improve drastically the results but requires more computation. So, this approach is not useful.

Solution of Exercise 3.4.

The following script (available in file APP1_M_FourierErr.m) calls two functions defined in files APP1_F_FourierTests.m (definition of the test functions) and APP1_F_FourierProj.m (projection of a function onto $\mathbb{S}_n$). The function APP1_F_FourierCoef.m called in APP1_F_FourierProj is available in file APP1_F_FourierCoef.m, it simply gives the (exact) Fourier coefficients of the test functions.

```
Slopes=[];
for f=[1,2,3]
        if f==2 | f==3,
```
```

```
            X=linspace(-pi,pi,1000)';
        else
            X=linspace(0,2*pi,1000)';
        end;
    F=APP1_F_FourierTests(f);Y=F(X);
    for k=1:15
        n(k)=10*k;
        Yn=APP1_F_FourierProj(f,n(k),X);
        e2(k,f)=norm(Yn-Y')/sqrt(2*n(k)+1);
    end;
    loglog(n,e2(:,f),'-s','Linewidth',4);grid on;hold on
    Slopes=[Slopes; polyfit(log(n),log(e2(:,f)),1)];
end
for s=[1,2,3]
    fprintf('Case %i: slope = %e \n',s,Slopes(s,1))
end;
```

Running the script produces the output

```
Case 1: slope = -1.910447e+00
Case 2: slope = -2.960324e+00
Case 3: slope = -3.901928e+00
```

It matches perfectly with the theory: the smoother the function, the faster the convergence.

Solution of Exercise 3.5.

It is enough to slightly change the previous script in order to compute the Fourier coefficients using Remark 3.2. See files APP1_M_FourierErr2.m, APP1_F_FourierProjDer.m and APP1_F_FourierTestDer.m. The output

```
For f2, slope = -1.818734e+00
For f3, slope = -2.960324e+00
```

matches perfectly with the theory since we lose one degree of smoothness for each function.

Solution of Exercise 3.6.

First, prove the series expansion $f_4(x) = \sum_{k=1}^{\infty} \frac{\cos(kx)}{2^{k-1}}$. The script in file APP1_M_FourierDErr.m generates plot on Fig. 3.7. It is clear, from the left plot, that the error related to f_4 decreases faster than the others. The right plot shows that this decrease is faster than any power of $1/n$ (exponential decreasing). If the error behaves like $C\varrho^n$, then $\log e_n$ behaves like a linear function of n. Hence, a plot $n \mapsto \log e_n$ will produce a straight line with slope

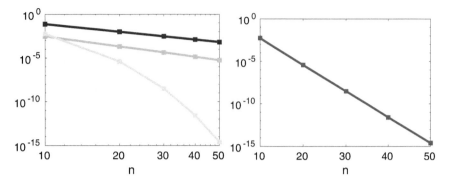

Fig. 3.7 Decreasing of the error. Left: Log–log plot of $\|f - \pi_n f\|_2$ versus n. Functions have different smoothness: a $\mathcal{C}^1$ function (top), a $\mathcal{C}^2$ function (center), and an analytic function (bottom). Right: $\log(\|f - \pi_n f\|_2)$ versus n for the analytic function.

ϱ. This is the case here with $\varrho \simeq 0.71$ which indicates clearly the exponential convergence of the error.

Solution of Exercise 3.7.

1. Straightforward computations give

$$
\widehat{(\varphi * \psi)}_k = \frac{1}{2\pi} \int_0^{2\pi} (\varphi * \psi)(x) e^{-ikx} dx
$$

$$
= \frac{1}{2\pi} \int_0^{2\pi} \varphi(y) \frac{1}{2\pi} \left[\int_0^{2\pi} \psi(x - y) e^{-ikx} dx \right] dy = \hat{\varphi}_k \hat{\psi}_k
$$

since the expression between square brackets is equal to

$$
\int_{-y}^{2\pi - y} \psi(z) e^{-ik(z+y)} dz = e^{-iky} \int_0^{2\pi} \psi(z) e^{-ikz} dz = 2\pi e^{-iky} \hat{\psi}_k.
$$

The identity $\pi_n f = D_n * f$ follows.
2. For $x \in I$, we have

$$
|\pi_n f(x)| = |\frac{1}{2\pi} \int_0^{2\pi} D_n(y) f(x - y) dy| \le \frac{\|f\|_\infty}{2\pi} \int_0^{2\pi} |D_n(y)| dy = \Lambda_n \|f\|_\infty.
$$

Hence, $\|\pi_n f\|_\infty \le \Lambda_n \|f\|_\infty$. We knew, from the Bessel inequality, that the L^2-norm of its projection is always less than the L^2-norm of the function itself. This is not true at all for the uniform norm since Λ_n goes to infinity with n as we will see in the following exercise.

Solution of Exercise 3.8.

Run of the script in file `APP1_M_FourierLebesgue.m` shows that the distance between Λ_n and $\frac{4}{\pi^2}\ln(2n+1)$ tends asymptotically to a constant $= 0.9894\cdots$

Solution of Exercise 3.9.

Using `fft` is much faster. See script in `APP1_M_FourierCPU.m`

Solution of Exercise 3.10.

The following script available in `APP1_M_FourierCPU2.m` produces a slope $\simeq 1.7$. With larger values of parameter n, the slope is closer to the expected value of 2.

```
f=inline('x.*exp(sin(2*x))');
time=[];
dim=5000:1000:10000;
for n=dim
    X0 = APP1_F_FourierColoc(n); Y0=f(X0);
    t0=cputime;Coef=APP1_F_FourierDCoef(X0,Y0);
    time=[time,cputime-t0];
end
figure (2);
logdim=log(dim);logtime=log(time);
plot(logdim,logtime,'r--o','Linewidth',2);grid on;
p=polyfit(logdim,logtime,1);
fprintf('Complexity of direct comput. slope = %e \n',p(1,1))
```

Chapter References

A. Cohen, *Numerical Analysis of Wavelet Methods. Studies in Mathematics and its Applications.* (North-Holland, Amsterdam, 2003)

D. Gottlieg, S.A. Orszag, *Numerical Analysis of Spectral Methods: Theory and Applications, CBMS-NSF Regional Conference Series in Applied Mathematics, No. 26.* (SIAM, Philadelphia, 1977)

L.N. Trefethen, *Spectral Methods in MATLAB, Software, Environments, and Tools,* vol. 10 (SIAM, Philadelphia, 2000)

A. Zygmund, *Trigonometric Series,* vol. I, II. (Cambridge University Press, Cambridge, 2002)

Chapter 4
Polynomial Approximation

Project Summary

Level of difficulty:	1
Keywords:	Polynomial approximation for nonperiodic functions, interpolation, best approximations, splines, series expansions
Application fields:	Approximation of functions

This project is devoted to the approximation of a given real function defined on an interval $I = [a, b] \subset \mathbb{R}$ by a simpler one that belongs to the set $\mathbb{P}_n$ of (algebraic) polynomials spanned by x^k for $k = 0, \cdots, n$ $(n \in \mathbb{N})$. Definitions and results of this chapter, given without proof, are widely used in the rest of the book. We refer the reader to monographs on analysis and applications of polynomial approximation, for instance, Crouzeix and Mignot (1989), Rivlin (1981), DeVore and Lorentz (1993), Trefethen (2000), and many others.

4.1 Introduction

The approximation of a given function by a polynomial is an efficient tool in many problems arising in applied mathematics. In the following examples, f is the function to be approximated by a polynomial p_n. *Approximation* means either *interpolation* or *best approximation*:

1. Visualization of some computational results. Given the values of a function f at some points x_i, the aim is to draw this function on the interval $[a, b]$. This is the interpolation problem if $[a, b] \subset [\min_i x_i, \max_i x_i]$; otherwise, it is an extrapolation problem. The following approximation is often made:

© The Author(s), under exclusive license to Springer Nature Switzerland AG 2023   75
I. Danaila et al., *An Introduction to Scientific Computing*,
https://doi.org/10.1007/978-3-031-35032-0_4

$$\forall x \in [a, b], \qquad f(x) \approx p_n(x).$$

2. Numerical quadrature: to compute an integral involving the function f, the following approximation is used:

$$\int_a^b f(x)dx \approx \int_a^b p_n(x)dx,$$

since the computation of the last integral is easy.

3. Differential equations: in spectral methods, the solution of an Ordinary Differential Equation or a Partial Differential Equation is approximated by a polynomial function, see Chap. 6.

Interpolation

The collocation or interpolation points $(x_i)_{i=0}^n$ are any $n+1$ distinct points in the interval I. Specific cases of equidistant points and Chebyshev points will be investigated. The problem is to find a polynomial $\mathcal{I}_n f$ in $\mathbb{P}_n$ that coincide with the continuous function f at collocation points x_i. Note that if $f \in \mathbb{P}_n$, then $\mathcal{I}_n f = f$. Associated to the collocation points are the Lagrange polynomials or fundamental polynomials defined by

$$\ell_i(x) = \prod_{j=0, j\neq i}^n \frac{x - x_j}{x_i - x_j}. \tag{4.1}$$

They form a basis of $\mathbb{P}_n$. Furthermore, $\ell_j(x_k) = \delta_{j,k}$, with δ the Kronecker symbol. Examples of Lagrange polynomials are displayed on Fig. 4.1. The Lagrange basis is mainly used to write in a very simple way the Lagrange polynomial interpolant as

$$\mathcal{I}_n f = \sum_{i=0}^n f(x_i)\ell_i. \tag{4.2}$$

We first address the stability of the Lagrange interpolation. The Lebesgue constant associated to points x_i is

$$\Lambda_n = \max_{x \in I} \sum_{i=0}^n |\ell_i(x)|. \tag{4.3}$$

It is important to notice that Λ_n does not depend on the function being interpolated, but only on interpolation points. From (4.2)-(4.3), we get $\|\mathcal{I}_n f\|_\infty \leq \Lambda_n \|f\|_\infty$, where $\|.\|_\infty$ denotes the supremum norm on I. Hence, for any continuous functions f and g, we have

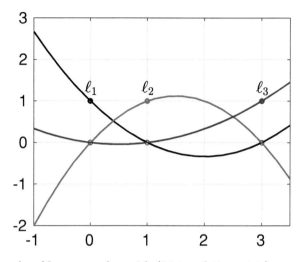

Fig. 4.1 Examples of Lagrange polynomials (3 interpolation points).

$$|\mathcal{I}_n f(x) - \mathcal{I}_n g(x)| \leq \Lambda_n \max_{0 \leq i \leq n} |f(x_i) - g(x_i)|. \qquad (4.4)$$

Suppose that g is very close to f at collocation points; $g(x_i) = f(x_i) + \varepsilon_i$, then

$$\max_{x \in I} |\mathcal{I}_n f(x) - \mathcal{I}_n g(x)| \leq \varepsilon \Lambda_n,$$

with $\varepsilon = \max_i |\varepsilon_i|$. The Lebesgue constant appears as a measure of the amplification of the error in the Lagrange interpolation, it is the *stability constant* of the Lagrange interpolation procedure. Unfortunately, the following result holds.

Proposition 4.1 *Whatever the interpolation points,*

$$\lim_{n \to +\infty} \Lambda_n = +\infty. \qquad (4.5)$$

Hence small perturbations (ε) on the data can lead to very large variations in the solution ($\mathcal{I}_n f$). This is the typical case of an *ill-conditioned* problem. We will also analyse the Hermite interpolation where not only the function is interpolated at collocation points but also some of its derivatives. We consider also the piecewise polynomial approximation (called spline approximation).

Best approximation

In this section, we seek a polynomial that is closest to f for a prescribed norm $\|.\|_{\mathcal{X}}$, $\mathcal{X}$ being a linear functional space that includes the polynomials. Notations and results presented in the Introduction section of Chap. 3 on Fourier approximation are valid here with slight modifications.

- Least square approximation (or Hilbertian approximation, or L^2 approximation). The functional space $\mathcal{X}$ is $L^2(I)$ equipped with the usual inner product and norm

$$\langle f, g \rangle = \int_a^b f(t)g(t)dt, \quad \|f\| = \sqrt{\langle f, f \rangle}, \quad f, g \in L^2(I).$$

- Uniform approximation. In this case, $\mathcal{X} = \mathcal{C}(I)$ the set of continuous functions over I. Weierstrass approximation theorem 3.1 and Proposition 3.2 are still valid.

4.2 Interpolation

In this section, $f : I = [a, b] \longrightarrow \mathbb{R}$ is a continuous function, $(x_i)_{i=0}^k$ a set of $k + 1$ *distinct* points in the interval $[a, b]$, and $(\alpha_i)_{i=0}^k$ a set of $(k + 1)$ integers. We set $n = k + \alpha_0 + \cdots + \alpha_k$ and consider the following problem: find a polynomial p that coincides with f and possibly with some derivatives of f at points x_i. The integer α_i indicates the highest derivative of f to be interpolated at the point x_i.

4.2.1 Lagrange Interpolation

Lagrange interpolation corresponds to situations where only the function f is interpolated and not its derivatives. In such a case, $\alpha_i = 0$ for all i and thus $k = n$. We have the following fundamental result.

Theorem 4.1 *Given $(n + 1)$ distinct points $x_0, x_1, \ldots, x_n$ and a continuous function f, there exists a unique polynomial $p_n \in \mathbb{P}_n$ such that $\forall i = 0, \ldots, n$,*

$$p_n(x_i) = f(x_i). \tag{4.6}$$

The polynomial p_n is called the Lagrange polynomial interpolant of f with respect to the points x_i. We denote it by $\mathcal{I}_n(f; x_0, \ldots, x_n)$ or simply $\mathcal{I}_n f$. The Lagrange polynomials form a basis of $\mathbb{P}_n$ and are explicitly given by (4.1). Since there are many other bases, a question arises naturally: what is the most appropriate basis of $\mathbb{P}_n$ for the computation of $\mathcal{I}_n f$? For the answer, we compare three bases.

- *Basis 1*. The canonical basis: $1, x, \ldots, x^n$.
- *Basis 2*. The basis given by the Lagrange polynomials.
- *Basis 3*. The Newton basis; it is the basis given by the polynomials

$$1, (x - x_0), (x - x_0)(x - x_1), \ldots, (x - x_0)(x - x_1) \cdots (x - x_{n-1}). \quad (4.7)$$

Exercise 4.1 Computations in the canonical basis.
Let $(a_k)_{k=0}^n$ be the coefficients of $\mathcal{I}_n f$ in the canonical basis,

$$\mathcal{I}_n f = \sum_{k=0}^n a_k x^k,$$

and $a = (a_0, \ldots, a_n)^T \in \mathbb{R}^{n+1}$.

1. Prove that the interpolation conditions (4.6) are equivalent to a linear system $Aa = b$, with matrix $A \in \mathbb{R}^{(n+1)\times(n+1)}$ and right-hand side $b \in \mathbb{R}^{n+1}$ to be determined.
2. For $n = 10$ (and 20) define an array x of $n + 1$ random numbers sorted in increasing order between 0 and 1. Write a program computing matrix A.
3. For $f(x) = \sin(10\, x \cos x)$, compute the coefficients of $\mathcal{I}_n f$ by solving the linear system $Aa = b$. Plot on the same figure $\mathcal{I}_n f$ and f evaluated at points x_i. Use the MATLAB function `polyval` (warning: handle carefully the ordering of the coefficients α_i).
4. For $n = 10$, compute $\|Aa - b\|_2$, then the condition number of the matrix A (use the function `cond`) and its rank (use the function `rank`). Same questions for $n = 20$. Comment on.

A solution to this exercise is proposed in Sect. 4.6 at page 95.

Exercise 4.2 For n going from 2 to 20 in steps of 2, compute the logarithm of the condition number of matrix A (see the previous exercise) for $n + 1$ points x_i uniformly chosen between 0 and 1. Plot the logarithm of the condition number of the matrix as a function of n. Comment on.
A solution of this exercise is proposed in Sect. 4.6 at page 95.

Exercise 4.3 Computations in the Lagrange basis.
For $n \in \{5, 10, 20\}$, define the points $x_i = i/n$ for $i = 0, \ldots, n$. Write a program (using n and k as input data) that computes the Lagrange polynomial ℓ_k. Use function `polyfit` of MATLAB . Evaluate $\ell_{[n/2]}$ at 0. Comment on.
A solution to this exercise is proposed in Sect. 4.6 at page 96.

Let us now consider Newton's basis. This basis is related to the so-called "divided differences". The divided difference of order $k \in \mathbb{N}$ of the function f with respect to $k + 1$ distinct points $x_0, \ldots, x_k$ is the real number denoted by $f[x_0, \ldots, x_k]$ and defined for $k = 0$ by $f[x_i] = f(x_i)$ and for $k \geq 1$ by

$$f\,[x_0,\ldots,x_k] = \frac{f\,[x_1,\ldots,x_k] - f\,[x_0,\ldots,x_{k-1}]}{x_k - x_0}.$$

The evaluation of the divided differences is computed by Newton's algorithm, described by the following "tree":

$f[x_0] = f(x_0)$

$\searrow f[x_0, x_1] = \frac{f(x_1)-f(x_0)}{x_1-x_0}$

$f[x_1] = f(x_1)$ $\searrow f[x_0, x_1, x_2] = \frac{f[x_1,x_2]-f[x_0,x_1]}{x_2-x_0}$ $\cdots$

$\nearrow$

$\searrow f[x_1, x_2] = \frac{f(x_2)-f(x_1)}{x_2-x_1}$ $\vdots$

$f[x_2] = f(x_2)$ $\vdots$ $\vdots$

$\vdots$ $\vdots$ $\vdots$

The first column of the tree contains the divided differences of order 0, the second column contains the divided differences of order 1, and so on. The following proposition shows that the divided differences are the coefficients of $\mathcal{I}_n f$ in Newton basis.

Proposition 4.2 *For all $x \in I$,*

$$\mathcal{I}_n f(x) = f[x_0] + \sum_{k=1}^{n} f[x_0,\ldots,x_k](x-x_0)(x-x_1)\cdots(x-x_{k-1}). \qquad (4.8)$$

Let c be an array that contains the divided differences $c_i = f[x_0,\ldots,x_i]$. To evaluate the polynomial $\mathcal{I}_n f$ at a point x, we write

$$\mathcal{I}_n f(x) = c_0 + (x-x_0)\Big\{ c_1 + (x-x_1)\Big\{ c_2 + c_3\,(x-x_2) + \cdots \Big. \Big..$$

This way of writing $\mathcal{I}_n f(x)$ is called the Horner form of the polynomial. It is a very efficient method since the computation of $\mathcal{I}_n f(x)$ in this form requires n multiplications, n subtractions, and n additions, while the form (4.8) requires $n(n+1)/2$ multiplications, $n(n+1)/2$ subtractions, and n additions. Here is the Horner's algorithm for the evaluation of $\mathcal{I}_n f(x)$:

$$y = c_n$$
$$\text{for } k = n-1 \searrow 0$$
$$\quad y = (x - x_k)y + c_k$$
$$\text{end}$$

Exercise 4.4 Divided differences.

1. Computation of the divided differences. Start from an array c that contains the $(n+1)$ values $f\,[x_i] = f(x_i)$. In the first step, $c_0 = f\,[x_0]$ is unchanged

and all the other values c_k ($k \geq 1$) are replaced by the divided differences of order 1. In the second step, $c_1 = f[x_0, x_1]$ is unchanged and the values c_k ($k \geq 2$) are replaced by new ones, and so on. Here is the algorithm to be implemented:

$$\text{for } k = 0 \nearrow n$$
$$\qquad c_k \leftarrow f(x_k)$$
$$\text{end}$$
$$\text{for } p = 1 \nearrow n$$
$$\qquad \text{for } k = n \searrow p$$
$$\qquad\qquad c_k \leftarrow (c_k - c_{k-1})/(x_k - x_{k-p})$$
$$\qquad \text{end}$$
$$\text{end}$$

2. Use Horner's algorithm to evaluate $\mathcal{I}_n f$ on a fine grid of points in $[0, 1]$. Draw $\mathcal{I}_n f$ and f on the same figure. In the same figure, mark the interpolation points x_i.

A solution of this exercise is proposed in Sect. 4.6 at page 96.

Actually, the error (local) error at a point x, $e_n(x) = f(x) - \mathcal{I}_n f(x)$ is precisely known through the following result.

Proposition 4.3 *Assume $f \in C^{n+1}([a, b])$. For all $x \in [a, b]$, there exists $\xi_x \in [a, b]$ such that*

$$e_n(x) = \frac{1}{(n+1)!} \Pi_n(x) f^{(n+1)}(\xi_x), \tag{4.9}$$

with $\Pi_n(x) = \prod_{i=0}^{n} (x - x_i)$.

We deduce from (4.9) the following upper bound

$$|e_n(x)| \leq \frac{1}{(n+1)!} \|\Pi_n\|_\infty \|f^{(n+1)}\|_\infty.$$

This suggests that a good way to choose the interpolation points consists in minimizing $\|\Pi_n\|_\infty$, since the term $\|f^{(n+1)}\|_\infty$ depends only on the function and not at all on the interpolation points. Let us first consider the case of equidistant interpolation points:

$$x_i = a + i \frac{b - a}{n}, \qquad 0 \leq i \leq n.$$

In this case, there exists a constant c independent of n such that for n large enough

$$\max_{a \leq x \leq b} |\Pi_n(x)| \geq c(b - a)^{n+1} e^{-n} n^{-5/2}. \tag{4.10}$$

We consider now the Chebyshev points. Chebyshev polynomial T_n $(n \in \mathbb{N})$ are defined on the interval $[-1, 1]$ by

$$T_n(t) = \cos(n\theta), \quad \text{with} \ \cos\theta = t. \tag{4.11}$$

The Chebyshev points on this interval are the n zeros of T_n

$$t_i = \cos(\theta_i), \quad \theta_i = \frac{\pi}{2n} + i\frac{\pi}{n}, \quad 0 \le i \le n - 1.$$

On an interval $[a, b]$, the Chebyshev points are defined as the image of the previous points by an affine transformation (see Fig. 4.2)

$$x_i = \frac{a+b}{2} + \frac{b-a}{2}\cos(\theta_i), \quad 0 \le i \le n - 1.$$

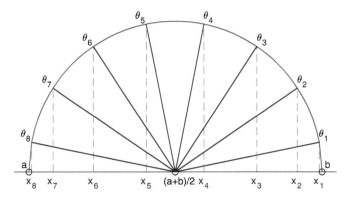

Fig. 4.2 The Chebyshev points on an interval $[a, b]$ $(n = 8)$.

Whatever the points x_i' in $[a, b]$, the following lower bound holds:

$$\max_{x \in [a,b]} \left| \prod_{i=0}^{n-1} (x - x_i') \right| \ge \max_{x \in [a,b]} \left| \prod_{i=0}^{n-1} (x - x_i) \right| = \frac{(b-a)^n}{2^{2n-1}}. \tag{4.12}$$

Comparing the bounds in (4.10) and (4.12) favors the Chebyshev points. We will see in the next paragraph another reason to prefer these points. The following exercise compares the Lebesgue constants in both cases: equidistant points and Chebyshev points.

Exercise 4.5 Computation of the Lebesgue constant.

1. Write a function that computes the Lebesgue constant associated with an array x of n real numbers (see (4.3)). Use MATLAB functions `polyval` and `polyfit` to evaluate ℓ_i. Compute the maximum in (4.3) on a uniform grid of 100 points between $\min_i x_i$ and $\max_i x_i$.

2. Uniform distribution. Compute for n going from 10 to 30 in steps of 5 the Lebesgue constant $\Lambda_{n,U}$ associated with $n+1$ equidistant points in the interval $[-1,1]$. Draw the curve $n \mapsto \ln(\Lambda_{n,U})$. Comment on.
3. Chebyshev points. Compute for n going from 10 to 20 in steps of 5 the Lebesgue constant $\Lambda_{n,T}$ associated with $n+1$ Chebyshev points on $[-1,1]$. Draw the curve $\ln n \mapsto \Lambda_{n,T}$. Comment on.

A solution to this exercise is proposed in Sect. 4.6 at page 97.

Exercise 4.6 Runge phenomenon. Draw on a uniform grid of 100 points the Lagrange polynomial interpolation of the function $f : x \mapsto 1/(x^2 + a^2)$ at the $n+1$ points $x_i = -1 + 2i/n$ $(i = 0, \ldots, n)$. Take $a = 2/5$ and $n = 5$, 10, then 15. Note that the function to be interpolated is very regular on $\mathbb{R}$, in contrast to the functions considered in the previous exercise. Comment on the results.

A solution of this exercise is proposed in Sect. 4.6 at page 98.

4.2.2 Hermite Interpolation

We assume in this section that function f has derivatives of order α_i at the point x_i. There exists a unique polynomial $p_n \in \mathbb{P}_n$ such that for all $i = 0, \ldots, k$ and $j = 0, \ldots, \alpha_i$

$$p_n^{(j)}(x_i) = f^{(j)}(x_i). \tag{4.13}$$

The polynomial p_n, which we denote by $\mathcal{I}_n(f; x_0, \ldots, x_k; \alpha_0, \ldots, \alpha_k)$, or simply $\mathcal{I}_n^H f$, is called the Hermite polynomial interpolation of f at points x_i with respect to indices α_i.

Theorem 4.2 *Suppose $f \in \mathcal{C}^{n+1}([a, b])$. For all $x \in [a, b]$, there exists $\xi_x \in [\min_i x_i, \max_i x_i]$ such that*

$$e_n^H(x) = f(x) - \mathcal{I}_n^H f(x) = \frac{1}{(n+1)!} \Pi_n^H(x) f^{(n+1)}(\xi_x), \tag{4.14}$$

with $\Pi_n^H(x) = \prod_{i=0}^{k} (x - x_i)^{1+\alpha_i}$.

Since the function f is of class $\mathcal{C}^{n+1}$ on the interval $[a, b]$, for all $n+1$ distinct points $x_0, \ldots, x_n$ in the interval, there exists $\xi \in [a, b]$ such that

$$f^{(n)}(\xi) = n! f[x_0, \ldots, x_n].$$

This relation defines a connection between the divided differences and the derivatives. More precisely, we make the following remark.

Remark 4.1 Letting each x_i go to x, we get an approximation of the nth derivative of f at x:

$$\frac{1}{n!} f^{(n)}(x) = \lim_{x_i \to x} f[x_0, \dots, x_n].$$

This remark combined with the Newton algorithm allows the evaluation of the Hermite polynomial interpolation, as in the following example.

Example 4.1 Computation of the polynomial interpolant p of minimal degree satisfying

$$p(0) = -1, \; p(1) = 0, \; p'(1) = \alpha \in \mathbb{R}. \tag{4.15}$$

First compute the divided differences with $x_2 = 1 + \varepsilon$, then use Remark 4.1 to obtain

$$\lim_{\varepsilon \to 0} f[x_1, x_2] = \lim_{\varepsilon \to 0} \frac{f(1+\varepsilon) - f(1)}{(1+\varepsilon) - 1} = f'(1).$$

We obtain

$x_0 = 0 \; f[x_0] = \boxed{-1}$

$\quad\quad\quad\quad f[x_0, x_1] = \boxed{1}$

$x_1 = 1 \quad f[x_1] = 0 \quad\quad\quad\quad f[x_0, x_1, x_2] = \frac{\alpha-1}{1-0} = \boxed{\alpha - 1}$

$\quad\quad\quad\quad \boxed{f'(1)} = \alpha$

$x_2 = 1 \quad f[x_1] = 0$

Then

$$p(x) = \boxed{-1} + \boxed{1}\,x + \boxed{(\alpha - 1)}\,x(x - 1).$$

Exercise 4.7 In this exercise, $f(x) = e^{-x} \cos(3\pi x)$.

1. Write a function based on the divided differences (as in Example 4.1) that computes the Hermite polynomial interpolant of a function (including the Lagrange case). The input parameters of this function are the interpolation points x_i, and for each point, the maximal derivative α_i to be interpolated at this point and the values $f^{(\ell)}(x_i)$ for $\ell = 0, \dots, \alpha_i$.
2. Compute the Lagrange interpolation of f at points $0, \frac{1}{4}, \frac{3}{4}$, and 1. Draw f and its polynomial interpolant on the interval $[0, 1]$.
3. Compute the Hermite interpolant of f at the same points (with $\alpha_i = 1$). Draw f and its Hermite polynomial on the interval $[0, 1]$. Compare to the previous results.
4. Answer the same questions in the case where f and f' are interpolated at the previous points and, in addition, point $\frac{1}{2}$.

A solution to this exercise is proposed in Sect. 4.6 at page 98.

Exercise 4.8 Draw on $[0, 1]$, and for several values of m, the polynomial of minimal degree p such that

$$p(0) = 0, \ p(1) = 1, \ \text{and} \ p^{(\ell)}(0) = p^{(\ell)}(1) = 0, \ \text{for} \ \ell = 1, \dots, m.$$

A solution of this exercise is proposed in Sect. 4.6 at page 100.

4.3 Best Polynomial Approximation

We consider two cases: uniform approximation and least square approximation.

4.3.1 Best Uniform Approximation

The following definition enables the characterization of the polynomial of best uniform approximation.

Definition 4.1 A continuous function φ is said to be equioscillatory on $n+1$ points of a real interval $[a, b]$ if φ takes alternately the values $\pm \|\varphi\|_\infty$ at $(n+1)$ points $x_0 < x_1 < \cdots < x_n$ of $[a, b]$ (see Fig. 4.3).

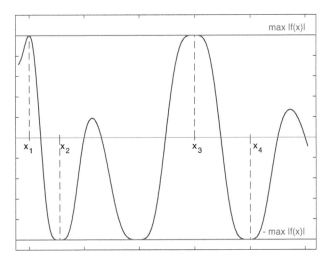

Fig. 4.3 Example of an equioscillatory function on 4 points.

The following theorem is known as the alternation theorem.

Theorem 4.3 *Let f be a continuous function defined on $I = [a, b]$. The polynomial p_n of best uniform approximation of f in $\mathbb{P}_n$ is the only polynomial* *in $\mathbb{P}_n$ for which the function $f - p_n$ is equioscillatory on (at least) $n+2$ distinct points of I.*

For example, the best uniform approximation of a continuous function f on $[a, b]$ by constants is $p_0 = \frac{1}{2} \{ \min_{x \in [a,b]} f(x) + \max_{x \in [a,b]} f(x) \}$, and there exist (at least) two points where the function $f - p_0$ equioscillates. These points are the two points where this function reaches its extremal values on $[a, b]$. To determine the best uniform approximation of a function f, it is sufficient to find a polynomial $p \in \mathbb{P}_n$ and $n + 2$ points such that $f - p$ equioscillates at these points. This is what the following algorithm (called the Remez algorithm) does.

The Remez algorithm

1. *Initialization.* Choose any $n + 2$ distinct points $x_0^0 < x_1^0 < \cdots < x_{n+1}^0$.
2. *Step k.* Suppose the $n + 2$ points $x_0^k < x_1^k < \cdots < x_{n+1}^k$ are known. Compute a polynomial $p_k \in \mathbb{P}_n$ (see Exercise 4.9) such that

$$f(x_i^k) - p_k(x_i^k) = (-1)^i \{ f(x_0^k) - p_k(x_0^k) \}, \qquad i = 1, \ldots, n+1.$$

(a) If

$$\| f - p_k \|_\infty = |f(x_i^k) - p_k(x_i^k)|, \qquad i = 0, \ldots, n+1, \qquad (4.16)$$

the algorithm stops, since the function $f - p_k$ equioscillates at these points. Hence p_k is the polynomial of best uniform approximation of f.

(b) Otherwise, there exists $y \in [a, b]$ such that for all $i = 0, \ldots, n+1$,

$$\| f - p_k \|_\infty = |f(y) - p_k(y)| > |f(x_i^k) - p_k(x_i^k)|. \qquad (4.17)$$

Design a new set of points $x_0^{k+1} < x_1^{k+1} < \cdots < x_{n+1}^{k+1}$ by replacing one of the points x_i^k by y in such a way that

$$\left(f(x_j^{k+1}) - p_k(x_j^{k+1}) \right) \left(f(x_{j-1}^{k+1}) - p_k(x_{j-1}^{k+1}) \right) \leq 0, \quad j = 1, \ldots, n+1.$$

Exercise 4.9 Prove the existence of a unique polynomial $p_k \in \mathbb{P}_n$ defined in step k of the Remez algorithm. Program a function that computes this polynomial (the input data are the $n + 2$ points x_i and a function f).
Hint: write $p_k(t) = \sum_{j=0}^{n} a_j t^j$ and use MATLAB to solve the linear system whose solution is $(a_0, \ldots, a_n)^T$.
A solution to this exercise is proposed in Sect. 4.6 at page 100.

Exercise 4.10 Remez algorithm.
The goal is to compute the best uniform approximation of function $x \mapsto \sin(2\pi \cos(\pi x))$ on $[0, 1]$ by the Remez algorithm. Discuss all the possible cases in point (b) (see the algorithm): $y < \min_i x_i$, $y > \max_i x_i$, $y \in]x_i^k, x_{i+1}^k[$,

and $(f(x_i^k) - p_k(x_i^k))(f(y) - p_k(y)) \geq 0$ or $(f(x_i^k) - p_k(x_i^k))(f(y) - p_k(y)) < 0$. To check the inequality (4.17):

- compute $\|f - p_k\|_\infty$ on a uniform grid of 100 points in the interval $[0, 1]$,
- The equality (4.16) in the algorithm is supposed to be true if the absolute value of the difference between the two quantities is larger than a prescribed tolerance (10^{-8} for example).

Compare the results (in terms of the number of iterations required for the convergence of the algorithm) for three choices of initialization points.

- Equidistant points: $x_i = \frac{i}{n+1}$, $i = 0, \ldots, n+1$.
- Chebyshev points: $x_i = \frac{1}{2}(1 - \cos(i\frac{\pi}{n+1}))$, $i = 0, \ldots, n+1$.
- Random points: $n + 1$ points given by the function rand then sorted out.

A solution to this exercise is proposed in Sect. 4.6 at page 101.

4.3.2 Best Hilbertian Approximation

Here $I =]-1, 1[$ since every interval $]a, b[$ can be mapped to I by a simple affine transformation. The Hilbertian structure of $\mathcal{X} = L^2(I)$ extends to this infinite-dimensional space some basic notions such as basis and orthogonal projection. See, for example, Schwartz (1980) for the definitions and results of this section. We are interested in the determination of the best approximation of a function in $L^2(I)$ by polynomials of the prescribed degree.

Hilbertian Basis

The Legendre polynomials[1] are defined by the recurrence relation (see Chap. 6)

$$(n + 1)L_{n+1}(x) = (2n + 1)xL_n(x) - nL_{n-1}(x) \qquad (\forall n \geq 1)$$

with $L_0(x) = 1$ and $L_1(x) = x$. The degree of L_n is n, and for all integers n and m,

$$\langle L_n, L_m \rangle = \begin{cases} 0 & \text{if } n \neq m, \\ 1/(n + 1/2) & \text{if } n = m. \end{cases}$$

These polynomials are said to be orthogonal. The family $L_n^* = L_n/\|L_n\|$ forms a Hilbertian basis of $L^2(I)$, that is, $(L_n^*)_{n \geq 0}$ is orthonormal and the set of all finite linear combinations of the L_n^* is dense in $L^2(I)$. As in finite dimension, we can expand every function in $L^2(I)$ in the (infinite) Legendre basis. The first five Legendre polynomials are displayed in Fig. 4.4.

---

[1] Sometimes denoted by $(P_n)_n$ in the literature.

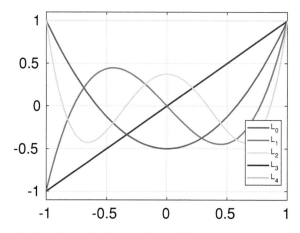

Fig. 4.4 Example of orthogonal polynomials: the first five Legendre polynomials.

Theorem 4.4 *Let $f \in L^2(I)$.*

1. *f has a Legendre series expansion: i.e., there exist real numbers $\hat{f}_k$ such that*

$$f = \sum_{k=0}^{\infty} \hat{f}_k L_k. \tag{4.18}$$

2. *For all $n \in \mathbb{N}$, there exists a unique polynomial in $\mathbb{P}_n$ (which we denote by $\pi_n f$) of best Hilbertian approximation of f in $\mathbb{P}_n$, i.e.,*

$$\|f - \pi_n f\| = \inf_{q \in \mathbb{P}_n} \|f - q\|.$$

Moreover, $\pi_n f$ is characterized by the orthogonality relations

$$\langle f - \pi_n f, p \rangle = 0, \qquad \forall p \in \mathbb{P}_n, \tag{4.19}$$

Therefore $\pi_n f$ is the orthogonal projection of f on $\mathbb{P}_n$ (see Fig. 4.11).

The real numbers $\hat{f}_k$ in (4.18) are called the Legendre (or Fourier–Legendre) coefficients of the function f. We deduce from the orthogonality of the Legendre polynomials that

$$\hat{f}_k = \frac{\langle f, L_k \rangle}{\|L_k\|^2} = (k + \frac{1}{2}) \int_{-1}^{1} f(t) L_k(t) dt \tag{4.20}$$

and $\pi_n f$ is the Legendre series of f, truncated to the order n:

$$\pi_n f = \sum_{k=0}^{n} \hat{f}_k L_k. \tag{4.21}$$

The computation of the best approximation of a function consists mainly in computing its Legendre coefficients. Since the integral in (4.20) can rarely be evaluated exactly, a numerical quadrature is required. See Chap. 6, where the Legendre polynomials are also used to solve a differential equation.

The convergence of the best Hilbertian approximation is stated in the following proposition.

Proposition 4.4 *For all $f \in L^2(I)$,*

$$\lim_{n \to +\infty} \|f - \pi_n f\| = 0. \tag{4.22}$$

Discrete Least Squares Approximation

In this section, we seek the best polynomial approximation of a function f with respect to a discrete norm. Given m distinct points $(x_i)_{i=1}^m$ and m values $(y_i)_{i=1}^m$, the goal is to determine a polynomial $p = \sum_{j=0}^{n-1} a_j x^j \in \mathbb{P}_{n-1}$ that minimizes the expression

$$E = \sum_{i=1}^m |y_i - p(x_i)|^2, \tag{4.23}$$

with m, in general, much larger than n. From a geometrical point of view, the problem is to find p such that its graph is as close as possible (in the Euclidean norm sense) to points (x_i, y_i). The function E defined by (4.23) is a function of n variables $(a_0, a_1, \dots, a_{n-1})$. To determine its minimum, we set to 0 its partial derivatives

$$\frac{\partial E}{\partial a_j} = 0 \iff -2 \sum_{i=1}^m (y_i - p(x_i)) \, x_i^j = 0 \iff \sum_{k=0}^{n-1} \left(\sum_{i=1}^m x_i^{k+j} \right) a_k = \sum_{i=1}^m x_i^j y_i.$$

Hence vector $a = (a_0, \dots, a_{n-1})^T$, whose components are the coefficients of the polynomial where the minimum of E is reached, is a solution of the linear system

$$\widetilde{A} a = \widetilde{b}, \tag{4.24}$$

with matrix $\widetilde{A}$ and right-hand side $\widetilde{b}$ defined by

$$\widetilde{A} = \begin{pmatrix} \sum_i 1 & \sum_i x_i & \cdots & \sum_i x_i^{n-1} \\ \sum_i x_i & \sum_i x_i^2 & \cdots & \sum_i x_i^n \\ \vdots & \vdots & \vdots & \vdots \\ \sum_i x_i^{n-1} & \sum_i x_i^n & \cdots & \sum_i x_i^{2n-1} \end{pmatrix} \in \mathbb{R}^{n \times n}, \quad \widetilde{b} = \begin{pmatrix} \sum_i y_i \\ \sum_i x_i y_i \\ \vdots \\ \sum_i x_i^{n-1} y_i \end{pmatrix}.$$

First of all, we consider the case $n = 2$, corresponding to the determination of a straight line called the *regression line*. In this case matrix $\widetilde{A}$ and vector

$\widetilde{b}$ are

$$\widetilde{A} = \begin{pmatrix} m & \sum_i x_i \\ \sum_i x_i & \sum_i x_i^2 \end{pmatrix}, \quad \widetilde{b} = \begin{pmatrix} \sum_i y_i \\ \sum_i x_i y_i \end{pmatrix}. \tag{4.25}$$

The determinant of $\widetilde{A}$,

$$\Delta = m \left(\sum_{i=1}^m x_i^2 \right) - \left(\sum_{i=1}^m x_i \right)^2 = m \sum_{i=1}^m \left(x_i - \frac{1}{m} \sum_{j=1}^m x_j \right)^2,$$

vanishes only if all the points x_i are identical. Hence the matrix $\widetilde{A}$ is invertible and the system (4.25) has a unique solution.

Let us go back to the general case. Noticing that the Vandermonde matrix

$$A = \begin{pmatrix} 1 & x_1 & \cdots & x_1^{n-1} \\ \vdots & \vdots & & \vdots \\ 1 & x_m & \cdots & x_m^{n-1} \end{pmatrix} \in \mathbb{R}^{m \times n}$$

is such that $\widetilde{A} = A^T A$ and $\widetilde{b} = A^T b$ with $b = (y_1, \ldots, y_m)^T$, we can write the system (4.24) as

$$A^T A a = A^T b. \tag{4.26}$$

These equations are called normal equations. The following theorem says that solutions of (4.26) are solutions of the minimization problem: find $a \in \mathbb{R}^n$ such that

$$\|Aa - b\| = \inf_{x \in \mathbb{R}^n} \|Ax - b\|. \tag{4.27}$$

Theorem 4.5 *A vector $a \in \mathbb{R}^n$ is solution of the normal equations (4.26) if and only if a is the solution to the minimization problem (4.27).*

Hence, to solve the least squares problem, one can either solve problem (4.27) by some optimization algorithms, or solve problem (4.26) by some linear system solvers. See Allaire and Kaber (2006), for instance. To compute a polynomial least squares approximation with MATLAB, use the instruction `polyfit(x,y,n)` with x a vector that contains the values x_i, y a vector that contains the y_i, and n the degree of the least squares polynomial.

Exercise 4.11 Compute the least squares approximation of the function $f(x) = \sin(2\pi \cos(\pi x))$ defined in Exercise 4.10. The optimal degree n could be determined in the following way. Starting from $n = 0$, one increases n in steps of 1 until the relative error $|e_n - e_{n-1}|/e_{n-1}$ becomes smaller than a pre-scribed value ($\frac{1}{2}$ for example). Here we consider $e_n = \|x - p_n(x)\|_2$ computed at a fixed set of points. For example, the set obtained by `linspace(0,1,m+1)` with $m = 10$.

A solution to this exercise is proposed in Sect. 4.6 at page 102.

4.4 Piecewise Polynomial Approximation

We display in Fig. 4.5 some Lagrange polynomial interpolants of the function f, defined on $I = [0, 1]$ by

$$f(x) = \begin{cases} 1 & \text{for } 0 \leq x \leq 0.25, \\ 2 - 4x & \text{for } 0.25 \leq x \leq 0.5, \\ 0 & \text{for } 0.5 \leq x \leq 1, \end{cases}$$

at respectively 4, 6, 8, and 10 points. Obviously, there is a problem although this function has a very simple structure; it is affine on each interval $[0, \frac{1}{4}]$, $[\frac{1}{4}, \frac{1}{2}]$ and $[1/2, 1]$. The problem is due to the lack of global regularity of f over the whole interval I.

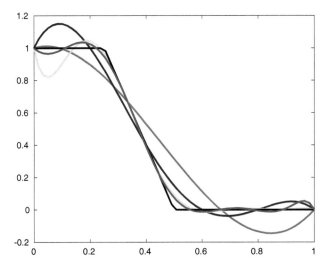

Fig. 4.5 Polynomial interpolation of a piecewise affine function.

The use of piecewise polynomials is a way to control problems related to this lack of global regularity. Another practical reason is the stability of the numerical computations: it is better to use several polynomials of low degree than one polynomial with high degree.

Let f be a continuous function defined on I. The goal is to approximate f by a piecewise polynomial function S. Such a function is called *spline*. The interval is divided into subintervals $I_i = [x_i, x_{i+1}]$ for $i = 1, \dots, n-1$. On each subinterval I_i, the function f is approximated by a polynomial $p_{k,i}$ of degree k. We denote by S_k the piecewise polynomial that coincides with $p_{k,i}$ on each interval I_i and satisfies some global regularity condition on the interval I: continuity, differentiability up to some order, etc.

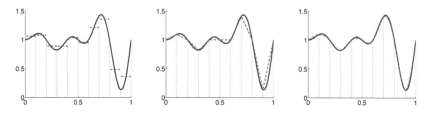

Fig. 4.6 From left to right: examples of piecewise constant, affine, and cubic approximations.

Piecewise Constant Approximation

Let S_0 be a function that is constant on each interval I_i and interpolates f at points $x_{i+1/2} = (x_i + x_{i+1})/2$:

$$S_{0|I_i}(x) = f(x_{i+1/2}).$$

Suppose that $f \in \mathcal{C}^1(I)$. According to Proposition 4.3, for all $x \in I_i$, there exists $\xi_{x,i} \in I_i$ such that

$$f(x) - S_0(x) = (x - x_{i+1/2}) f'(\xi_{x,i}).$$

We deduce from this that if the points x_i are equidistant $(x_{i+1} - x_i = h = 1/n)$ then

$$\|f - S_0\|_\infty \le \frac{h}{2} M_1, \tag{4.28}$$

with M_1 an upper bound of f' on I. Hence, as h goes to 0, S_0 converges uniformly toward f.

Remark 4.2 The power of h in (4.28) indicates that if the discretization parameter h is divided by a constant $c > 0$, the bound on the error $\|f - S_0\|_\infty$ is divided by the same constant c.

Exercise 4.12 Let $f : [0,1] \mapsto f(x) = \sin(4\pi x)$. Draw the curve $\ln n \mapsto \ln \|f - S_0\|_\infty$ and check (an approximation of) the estimate (4.28). Take the values $n = 10\,k$ with $k = 1, \ldots, 10$.
A solution of this exercise is proposed in Sect. 4.6 at page 102.

Piecewise Affine Approximation

We seek a spline S_1 that is affine on each interval I_i and coincides with f at the endpoints x_i and x_{i+1}:

$$S_{1|I_i}(x) = \frac{f(x_{i+1}) - f(x_i)}{h}(x - x_i) + f(x_i).$$

First suppose $f \in C^2(I)$. According to Proposition 4.3, for all $x \in I_i$, there exists $\xi_{x,i} \in I_i$ such that

$$f(x) - S_1(x) = \frac{(x - x_i)(x - x_{i+1})}{2} f''(\xi_{x,i}).$$

We deduce from this

$$\|f - S_1\|_\infty \leq \frac{h^2}{8} M_2, \tag{4.29}$$

with M_2 an upper bound of f'' on I. Hence the uniform convergence of S_1 toward f.

Remark 4.3 The power of h in (4.29) indicates that if the discretization parameter h is divided by a constant $c > 0$, the bound on the error $\|f - S_1\|_\infty$ is divided by c^2. For example, changing h into $h/2$ divides the bound on the error by 4.

Exercise 4.13 Same questions as in the previous exercise to check the estimate (4.29).
A solution of this exercise is proposed in Sect.4.6 at page 104.

If f is only in C^1, convergence holds too. To prove it, write $f(x)$ as an integral,

$$f(x) = f(x_i) + \int_{x_i}^x f'(t)dt, \quad \text{and} \quad S_1(x) = f(x_i) + \frac{x - x_i}{h} \int_{x_i}^{x_{i+1}} f'(t)dt,$$

and use the assumed bound on f',

$$\|f - S_1\|_\infty \leq 2hM_1.$$

That implies convergence. Note that this estimate is less accurate than (4.29), but it requires less regularity on function f.

Piecewise Cubic Approximation

Now we seek a cubic spline, that is $S_3 \in C^2(I)$ is a cubic polynomial on each interval I_i and coincides with f at the endpoints x_i and x_{i+1}. Let p_i be the restriction of S_3 to the interval I_i, for $i = 0, \ldots, n - 1$:

$$p_i(x) = a_i(x - x_i)^3 + b_i(x - x_i)^2 + c_i(x - x_i) + d_i.$$

Obviously $d_i = f(x_i)$. The unknowns a_i, b_i, and c_i can be expressed in terms of the values of f and its second derivative at points x_i. Setting $\alpha_i = p_i''(x_i)$ and using the continuity of the first and second derivatives of the approximation at the points x_i, we get for $i = 0, \ldots, n - 1$,

$$b_i = \frac{1}{2}\alpha_i, \quad a_i = \frac{\alpha_{i+1} - \alpha_i}{6h}, \quad c_i = \frac{f_{i+1} - f_i}{h} - \frac{2\alpha_i + \alpha_{i+1}}{6}h,$$

and a recurrence relation between consecutive values α_{i-1}, α_i, and α_{i+1}:

$$h(\alpha_{i-1} + 4\alpha_i + \alpha_{i+1}) = \frac{6}{h}(f_{i-1} - 2f_i + f_{i+1}).$$

We have to add to these $n-1$ equations two other equations in order to close the system and compute the $n+1$ unknowns α_i. Several choices of these two equations exist. If α_0 and α_n are fixed, say

$$\alpha_0 = \alpha_n = 0, \tag{4.30}$$

the vector $\alpha = (\alpha_1, \ldots, \alpha_{n-1})^T$ is a solution of the linear tridiagonal system $Ax = b$, with

$$A = h \begin{pmatrix} 4 & 1 & 0 & \ldots & 0 \\ 1 & 4 & 1 & & \vdots \\ 0 & \ddots & \ddots & \ddots & 0 \\ \vdots & & 1 & 4 & 1 \\ 0 & \ldots & 0 & 1 & 4 \end{pmatrix} \quad \text{and} \quad b = \frac{6}{h} \begin{pmatrix} f_0 - 2f_1 + f_2 \\ \vdots \\ f_{i-1} - 2f_i + f_{i+1} \\ \vdots \\ f_{n-2} - 2f_{n-1} + f_n \end{pmatrix}. \tag{4.31}$$

The matrix A is invertible (nonsingular) since its diagonal is strictly dominant.

Exercise 4.14 Write a program that computes the cubic spline with conditions (4.30) and $n+1$ points $(i/n)_{i=0}^n$. Test your program with the function $f(x) = \sin(4\pi x)$. Take $n = 5$, then $n = 10$. Draw on the same plot the function f and the spline. In order to see the behavior of the spline between two interpolation points, add ten or twenty more points in each interval I_i to get a smooth graphical representation.
A solution to this exercise is proposed in Sect. 4.6 at page 104.

4.5 Further Reading

For the general theory of polynomial approximation, we refer the reader to Rivlin (1981) and DeVore and Lorentz (1993). The Legendre polynomials are used in Chap. 6 to solve a differential equation. See Bernardi and Maday (1997) for the use of spectral methods in numerical analysis. Related to the splines are the Bézier curves, which have many applications in computer-aided geometric design, see Chap. 11.

4.6 Solutions and Programs

Solution of Exercise 4.1.

1. $b = (f(x_0), \ldots, f(x_n))^T$ and $A = \begin{pmatrix} 1 & x_0 & x_0^2 & \cdots & x_0^n \\ 1 & x_1 & x_1^2 & \cdots & x_1^n \\ \vdots & \vdots & \vdots & & \vdots \\ 1 & x_n & x_n^2 & \cdots & x_n^n \end{pmatrix}$.

2.
```
n=10;x=sort(rand(n+1,1));
A=ones(length(x),1);
for k=1:length(x)-1
    A=[A x.^k];
end;
```

3.
```
cf=A\test1(x);
%reordering of the coefficients
cf=cf(end:-1:1);
y=polyval(cf,x);
plot(x,test1(x),x,y,'r+');
```
the function test1 is defined by

```
test1=inline('sin(10.*x.*cos(x))');
```

4. A is a Vandermonde matrix, it is invertible if all the points are distinct. That is the case in this experiment. However, for MATLAB, $A\alpha - b$ is not zero. This is due to the very bad condition number of matrix A. Recall that the condition number of a matrix A measures the sensitivity of linear system $Ax = b$ to perturbations of the data A or b. For large values of n (say $n = 10$), matrix A becomes a singular matrix for MATLAB: the numerical rank of matrix A computed by MATLAB is e^{18}, while the correct one is $n + 1$.

See the script in the file APP2_M_PolyCanonic.m.

Solution of Exercise 4.2.

The following script is written in file APP2_M_PolyCanonicCond.m..

```
%Condition number of a Vandermonde matrix
N=2:2:20;cd=[ ];
for n=N
    cd=[cd APP2_F_PolycondVdM(n)];
end;
plot(N,log(cd),'+-')
```

It uses function APP2_F_PolycondVdM defined as follows:

```
function y=APP2_F_PolycondVdM(n)
% compute the condition number of a Vandermonde matrix
% The n+1 points are uniformly chosen between 0 and 1.
x=(0:n)'/n;
A=ones(length(x),1);
for k=1:length(x)-1
    A=[A x.^k];
end;
y=cond(A);
```

We deduce from the plot that $\ln(\text{cond}(A))$, as a function of n, is a straight line. Hence cond(A) grows exponentially with n. The reader is asked to compare function APP2_F_PolycondVdM to the following function, in terms of numerical complexity

```
function c=APP2_F_PolycondVdMBis(n)
%compute the condition number of a Vandermonde matrix
%The n+1 points are uniformly chosen between 0 and 1.
x=(0:n)'/n;
A=ones(length(x),1);y=x;
for k=1:length(x)-1
    A=[A y];y=y.*x;
end;
c=cond(A);
```

Solution of Exercise 4.3.

```
%The Lagrange basis
n=10;x=(0:n)'/n;i=round(n/2);
twon=n;g=(0:twon)'/twon;
y=zeros(size(x));y(i)=1;cf=polyfit(x,y,n);
y0=polyval(cf,0);
```

For $n = 5$ or 10 everything goes well, since the program computes a value for $\ell_{n/2}(x_0)$ close to the exact value 0. But for $n = 20$, the computation of $\ell_{n/2}(x_0)$ gives -1.0460. This is again a consequence of the ill conditioning of the matrix. Note that in that case, MATLAB displays a warning message. See the script in APP2_M_PolyLagrange.m.

Solution of Exercise 4.4.

1. The function APP2_F_PolyDD defined below computes the divided differences.

```
function c=APP2_F_PolyDD(x)
```

```
% x contains the points xi
% c contains the divided differences
c=f(x); % f is defined either in another file or "inline"
n=length(x);
for p=1:n-1
    for k=n:-1:p+1
        c(k)=(c(k)-c(k-1))/(x(k)-x(k-p));
    end;
end;
```

It is sometimes useful to send the name of a function as an input parameter of APP2_F_PolyDD.

```
function c=APP2_F_PolyDD(x,f)
c=feval(f,x);
...
```

2. This script is available in file APP2_M_PolyDD.m as well as functions APP2_F_PolyDD and APP2_F_Polyinterpol.

```
function y=APP2_F_Polyinterpol(c,x,g)
%compute the interpolation of the function f on the grid g
%knowing the divided differences c computed at the points x
n=length(c);
y=c(n)*ones(size(g));
for k=n-1:-1:1
    y=c(k)+y.*(g-x(k));
end;

n=20;x=(0:n)'/n;g=0:0.01:1;
c=APP2_F_PolyDD(x);y=APP2_F_PolyInterpol(c,x,g);
yg=test1(g);plot(g,yg,g,y,'r+')
hold on;yx=test1(x);plot(x,yx,'O');hold off
```

Solution of Exercise 4.5.

The script of this exercise is available in file APP2_M_PolyLebesgue.m as well as function APP2_F_PolyLebesgue.

1. Computation of the Lebesgue constant:

```
function leb=APP2_F_PolyLebesgue(x)
%Computation of the Lebesgue constant related
%to the points in the array x
n=length(x)-1;
xx=linspace(min(x),max(x),100);
%fine grid of 100 points
```

```
y=zeros(size(xx));
for i=1:n+1;
   %computation of l_i(x)
   l=zeros(size(x));l(i)=1;cf=polyfit(x,l,n);
   y=y+abs(polyval(cf,xx));
end;
leb=max(y);
```

2. The uniform case.

```
ind=[ ];lebE=[ ];
for n=10:5:30
    x=(-n/2:n/2)/n*2;    %equidistant points
    l=APP2_F_PolyLebesgue(x)
    ind=[ind;n];lebE=[lebE;l];
end;
figure(1);plot(ind,log(lebE),'+-')
```

We note from the plot $\ln(\Lambda(n))$, as a function of n, is a straight line with slope (approximately) $\frac{1}{2}$. Hence $\Lambda(n) \approx e^{n/2}$.

3. The Chebyshev case.

```
ind=[ ];lebT=[ ];
for n=10:5:30
    x=cos(pi*(.5+n:-1:0)/(n+1)); %Chebyshev points
    l=APP2_F_PolyLebesgue(x);
    ind=[ind;n];lebT=[lebT;l];
end;
figure(2);plot(log(ind),lebT,'+-')
```

We note from the plot that $e^{\Lambda_{n,T}}$, as a function of n, is a straight line with slope (approximately) 0.6, hence $\Lambda_{n,T} \approx 0.6 \ln(n)$. Indeed, one can prove rigorously that $\Lambda_{n,T} \approx \frac{2}{\pi} \ln n$.

Solution of Exercise 4.6.

See the script in APP2_M_PolyRunge.m. The results for $n = 8, 10, 12$ are shown in Fig. 4.7: the interpolation diverges at the boundaries. Note that in this case the function to be interpolated is very smooth, but its Lebesgue constant "explodes" (see Proposition 4.1).

Solution of Exercise 4.7.

1. Computation of the Hermite interpolant using divided differences. See function APP2_F_PolyHermite. The input data of this function is an array Tab whose first column contains points x_i, and for each i:

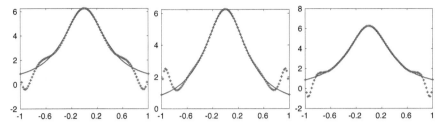

Fig. 4.7 Runge phenomenon. From left to right: interpolation at 9, 11, and 13 equidistant points.

- `Tab(i,1)` contains the points x_i.
- `Tab(i,2)` contains an integer α_i: the function and its derivatives up to α_i are interpolated.
- `Tab(i,3:Tab(i,2)+3)` contains the values of the function at point x_i and its derivatives up to order α_i.

The call `[xx,dd]=APP2_F_PolyHermite(Tab)` returns two vectors:

- the first vector contains points x_i taking into account their "multiplicity": if the function and its α_i derivatives have to be interpolated at the point x_i, this point is copied $\alpha_i + 1$ times in `xx`.
- the vector `dd` contains the divided differences.

With the help of these two vectors, we can implement the Horner algorithm (see page 80) to evaluate $\mathcal{I}_n^H f(x)$.

2. This is done vector-wise as follows:

```
f=inline('cos(3*pi*x).*exp(-x)');
coll=[0 1/4 3/4 1]';
T=[coll zeros(size(coll)) f(coll)];
[xx,dd]=APP2_F_PolyHermite(T);
%plot the function on a fine grid
x=linspace(0,1,100);n=length(dd);
y=dd(n)*ones(size(x));
for k=n-1:-1:1
    y=dd(k)+y.*(x-xx(k));
end;
plot(x,y,x,f(x),'r');hold on;plot(coll,f(coll),'+')
```

See script in `APP2_M_PolyHermite.m`. The left panel of Fig. 4.8 shows the polynomial interpolation (see script `APP2_M_PolyDD.m`. We see that the curves intersect at only the four interpolation points; they do not match at all, except at these points.

3. Imposing the matching of the derivatives forces the polynomial to fit the function more closely (see center panel of Fig. 4.8). However, there is still

a region of the interval $[0, 1]$ where the two curves are not close to each other.

4. By adding another interpolation point in this region, the approximation is much sharper, as shown in the right panel of Fig. 4.8.

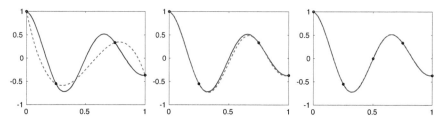

Fig. 4.8 Polynomial approximations (dashed line) of the function f (solid line). Left: the Lagrange interpolation polynomial p_3 at points $0, \frac{1}{4}, \frac{3}{4}$, and 1. Center: the Hermite interpolation polynomial p_7 at the same points. Right: the Hermite interpolation polynomial p_9 at points $0, \frac{1}{4}, \frac{1}{2}, \frac{3}{4}$, and 1. The interpolation points are marked by circles.

Solution of Exercise 4.8.

The implementation is done in the script in `APP2_M_PolyHermiteBis.m`. For several values $m = 5(m' - 1)$ with $m' \in \{1, 2, 3, 4, 5\}$, we compute the corresponding polynomial p_m with the help of the function `APP2_F_PolyHermite`. The column m' of the matrix `Y` contains the values of p_m on a uniform fine grid of 100 points in $[0, 1]$. Fig. 4.9 is generated by the script.

Solution of Exercise 4.9.

The equations form a linear system of $n + 1$ equations and $n + 1$ unknowns (the coefficients of the polynomial). This system has a unique solution if and only if the unique solution of the homogeneous problem (i.e., $f = 0$) is the null polynomial. This is the case since:

- The null polynomial is a trivial solution.
- If $p \in \mathbb{P}_n$ is a solution of the problem and if $p(x_0) = 0$, we deduce that $p(x_i) = 0$ for all $i = 1, \ldots, n$. Hence p is the null polynomial (a polynomial of $\mathbb{P}_n$ cannot have more than n distinct zeros). If $p(x_0) \neq 0$ then p has alternating signs between two successive x_i; hence it vanishes at $n + 1$ distinct points and is again the null polynomial.

Writing $p(t) = \sum_{j=0}^{n} p_j t^j$, the equations become

$$p(x_i) - (-1)^i p(x_0) = f(x_i) - (-1)^i f(x_0)\}, \qquad i = 1, \ldots, n + 1.$$

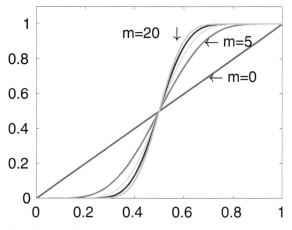

Fig. 4.9 Hermite polynomial interpolants.

Vector $a = (p_0, \ldots, p_n)^T$ solves the system $Aa = b$ with

$$A_{i,j} = x_i^j - (-1)^i x_0^j, \quad b_i = f(x_i) - (-1)^i f(x_0) \qquad (1 \le i \le n+1, 0 \le j \le n).$$

The implementation is done in the function APP2_F_PolyEquiosc. The input data of this function is a vector containing the values x_i. The function computes matrix A and vector b defined above and returns the coefficients p_j of the polynomial.

Solution of Exercise 4.10.

Function APP2_F_PolyRemez computes, for a given integer n, the polynomial of best uniform approximation of a function f. Three cases are considered for the initialization of the algorithm: the Chebyshev points, equidistant points, and randomly chosen points. The parameter Tol relaxes the equality constraint in step k of the algorithm (that is, a test of the form $a = b$ is replaced by the test $|a - b| < $ Tol). For $n = 5, 10, 15$, the best uniform approximations on $[0, 1]$ of the function $x \mapsto \sin(2\pi \cos(\pi x))$ are displayed in Fig. 4.10.
For $n = 15$, the algorithm initialized with the Chebyshev points converges in 21 iterations. Initialized with equidistant points, it converges in 28 iterations. Initialized with random points, it does not converge (in general) after 100 iterations.

The good result obtained with the Chebyshev points can be explained: since the function f has a Chebyshev expansion, analogous to the Legendre series in (4.18),

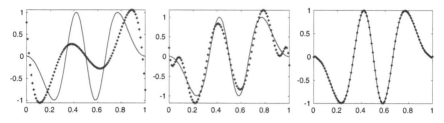

Fig. 4.10 Best uniform approximation in $\mathbb{P}_n$ (crosses) of $f(x) = \sin(2\pi\cos(\pi x))$ (solid line). From left to right: $n = 5$, $n = 10$, and $n = 15$.

$$f = \sum_{k=0}^{\infty} \hat{f}_k T_k,$$

we can use the approximation

$$f - \sum_{k=0}^{n} \hat{f}_k T_k \approx \hat{f}_{n+1} T_{n+1}$$

by neglecting the remainder of the expansion. We remark that T_{n+1} equioscillates over $[-1,1]$ at the $n+2$ points $t_i = \cos(i\frac{\pi}{n+1})$, $(i = 0,\ldots,n+1)$ since $T_{n+1}(t_i) = \cos(i\pi) = (-1)^i$. Hence $\sum_{k=0}^{n} \hat{f}_k T_k$ is close to the best uniform approximation of f in $\mathbb{P}_n$ (see Theorem 4.3). This is the reason why the Chebyshev points are good candidates for the initialization of the Remez algorithm.

Solution of Exercise 4.11.

The script in APP2_M_PolyBpa2.m computes the least squares approximation on $[0,1]$ of a given function. The instruction p=polyfit(x,y,n) returns an array p containing the coefficients of the polynomial with degree less than or equal to n that interpolates the values y(i) at the points x(i). In order to evaluate this polynomial on a grid of points, we use the MATLAB function polyval. Running the script in APP2_M_PolyBpa2.m returns the value $n = 10$. The polynomial is displayed in Fig. 4.11 with the points (x_i, y_i) marked by the symbol 'o'.

Solution of Exercise 4.12.

Running this following script (available in APP2_M_PolySpline0.m.) produces a plot. The slope of the displayed straight line is about -0.971325, which is a good approximation of the exact value -1 given by (4.28).

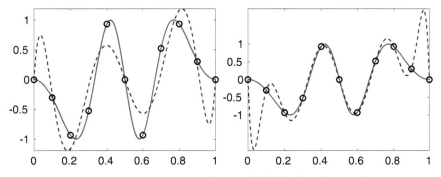

Fig. 4.11 Least squares approximation: fonction (solid line), polynomial approximation (dashed line), and the points $(x_i, f(x_i))$. Left: $n = 9$, right: $n = 10$.

```
n0=10;E=[];N=[];
for i=1:10,
    n=i*n0;E=[E;errorS0(n)];N=[N;n];
end;
%
loglog(N,E,'-+','MarkerSize',10,'LineWidth',2);
set(gca,'FontSize',20);
xlabel('log n');ylabel('log Error');
r=(log(E(end))-log(E(1)))/(log(N(end))-log(N(1)));
fprintf('slope of the straight line  = %g\n ',r);

function y=errorS0(n)
x=(0:n)'/n;h=1/n;fx=f(x);
%Evaluation of $p_i$ on each interval $[x_i,x_i+1]$
y=[];
for i=1:n
    Ii=linspace(x(i),x(i+1),20);
    fi=f(Ii);
    Si=f(.5*(x(i)+x(i+1)));
    y=[y norm(Si-fi,'inf')];
end
y=max(y);
end

function y=f(x)
y=sin(4*pi*x);
end
```

Solution of Exercise 4.13.

The script is available in file `APP2_M_PolySpline1.m`). The slope of the straight line is now about -1.965, which is a good approximation of the exact value -2 given by (4.29).

Solution of Exercise 4.14.

See the script in file `APP2_M_PolySpline3.m`. The results obtained with $n = 5$ and $n = 10$ are displayed in Fig. 4.12.

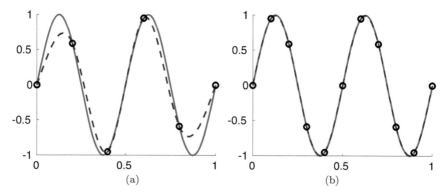

(a)   (b)

Fig. 4.12 Cubic splines: approximation (dashed line) of f (solid line) on 5 intervals (a) and 10 intervals (b).

Chapter References

G. Allaire, S.M. Kaber, *Numerical Linear Algebra.* (Springer, New York, forthcoming, 2007)

C. Bernardi, Y. Maday, *Spectral Methods, in Handbook of Numerical Analysis,* vol. V (North-Holland, Amsterdam, 1997)

M. Crouzeix, A. Mignot, *Analyse Numérique des équations différentielles* (Masson, Paris, 1989)

R.A. DeVore, G.G. Lorentz, *Constructive Approximation* (Springer-Verlag, Berlin, 1993)

T.J. Rivlin, *An Introduction to the Approximation of Functions* (Dover Publications Inc., New York, 1981)

L. Schwartz, *Analyse, topologie générale et analyse fonctionnelle* (Hermann, Paris, 1980)

L.N. Trefethen, *Spectral Methods in MATLAB, Software, Environments, and Tools,* vol. 10 (SIAM, Philadelphia, 2000)

Chapter 5
Solving an Advection–Diffusion Equation by a Finite Element Method

Project Summary

Level of difficulty:	1
Keywords:	Advection–diffusion equation, finite element method, stabilization of a numerical scheme
Application fields:	Advection and diffusion phenomena

In this project, we seek a numerical approximation of the solution $u : [0,1] \longrightarrow \mathbb{R}$ of the following problem:

$$\begin{cases} -\varepsilon u''(x) + \lambda u'(x) = f(x), & x \in \,]0,1[\,, \\ u(0) = 0, \\ u(1) = 0. \end{cases} \tag{5.1}$$

The function f and the real numbers $\varepsilon > 0$ and λ are chosen in such a way that there exists a unique twice differentiable solution of this problem. Our aim is to approximate the solution using a continuous piecewise polynomial function.

The differential equation in the problem (5.1) is an advection–diffusion equation. It models several phenomena, as, for example, the concentration of some chemical species transported in a fluid with speed λ; the parameter ε is the diffusivity of the chemical species. The ratio $\theta = \lambda/\varepsilon$ measures the importance of the advection compared to the diffusion. For large values of this ratio, the numerical solution of the problem (5.1) is delicate. The production and the vanishing of the chemical species are modeled by function f, which in the general case depends on the unknown u. In this problem, we assume that f depends only on the position x and we consider λ and ε as constants.

© The Author(s), under exclusive license to Springer Nature Switzerland AG 2023   105
I. Danaila et al., *An Introduction to Scientific Computing*,
https://doi.org/10.1007/978-3-031-35032-0_5

5.1 Variational Formulation of the Problem

A solution u of the boundary value problem (5.1) is also a solution of the following problem:

$$\begin{cases} \text{Find } u \in V \text{ such that} \\ \text{for all } v \in V : \quad a(u, v) = \int_0^1 f(x)v(x)dx. \end{cases} \tag{5.2}$$

Here, V denotes $H_0^1(]0,1[)$, the space of functions $v : [0,1] \to \mathbb{R}$ such that the integrals $\int_0^1 |v|^2$ and $\int_0^1 |v'|^2$ are bounded and $v(0) = v(1) = 0$. The bilinear form a is defined on $V \times V$ by

$$a(u, v) = \varepsilon \int_0^1 u'(x)v'(x)dx + \lambda \int_0^1 u'(x)v(x)dx. \tag{5.3}$$

Conversely, every regular solution of (5.2) is also a solution of (5.1). The problem (5.2) is called a **variational formulation** of (5.1). The finite element method is based on the computation of the solution of the variational problem while finite difference methods are based on a direct discretization of equation (5.1) (see Chaps. 1, 2, and 7).

Given a strictly positive integer n, we divide the interval $[0,1]$ into $n+1$ subintervals I_i. For a positive integer ℓ, we denote by $\mathbb{P}_\ell(I_i)$ the set of algebraic polynomials of degree less than or equal to ℓ on I_i and $\mathcal{V}_\ell^h$ the set of continuous functions defined on $[0,1]$ whose restriction to each interval I_i belongs to $\mathbb{P}_\ell(I_i)$. Figure 5.1 displays two examples of functions of $\mathcal{V}_\ell^h$.

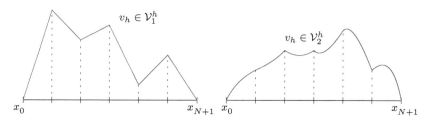

Fig. 5.1 Examples of functions in $\mathcal{V}_1^h$ (left) and in $\mathcal{V}_2^h$ (right).

The finite element method (Atkinson and Han (2001), Bernardi et al. (2004), Danaila et al. (2003), Joly (1990), Le Dret and Lucquin (2016), Lucquin and Pironneau (1996)) consists in searching for an approximation $u_h \in \mathcal{V}_\ell^h$ of the function u, defined as the solution of the following problem (compared to (5.2)):

$$\begin{cases} \text{Find } u_h \in \mathcal{V}_\ell^h \text{ such that} \\ \text{for all } v_h \in \mathcal{V}_\ell^h : \quad a(u_h, v_h) = \int_0^1 f(x)v_h(x)dx. \end{cases} \tag{5.4}$$

The integrals on the left-hand side of (5.4) are easily computed since they involve products of polynomials. More difficult is the computation of the integral $\int_0^1 f(x)v_h(x)dx$, whose explicit calculation is rarely possible. In this case, some quadrature rules are necessary. We will make use of two rules:

- The trapezoidal quadrature rule

$$\int_\alpha^\beta g(x)dx \approx (\beta - \alpha)\frac{g(\alpha) + g(\beta)}{2}.$$

This method is of order 1, that is, it is exact for all $g \in \mathbb{P}_1([\alpha, \beta])$.
- The Simpson quadrature rule

$$\int_\alpha^\beta g(x)dx \approx \frac{\beta - \alpha}{6}\left(g(\alpha) + 4g(\frac{\alpha + \beta}{2}) + g(\beta)\right). \tag{5.5}$$

This method is of order 3; it is exact for all $g \in \mathbb{P}_3([\alpha, \beta])$.

Using one of these basic quadrature formulas on each subinterval I_i, we get a quadrature formula on the whole interval $[0, 1]$.

In this project, we compare two finite element methods (FEM) to solve the advection–diffusion problem. The first method is called $P1$ since it uses functions in $\mathcal{V}_1^h$; the second method is a $P2$ method since the approximation space for this method is $\mathcal{V}_2^h$.

To validate the computations, we have to compare the computed solution to the exact one. Generally, this can be done only in some special simple cases. The aim of the first exercise is to compute the exact solution of (5.1) in the case of a constant source term.

Exercise 5.1 The exact solution of (5.1) in the case of constant f.

1. Derive explicit formulas for the solution u of problem (5.1).
2. Prove the existence of $x_\theta \in]0, 1[$ depending only on the ratio $\theta = \lambda/\varepsilon$ such that the function u is strictly increasing (respectively decreasing) over $]0, x_\theta[$ (respectively $]x_\theta, 1[$). Determine $\lim_{|\theta|\to+\infty} x_\theta$.
3. For $\lambda > 0$ fixed, we are interested in the behavior of the solution u for ε going to 0^+ (and thus $\theta \to +\infty$). Determine $u(x_\theta)$ and $\lim_{\varepsilon\to0+} u(x_\theta)$. Show that

$$\lim_{\varepsilon\to0+} \lim_{x\to1} u(x) \neq \lim_{x\to1} \lim_{\varepsilon\to0+} u(x).$$

Explain the meaning of the sentence, "*for small values of ε, the solution of the differential problem (5.1) contains a thin boundary layer in a neighborhood of point $x = 1$*".
4. Write a program that computes the values of the solution u on a given set of points (an array). Plot u for $f = 1$, $\lambda \in \{-1, 1\}$, and $\varepsilon \in \{1, 0.5, 10^{-1}, 10^{-2}\}$. Comment on the results.

A solution of this exercise is proposed in Sect. 5.6 at page 118.

5.2 A $P1$ Finite Element Method

For $n \in \mathbb{N}^*$, we define the points

$$x_k^{(1)} = kh, \qquad k = 0, \dots, n+1,$$

and the intervals

$$I_k = \left] x_k^{(1)}, x_{k+1}^{(1)} \right[, \qquad k = 0, \dots, n,$$

with grid size $h = 1/(n+1)$. We also define k "hat functions" $\varphi_{h,k}^{(1)}$ ($k = 1, \dots, n$) (see Fig. 5.2), such that

$$\varphi_{h,k}^{(1)} \in \mathcal{V}_1^h \quad \text{and} \quad \varphi_{h,k}^{(1)}(x_j^{(1)}) = \delta_{j,k}, \quad \forall j = 1, \dots, n,$$

with $\delta_{j,k}$ the Kronecker symbol. Note that the support of the function $\varphi_{h,k}^{(1)}$ is the union of two intervals I_{k-1} and I_k.

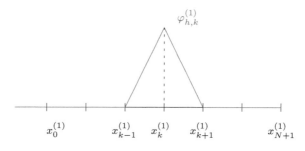

Fig. 5.2 A hat function $\varphi_{h,k}^{(1)}$ of $\mathcal{V}_1^h$.

In the finite element methods, the points $x_k^{(1)}$ are called *nodes* and the intervals I_k are called *cells*. We seek an approximation $u_h^{(1)} \in \mathcal{V}_1^h$ of the function u, a solution of the problem (5.4) with $\ell = 1$:

$$\begin{cases} \text{Find } u_h^{(1)} \in \mathcal{V}_1^h \text{ such that} \\ \text{for all } v_h \in \mathcal{V}_1^h : \quad a\left(u_h^{(1)}, v_h\right) = \int_0^1 f(x) v_h(x) dx. \end{cases} \qquad (5.6)$$

Exercise 5.2

1. Prove that the functions $(\varphi_{h,k}^{(1)})_{k=1}^n$ form a basis of $\mathcal{V}_1^h$.
2. Deduce that problem (5.6) is equivalent to the following problem:

$$\begin{cases} \text{Find } u_h^{(1)} \in V_1^h \text{ such that} \\ \text{for all } k = 1, \dots, n : \quad a\left(u_h^{(1)}, \varphi_{h,k}^{(1)}\right) = \int_0^1 f(x)\varphi_{h,k}^{(1)}(x)dx. \end{cases} \tag{5.7}$$

3. By expanding $u_h^{(1)}$ in the basis $(\varphi_{h,k}^{(1)})_{k=1}^n$,

$$u_h^{(1)} = \sum_{m=1}^n \alpha_m \varphi_{h,m}^{(1)},$$

show that $\alpha_k = u_h(x_k^{(1)})$. Show that $\tilde{u}_h^{(1)} = (u_h^{(1)}(x_1^{(1)}), \dots, u_h^{(1)}(x_n^{(1)}))^T$ is a solution of the linear system

$$A_h^{(1)} \tilde{u}_h^{(1)} = b_h^{(1)}, \tag{5.8}$$

where $A_h^{(1)}$ is the real matrix of size $n \times n$ defined by

$$(A_h^{(1)})_{k,m} = a(\varphi_{h,m}^{(1)}, \varphi_{h,k}^{(1)}), \quad 1 \le m, k \le n,$$

and $b_h^{(1)}$ is the vector of $\mathbb{R}^n$ with elements

$$(b_h^{(1)})_k = \int_0^1 f(x)\varphi_{h,k}^{(1)}(x)dx, \quad 1 \le k \le n.$$

Show that $A_h^{(1)} = \varepsilon B_h^{(1)} + \lambda C_h^{(1)}$, where $B_h^{(1)}$ is a tridiagonal symmetric matrix and $C_h^{(1)}$ a tridiagonal skew-symmetric matrix.

4. Prove that the symmetric matrix $B_h^{(1)}$ is positive definite, i.e., for all $x \in \mathbb{R}^n$, $\langle B_h^{(1)} x, x \rangle \ge 0$, with equality if and only if x is the null vector ($\langle ., . \rangle$ denotes the usual inner product). This property is very useful in the numerical analysis of linear problems. It implies, in particular, invertiblity of the matrix.

5. Show that $\langle A_h^{(1)} x, x \rangle = \langle B_h^{(1)} x, x \rangle$. Conclude that $A_h^{(1)}$ is invertible.

A solution of this exercise is proposed in Sect. 5.6 at page 119.

System (5.7) has a unique solution that will be computed now by solving the linear system (5.8).

Exercise 5.3 Computation of the *P*1 solution by solving (5.8).

1. Derive the following explicit formulas for $B_h^{(1)}$ and $C_h^{(1)}$:

$$B_h^{(1)} = \frac{1}{h} \begin{pmatrix} 2 & -1 & 0 & \cdots & \cdots & 0 \\ -1 & 2 & -1 & 0 & & \vdots \\ 0 & & \ddots & \ddots & \ddots & \vdots \\ \vdots & & & \ddots & \ddots & \vdots \\ \vdots & & 0 & -1 & 2 & -1 \\ 0 & \cdots & \cdots & 0 & -1 & 2 \end{pmatrix}, \quad C_h^{(1)} = \frac{1}{2} \begin{pmatrix} 0 & 1 & 0 & \cdots & \cdots & 0 \\ -1 & 0 & 1 & 0 & & \vdots \\ 0 & & \ddots & \ddots & \ddots & \vdots \\ \vdots & & & \ddots & \ddots & \vdots \\ \vdots & & 0 & -1 & 0 & 1 \\ 0 & \cdots & \cdots & 0 & -1 & 0 \end{pmatrix}.$$

2. Write a program that computes matrix $A_h^{(1)}$ with input data n: ε and λ.
3. Write a program that computes the right-hand side $b_h^{(1)}$ (input data: n and f). Use the trapezoidal rule to compute the components of the vector $b_h^{(1)}$.
4. For $\varepsilon = 0.1$, $\lambda = 1$, $f = 1$, and $n \in \{10, 20\}$, compute the solution $\tilde{u}_h$ of (5.8) and compare it to the exact solution.
5. Error analysis. Let $e_h^{(1)}(x) = u_h^{(1)}(x) - u(x)$ denote the pointwise error. Its L^2 norm is $\|e_h^{(1)}\|_2^2 = \int_I (e_h^{(1)}(x))^2 dx = \sum_k \int_{I_k} (e_h^{(1)}(x))^2 dx$. We could either compute exactly the integrals or make the approximation $\|e_h^{(1)}\|_2^2 \simeq h \sum_k (e_h^{(1)}(x_k^{(1)}))^2$. So, if $\mathbf{e}_n \in \mathbb{R}^n$ is the vector defined by $(\mathbf{e}_n)_k = \tilde{u}_h^{(1)}(k) - u(x_k^{(1)})$, then $\|e_h^{(1)}\|_2 \simeq \sqrt{h}\|\mathbf{e}_n\|_2$. Fix the parameters $\varepsilon = 0.1$, $\lambda = 1$, and $f = 1$. For n going from 10 to 200 in steps of 10, draw the curves $\log n \longmapsto \log \|e_n^{(1)}\|_2$. Deduce a decreasing law for $\|e_n^{(1)}\|_2$ of the form $\|e_n^{(1)}\|_2 \approx \text{constant}/n^s$ with $s > 0$ to be determined (use MATLAB function polyfit).

A solution of this exercise is proposed in Sect. 5.6 at page 121.

The next exercise answers the following question: for a fixed ε, what is the minimum number of subintervals required to resolve the boundary layer?

Exercise 5.4 Fix $\lambda = 1$ and $f = 1$. For various "small" values of ε (for example in the range $[0.005, 0.02]$), determine the integer $n \equiv n(\varepsilon)$ from which the numerical solution seems to be a reasonable approximation of the exact solution in the boundary layer (i.e., the numerical approximation is not oscillating and is close to the exact solution in the boundary layer).
Hint: Use the following strategy: for fixed ε, run the program for $n = 10, 20, 30$, etc. For each value of n, plot the exact solution and the numerical approximation. From these graphs, decide whether the approximation is good or not. For each value of n, compute $P = \frac{|\lambda| h}{2\varepsilon}$, the Peclet number of the grid. What conclusion can be drawn from this?
A solution of this exercise is proposed in Sect. 5.6 at page 122.

The $P1$ finite element method seems to be well suited to solve the advection–diffusion problem. Unfortunately, things are not so simple. For $\lambda = 1$, $f = 1$, $\varepsilon = 0.01$, and $n = 10$, we obtain the results displayed in Fig. 5.3. The oscillatory behavior of $u_h^{(1)}$ shows that it is clearly not a good approximation of u, especially in the boundary layer. We investigate in the next section whether a high-order finite element method is able to suppress these oscillations.

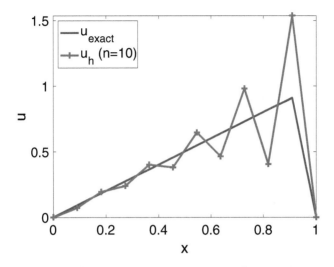

Fig. 5.3 Approximation of the solution of the advection–diffusion problem by a *P*1 finite element method, $\varepsilon = 0.01$, $\lambda = 1$, and $n = 10$.

5.3 A *P*2 Finite Element Method

For $n \in \mathbb{N}^*$, we set $h = 1/(n + 1)$ and define points $x_k^{(2)} = kh/2$ ($k = 0, \ldots, 2(n + 1)$). Note that $x_{2k}^{(2)} = x_k^{(1)}$ and intervals $I_k =]x_{2k}^{(2)}, x_{2k+2}^{(2)}[$ ($k = 0, \ldots, n$) are those used in the previous section. In other words, we keep the same number of intervals and add to each interval a new node, namely the center of the interval. To get a better approximation, we shall associate with each node of the mesh a piecewise quadratic function rather than a piecewise affine one.

We seek an approximation $u_h \in \mathcal{V}_2^h$ of the function u, a solution of problem (5.4) with $\ell = 2$:

$$\begin{cases} \text{Find } u_h^{(2)} \in \mathcal{V}_2^h \text{ such that} \\ \text{for all } v_h \in \mathcal{V}_2^h : \quad a(u_h^{(2)}, v_h) = \int_0^1 f(x)v_h(x)dx. \end{cases} \qquad (5.9)$$

As in the previous section, we begin by building a simple basis of $\mathcal{V}_2^h$. On each interval I_k, we define three quadratic Lagrange polynomials associated with points $x_{2k}^{(2)}$, $x_{2k+1}^{(2)}$, and $x_{2k+2}^{(2)}$:

$$\begin{cases} \psi_{h,k}^{(-)}(x) = 2(x - x_{2k+1}^{(2)})(x - x_{2k+2}^{(2)})/h^2, \\ \psi_{h,k}^{(0)}(x) = -4(x - x_{2k}^{(2)})(x - x_{2k+2}^{(2)})/h^2, \\ \psi_{h,k}^{(+)}(x) = 2(x - x_{2k}^{(2)})(x - x_{2k+1}^{(2)})/h^2. \end{cases}$$

To each node $x_k^{(2)}$ of the mesh, we associate the function $\varphi_{h,k}^{(2)} \in V_2^h$ defined by (see also Fig. 5.4)

$$\varphi_{h,2k+1}^{(2)}(x) = \begin{cases} \psi_{h,k}^{(0)}(x) \text{ for } x \in I_k, \\ 0 \qquad\quad \text{otherwise,} \end{cases} \text{ and } \varphi_{h,2k}^{(2)}(x) = \begin{cases} \psi_{h,k}^{(-)}(x) & \text{for } x \in I_k, \\ \psi_{h,k-1}^{(+)}(x) & \text{for } x \in I_{k-1}, \\ 0 & \text{otherwise.} \end{cases}$$

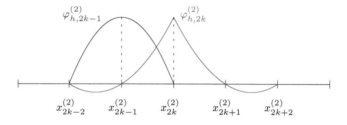

Fig. 5.4 Generic functions forming a basis of V_2^h.

Note that the support of function $\varphi_{h,k}^{(2)}$ is either the interval I_k or the union of intervals I_{k-1} and I_k, according to the parity of k. Since functions $(\varphi_{h,k}^{(2)})_{k=1}^{2n+1}$ form a basis of V_2^h, problem (5.9) is equivalent to the problem

$$\begin{cases} \text{Find } u_h^{(2)} \in V_2^h \text{ such that} \\ \text{for all } k = 1, \ldots, 2n+1: \quad a(u_h^{(2)}, \varphi_{h,k}^{(2)}) = \int_0^1 f(x)\varphi_{h,k}^{(2)}(x)dx. \end{cases} \quad (5.10)$$

Expanding $u_h^{(2)}$ in the basis $\varphi_{h,k}^{(2)}$

$$u_h^{(2)} = \sum_{m=1}^{2n+1} \alpha_m \varphi_{h,m},$$

we prove, as in the $P1$ case, that $\alpha_m = u_h^{(2)}(x_m^{(2)})$ and the vector $\tilde{u}_h^{(2)} = (\alpha_1, \ldots, \alpha_{2n+1})^T$ solves a linear system

$$A_h^{(2)} \tilde{u}_h^{(2)} = b_h^{(2)}, \quad (5.11)$$

with $A_h^{(2)}$ the $(2n+1) \times (2n+1)$ matrix defined by

$$(A_h^{(2)})_{k,m} = a(\varphi_{h,m}^{(2)}, \varphi_{h,k}^{(2)}) \quad 1 \leq m, k \leq 2n+1$$

and b_h the vector of $\mathbb{R}^{2n+1}$ defined by

$$(b_h^{(2)})_k = \int_0^1 f(x)\varphi_{h,k}^{(2)}(x)dx \qquad 1 \le k \le 2n+1.$$

Exercise 5.5 Split matrix $A_h^{(2)}$ as $A_h^{(2)} = \varepsilon B_h^{(2)} + \lambda C_h^{(2)}$. Is matrix $B_h^{(2)}$ symmetric or tridiagonal? Is matrix $C_h^{(2)}$ skew-symmetric or tridiagonal? Prove that matrix $A_h^{(2)}$ is invertible (nonsingular).
A solution of this exercise is proposed in Sect. 5.6 at page 123.

System (5.10) has thus a unique solution, which we compute by solving the linear system (5.11).

Exercise 5.6 Computation of the $P2$ solution.

1. Prove that $B_h^{(2)}$ and $C_h^{(2)}$ have the following patterns (presented here for $n = 3$):

$$B_h^{(2)} = \frac{1}{3h}\begin{pmatrix} 16 & -8 & 0 & 0 & 0 & 0 & 0 \\ -8 & 14 & -8 & 1 & 0 & 0 & 0 \\ 0 & -8 & 16 & -8 & 0 & 0 & 0 \\ 0 & 1 & -8 & 14 & -8 & 1 & 0 \\ 0 & 0 & 0 & -8 & 16 & -8 & 0 \\ 0 & 0 & 0 & 1 & -8 & 14 & -8 \\ 0 & 0 & 0 & 0 & 0 & -8 & 16 \end{pmatrix}, \quad C_h^{(2)} = \frac{1}{6}\begin{pmatrix} 0 & 4 & 0 & 0 & 0 & 0 & 0 \\ -4 & 0 & 4 & -1 & 0 & 0 & 0 \\ 0 & -4 & 0 & 4 & 0 & 0 & 0 \\ 0 & 1 & -4 & 0 & 4 & -1 & 0 \\ 0 & 0 & 0 & -4 & 0 & 4 & 0 \\ 0 & 0 & 0 & 1 & -4 & 0 & 4 \\ 0 & 0 & 0 & 0 & 0 & -4 & 0 \end{pmatrix}.$$

2. Write a program computing matrix $A_h^{(2)}$ (input data: n, ε, and λ).
3. Write a program computing the right-hand side $b_h^{(2)}$ (input data: n and f). Use Simpson's rule to compute the components of the vector $b_h^{(2)}$.
4. For $\varepsilon = 0.1$, $\lambda = 1$, and $f = 1$, compare the error with the $P1$ case.

A solution of this exercise is proposed in Sect. 5.6 at page 123.

Running the $P2$ program with parameters $\lambda = 1$, $f = 1$, $\varepsilon = 0.01$, and $n = 10$, we obtain Fig. . 5.5 (left). At first glance, the oscillations persist as in Fig. 5.3 for the $P1$ method. However, if we consider only the values of the solution at the endpoints of the intervals (the points $x_{2k}^{(2)}$), we observe a nonoscillatory approximation of the exact solution, see Fig. 5.5 (right).

5.4 A Stabilization Method

In this section, we propose a method (Brezzi and Russo, 1994) for removing the oscillations that we have observed in Fig. 5.3.
First of all, we define a way to compute the values of $u_h^{(2)}$ at the endpoints of the intervals, without computing the values at midpoints.

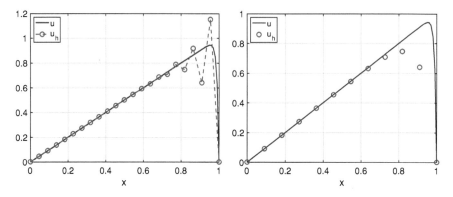

Fig. 5.5 Approximation of the solution of the advection–diffusion problem for $\varepsilon = 0.01$, $\lambda = 1$, and $n = 10$. Left: $P2$ finite element solution; right: the same solution displayed only at the endpoints of the intervals.

5.4.1 Computation of the Solution at the Endpoints of the Intervals

Let $\widehat{A}_h^{(2)}$ denote the matrix obtained after permutation of the rows of the matrix $A_h^{(2)}$ in order to put in the first places the rows with even indices (the same operation is performed for the columns). In the same way, we define the vector $\widehat{b}_h^{(2)}$ from the vector $b_h^{(2)}$. We also define the vector $\widehat{u}_h^{(2)}$ from the vector of unknowns $\tilde{u}_h^{(2)}$. The system (5.11) is equivalent to system

$$\widehat{A}_h^{(2)}\widehat{u}_h^{(2)} = \widehat{b}_h^{(2)},$$

which one could have obtained directly from the variational formulation by changing the numbering of the unknowns. Finally, we split $\widehat{A}_h^{(2)}$, $\widehat{b}_h^{(2)}$, and $\widehat{u}_h^{(2)}$ as follows:

$$\widehat{A}_h^{(2)} = \begin{bmatrix} A & B \\ C & D \end{bmatrix}, \quad \widehat{b}_h^{(2)} = \begin{bmatrix} c \\ d \end{bmatrix}, \quad \widehat{u}_h^{(2)} = \begin{bmatrix} v \\ w \end{bmatrix},$$

where $A \in \mathbb{R}^{n \times n}$, $B \in \mathbb{R}^{n \times (n+1)}$, $C \in \mathbb{R}^{(n+1) \times n}$, $D \in \mathbb{R}^{(n+1) \times (n+1)}$, $c \in \mathbb{R}^n$, $d \in \mathbb{R}^{n+1}$, $v \in \mathbb{R}^n$, $w \in \mathbb{R}^{n+1}$, and

$$
\begin{aligned}
A_{i,j} &= a(\varphi_{h,2j}^{(2)}, \varphi_{h,2i}^{(2)}), & c_i &= \int_0^1 f(x)\varphi_{h,2i}^{(2)}(x)dx, \\
B_{i,j} &= a(\varphi_{h,2j}^{(2)}, \varphi_{h,2i-1}^{(2)}), & d_i &= \int_0^1 f(x)\varphi_{h,2i-1}^{(2)}(x)dx, \\
C_{i,j} &= a(\varphi_{h,2j-1}^{(2)}, \varphi_{h,2i}^{(2)}), & v_i &= u_h^{(2)}(x_{h,2i}^{(2)}), \\
D_{i,j} &= a(\varphi_{h,2j-1}^{(2)}, \varphi_{h,2i-1}^{(2)}), & w_i &= u_h^{(2)}(x_{h,2i-1}^{(2)}).
\end{aligned}
$$

The matrix A is tridiagonal, B is upper triangular and bidiagonal, C is lower triangular and bidiagonal, and D is diagonal with diagonal components

$$D_{k,k} = a(\varphi_{h,2k-1}^{(2)}, \varphi_{h,2k-1}^{(2)}) = \varepsilon \int_{x_{h,2k-2}^{(2)}}^{x_{h,2k}^{(2)}} [\varphi_{h,2k-1}^{(2)}{}'(x)]^2 dx > 0,$$

and consequently it is invertible. The unknowns v and w are solutions of the linear system

$$\begin{cases} Av + Bw = c, \\ Cv + Dw = d. \end{cases} \tag{5.12}$$

From the second equation, we get $w = D^{-1}(d - Cv)$. Plugging this expression for w into the first equation, we obtain

$$(A - BD^{-1}C)v = c - BD^{-1}d. \tag{5.13}$$

This equation allows us to compute the vector v (i.e., the components of the $P2$ solution at the endpoints of the subintervals). Note that matrix $A - BD^{-1}C$ is tridiagonal, whereas matrix $A_h^{(2)}$ is pentadiagonal. This method, which consists in isolating, in a linear system, some of the unknowns that solve another simpler system, is called *condensation*. It is simply a Gaussian elimination of the unknowns associated with the centers of the subintervals.

Exercise 5.7

1. Show that matrix $A - BD^{-1}C$ is invertible.
2. Compute matrices A, B, C, and D by extraction of rows and columns of matrix $A_h^{(2)}$ computed in Exercise 5.6.
3. Fix $\lambda = 1$, $f = 1$, and $\varepsilon = 0.01$. For $n = 10$ and $n = 20$, solve problem (5.13). In the same figure, plot (using solid line) the exact solution computed in 100 points in $[0, 1]$ and the numerical solution (using symbols), i.e., the components of the vector v.
4. For $n = 20$, compare the results with the $P1$ case. What is the minimum value n_0 starting from which the $P1$ method gives the same quality of approximation? (A visual appreciation is enough.)

A solution of this exercise is proposed in Sect. 5.6 at page 125.

In conclusion, the $P2$ method, used only for the endpoints of the subintervals, provides good results. In the next section, we justify these observations.

5.4.2 Analysis of the Stabilized Method

Let us consider only the values $u_h^{(2)}(x_{h,2k}^{(2)})$, i.e., the values of $u_h^{(2)}$ on the grid of the $P1$ method. We prove that these values can be computed by a slightly modified $P1$ scheme. More precisely, setting

$$A_h^{(1S)} = A - BD^{-1}C,$$

the following result holds.

Proposition 5.1 *Matrix* $A_h^{(1S)}$ *equals matrix* $A_h^{(1)}$ *in which* ε *is replaced by* $\varepsilon' = \varepsilon + \lambda^2 h^2/(12\varepsilon)$.

One could understand this result as an addition of a viscosity term to the original scheme that makes the solution smoother (less oscillatory). The remainder of the section is devoted to prove Proposition 5.1. In order to compute the matrix $A_h^{(1S)}$, we denote by $X = \text{Tridiag}(a, b, c; n, m)$ the tridiagonal $n \times m$ matrix X defined by

$$X_{i-1,i} = a, X_{i,i} = b, X_{i,i+1} = c, \quad \text{if these indices are defined.}$$

Note that $\text{Tridiag}(a, b, c; n, m)$ is not necessarily a square matrix. The reader who enjoys long calculations will prove that

$$A = \frac{\varepsilon}{3h} \text{Tridiag}(1, 14, 1; n, n) + \frac{\lambda}{6} \text{Tridiag}(1, 0, -1; n, n),$$

$$B = -\frac{8\varepsilon}{3h} \text{Tridiag}(0, 1, 1; n, n+1) - \frac{2\lambda}{3} \text{Tridiag}(0, 1, -1; n, n+1),$$

$$C = -\frac{8\varepsilon}{3h} \text{Tridiag}(1, 1, 0; n+1, n) + \frac{2\lambda}{3} \text{Tridiag}(-1, 1, 0; n+1, n),$$

$$D = \frac{16\varepsilon}{3h} \text{Tridiag}(0, 1, 0; n+1, n+1),$$

and that

$$BC = \frac{\sigma^2}{4} \text{Tridiag}(1, 2, 1; n, n) + \frac{\lambda\sigma}{3} \text{Tridiag}(2, 0, -2; n, n),$$
$$- \frac{4}{9}\lambda^2 \text{Tridiag}(-1, 2, -1; n, n),$$

where $\sigma = \frac{16\varepsilon}{3h}$. The reader may also calculate

$$A_h^{(1S)} = A - \frac{1}{\sigma}BC = \text{Tridiag}(\alpha, \beta, \gamma; n, n),$$

where

$$\alpha = -\frac{\varepsilon}{h} - \frac{\lambda}{2} - \frac{\lambda^2 h}{12\varepsilon} = -\frac{1}{h}\varepsilon' - \frac{\lambda}{2},$$

$$\beta = 2\frac{\varepsilon}{h} + \frac{\lambda^2 h}{6\varepsilon} = \frac{2}{h}\varepsilon',$$

$$\gamma = -\frac{\varepsilon}{h} + \frac{\lambda}{2} - \frac{\lambda^2 h}{12\varepsilon} = -\frac{1}{h}\varepsilon' + \frac{\lambda}{2}.$$

Finally,

$$A_h^{(1S)} = \frac{\varepsilon'}{h}\,\mathrm{Tridiag}(-1, 2, -1; n, n) + \frac{\lambda}{2}\,\mathrm{Tridiag}(-1, 0, 1; n, n),$$

and the proposition follows since (see Exercise 5.3)

$$A_h^{(1)} = \frac{\varepsilon}{h}\,\mathrm{Tridiag}(-1, 2, -1; n, n) + \frac{\lambda}{2}\,\mathrm{Tridiag}(-1, 0, 1; n, n).$$

5.5 The Case of a Variable Source Term

We consider in the last section the advection–diffusion equation (5.1) with a nonconstant source term f in order to understand the effect of this term on the existence of a boundary layer.

Exercise 5.8

1. Fix $\varepsilon = 0,01$, $\lambda = 1$, and $f(x) = \cos(a\pi x)$, $a \in \mathbb{R}$. For $a = 0$, we already know the existence of a boundary layer near the endpoint $x = 1$. We assume in this exercise that $a > 0$. For several values of $a = 1, 2, 3, \ldots$, compute (by any of the previous numerical schemes) the solution of equation (5.1). (Hint: take n large enough to avoid oscillations in the numerical solution.) Answer the same questions for $a = \frac{3}{2}$. Comment on the results.
2. Exact solution. For fixed θ, let F_θ be the function defined for $x \in [0, 1]$ by

$$F_\theta(x) = \int_0^x e^{\theta z} \left[\int_0^z f(y) e^{-\theta y} dy \right] dz.$$

   (a) Show that for all real numbers α and β, function $\alpha + \beta e^{\theta x} - \frac{1}{\varepsilon} F_\theta$ is a solution of the differential equation (5.1).
   (b) Determine α and β so that $u = \alpha + \beta e^{\theta x} - \frac{1}{\varepsilon} F_\theta$ is the solution of problem (5.1), i.e., it satisfies the differential equation and the boundary conditions.
   (c) For $f(x) = \cos(a\pi x)$ with $a \in \mathbb{R}^*$, prove that

$$\lim_{\varepsilon \to 0^+} u(x) = \frac{1}{\lambda a \pi} \sin(a\pi x).$$

3. Explain the results obtained in question 1.

A solution of this exercise is proposed in Sect. 5.6 at page 125.

5.6 Solutions and Programs

Solution of Exercise 5.1.

Computation of the exact solution.

1. For a constant function f, the solution of problem (5.1) is

$$u(x) = \frac{f}{\lambda}\left(x - \frac{e^{\lambda x/\varepsilon} - 1}{e^{\lambda/\varepsilon} - 1}\right).$$

2. With $\theta = \frac{\lambda}{\varepsilon}$, we rewrite u as

$$u(x) = \frac{f}{\lambda}\left(x - \frac{e^{\theta x} - 1}{e^{\theta} - 1}\right).$$

Hence

$$\frac{\lambda}{f}u'(x) = 0 \Longleftrightarrow 1 - \frac{\theta e^{\theta x}}{e^{\theta} - 1} = 0 \Longleftrightarrow \theta e^{\theta x} = e^{\theta} - 1 \Longleftrightarrow x = \frac{1}{\theta}\ln\frac{e^{\theta} - 1}{\theta}.$$

From this, we deduce that $x_\theta = \frac{1}{\theta}\ln\frac{e^{\theta}-1}{\theta} = 1 + \frac{1}{\theta}\ln\frac{1-e^{-\theta}}{\theta} \in\ ]0, 1[$ and

$$u'(x) > 0 \Longleftrightarrow e^{\theta} - 1 - \theta e^{\theta x} > 0 \Longleftrightarrow \theta e^{\theta x} < \theta e^{\theta x_\theta} \Longleftrightarrow x < x_\theta.$$

The limits are $\lim_{\theta\to+\infty} x_\theta = 1$ and $\lim_{\theta\to-\infty} x_\theta = 0$.
3. It is easy to check that

$$\lim_{\varepsilon\to 0+} u(x_\theta) = \frac{f}{\lambda}, \quad \lim_{\varepsilon\to 0+}\lim_{x\to 1} u(x) = 0, \quad \text{and} \quad \lim_{x\to 1}\lim_{\varepsilon\to 0+} u(x) = \frac{f}{\lambda}.$$

The "boundary layer" is due to the strong variation of the solution (from f/λ to 0) over a small interval $[x_\theta, 1]$ whose length $1 - x_\theta = \frac{1}{\theta}\ln\frac{1-e^{-\theta}}{\theta}$ goes to 0 as θ goes to $+\infty$.
4. The following function FEM_F_ConvecDiffSolExa computes the exact solution of the problem:

```
function y=FEM_F_ConvecDiffSolExa(e,lambda,fc,x)
% solution of the convection--diffusion problem
% case ε,λ, and f constant
y=fc/lambda*(x-(1-exp(lambda*x/e))./(1-exp(lambda/e)));
```

We display in Fig. 5.6 the solutions for $f = 1$, $\varepsilon \in \{1, 1/2, 10^{-1}, 10^{-2}\}$, and (a) $\lambda = -1$ and (b) $\lambda = 1$. For small values of ε, we observe a small interval near $x = 1$ or $x = 0$ (according to the sign of λ) in which the solution suddenly changes from a value close to 1 to zero; this is the boundary layer.

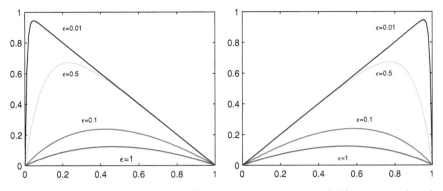

Fig. 5.6 Solution of the convection–diffusion problem for $\lambda = -1$ (left) and $\lambda = 1$ (right).

Solution of Exercise 5.2.

1. The linear space $\mathcal{V}_1^h$ has dimension n since there is an obvious isomorphism from this space onto $\mathbb{R}^n$: to each $u = (u_1, \ldots, u_n)^T \in \mathbb{R}^n$ corresponds a function $v \in \mathcal{V}_1^h$ defined by $v(x_i) = u_i$. From the identities $\varphi_{h,k}^{(1)}(x_j) = \delta_{j,k}$, we deduce that functions $(\varphi_{h,k}^{(1)})_{k=1}^n$ are linearly independent:

$$\left(\sum_{k=1}^n c_k \varphi_{h,k}^{(1)}(x) = 0, \forall x\right) \Longrightarrow \left(\sum_{k=1}^n c_k \varphi_{h,k}^{(1)}(x_j) = 0, \forall j\right) \Longrightarrow (c_j = 0, \forall j).$$

Since $\mathcal{V}_1^h$ has dimension n, $(\varphi_{h,k}^{(1)})_{k=1}^n$ form a basis of this space. Here are analytical expressions for $\varphi_{h,k}^{(1)}$ and its derivative

- for $x \notin I_{k-1} \cup I_k$, $\varphi_{h,k}^{(1)}(x) = \varphi_{h,k}^{(1)'}(x) = 0$,
- for $x \in I_{k-1}$, $\varphi_{h,k}^{(1)}(x) = (x - x_{k-1})/h$ and $\varphi_{h,k}^{(1)'}(x) = 1/h$,
- for $x \in I_k$, $\varphi_{h,k}^{(1)}(x) = (x_{k+1} - x)/h$ and $\varphi_{h,k}^{(1)'}(x) = -1/h$.

2. Since $(\varphi_{h,j}^{(1)})_{j=1}^n$ form a basis of $\mathcal{V}_1^h$, we can replace in (5.6) $v_h \in \mathcal{V}_1^h$ by any element of the basis $(\varphi_{h,j}^{(1)})_{j=1}^n$.

3. Using again the identities $\varphi_{h,k}^{(1)}(x_j) = \delta_{j,k}$, we obtain

$$u_h^{(1)}(x_j) = \sum_{k=1}^n \alpha f_k \varphi_{h,k}^{(1)}(x_j) = \alpha_j.$$

Replacing in (5.7) u_h by its expansion in the basis $\varphi_{h,k}^{(1)}$, we get

$$\sum_{k=1}^{n} a(\varphi_{h,k}^{(1)}, \varphi_{h,j}^{(1)})\alpha_k = \int_0^1 f\varphi_{h,j}dx, \quad \forall j = 1, \ldots, n.$$

Let $A_h^{(1)}$ be the $n \times n$ matrix and $b_h^{(1)}$ the vector of $\mathbb{R}^n$ defined by

$$(A_h^{(1)})_{j,k} = a(\varphi_{h,k}^{(1)}, \varphi_{h,j}^{(1)}), \quad (b_h^{(1)})_j = \int_0^1 f\varphi_{h,j}^{(1)}dx.$$

The vector $\tilde{u}_h = (\alpha_1, \ldots, \alpha_n)^T$ solves the linear system $A_h^{(1)}\tilde{u}_h^{(1)} = b_h^{(1)}$. The matrix of this system can be split as $A_h^{(1)} = \varepsilon B_h^{(1)} + \lambda C_h^{(1)}$, where

$$(B_h^{(1)})_{j,k} = \int_0^1 \varphi_{h,k}^{(1)}{}' \varphi_{h,j}^{(1)}{}' dx, \quad \text{and} \quad (C_h^{(1)})_{j,k} = \int_0^1 \varphi_{h,k}^{(1)}{}' \varphi_{h,j}^{(1)} dx.$$

Note the symmetry of the first matrix $(B_h^{(1)})_{j,k} = (B_h^{(1)})_{k,j}$. Using an integration by parts and the fact that the basis functions are null for $x = 0$ and $x = 1$ yields $(C_h^{(1)})_{j,k} = -(C_h^{(1)})_{k,j}$. The matrices are tridiagonal since the supports of any two functions $\varphi_{h,j}^{(1)}$ and $\varphi_{h,k}^{(1)}$ are disjoint for $|j - k| > 1$.

4. For $x \in \mathbb{R}^n$ with components $(x_k)_{k=1}^n$, we have

$$\langle B_h^{(1)}x, x \rangle = \sum_{k=1}^{n}(B_h^{(1)}x)_k x_k = \sum_{k=1}^{n}\sum_{j=1}^{n}(B_h^{(1)})_{k,j}x_j x_k$$

$$= \sum_{k=1}^{n} x_k \sum_{j=1}^{n} x_j \int_0^1 \varphi_{h,k}^{(1)}{}'(x)\varphi_{h,j}^{(1)}{}'(x)dx$$

$$= \int_0^1 \left(\sum_{k=1}^{n} x_k\varphi_{h,k}^{(1)}{}'(x)\right)^2 dx \geq 0.$$

Moreover, $\langle B_h^{(1)}x, x \rangle = 0$ implies that

$$\sum_{k=1}^{n} x_k\varphi_{h,k}^{(1)}{}'(x) = 0$$

for all $x \in [0, 1]$; thus the x_k are all zero. Consequently, the symmetric matrix $B_h^{(1)}$ is positive definite.

5. For matrix $C_h^{(1)}$, we have

$$\langle C_h^{(1)}x, x \rangle = \langle x, C_h^{(1)}{}^T x \rangle = -\langle x, C_h^{(1)}x \rangle = -\langle C_h^{(1)}x, x \rangle,$$

that is, $\langle C_h^{(1)}x, x \rangle = 0$, and the result follows.

Solution of Exercise 5.3.

Numerical computation of the $P1$ solution.

1. Computation of A_h. Recall that the supports of two sufficiently distant basis functions $\varphi_{h,j}^{(1)}$ and $\varphi_{h,k}^{(1)}$ are disjoint. More precisely, defining $b_{k,j} = \int_0^1 \varphi_{h,k}^{(1)'} \varphi_{h,j}^{(1)'}$ and $c_{j,k} = \int_0^1 \varphi_{h,k}^{(1)'} \varphi_{h,j}$, we get

   - for $|k - j| > 1$, $b_{k,j} = c_{k,j} = 0$,
   - for $k = j$,

$$b_{k,k} = \int_0^1 (\varphi_{h,k}^{(1)'})^2 = \int_{I_{k-1}} (\varphi_{h,k}^{(1)'})^2 + \int_{I_k} (\varphi_{h,k}^{(1)'})^2 = \frac{2}{h},$$

$$c_{k,k} = \int_0^1 \varphi_{h,k}^{(1)'} \varphi_{h,k} = \int_{I_{k-1}} \varphi_{h,k}^{(1)'} \varphi_{h,k} + \int_{I_k} \varphi_{h,k}^{(1)'} \varphi_{h,k} = 0,$$

   - for $k = j + 1$,

$$b_{j+1,j} = b_{j,j+1} = \int_0^1 \varphi_{h,j+1}^{(1)'} \varphi_{h,j}^{(1)'} = -\frac{1}{h},$$

$$c_{j+1,j} = -c_{j,j+1} = \int_0^1 \varphi_{h,j+1}^{(1)'} \varphi_{h,j} = -\frac{1}{2}.$$

2. See function FEM_F_ConvecDiffAP1.
3. Using the trapezoidal rule, we have

$$(b_h^{(1)})_k = \int_0^1 f\varphi_{h,k}^{(1)} dx = \int_{x_{k-1}^{(1)}}^{x_k^{(1)}} f\varphi_{h,k}^{(1)} dx + \int_{x_k^{(1)}}^{x_{k+1}^{(1)}} f\varphi_{h,k}^{(1)} dx$$

$$\approx \frac{h}{2} \left[f(x_{k-1}^{(1)})\varphi_{h,k}^{(1)}(x_{k-1}^{(1)}) + 2f(x_k^{(1)})\varphi_{h,k}^{(1)}(x_k^{(1)}) + f(x_{k+1}^{(1)})\varphi_{h,k}^{(1)}(x_{k+1}^{(1)}) \right]$$

$$= hf(x_k^{(1)}).$$

   See the function FEM_F_ConvecDiffbP1.
4. The following MATLAB script produces the left panel in Fig. 5.7.

```
eps=0.1;lambda=1;              %physical parameters
f = inline('ones(size(x))');  %right-hand side of the equation
n=10;
A=FEM_F_ConvecDiffAP1(eps,lambda,n);%matrix of the linear system
b=FEM_F_ConvecDiffbP1(n,f);%right-hand side of the linear system
u=A\b;                        %FEM solution
u=[0;u;0];                    %add to u the boundary values
x=(0:n+1)/(n+1);              %the mesh
uexa=FEM_F_ConvecDiffSolExa(eps,lambda,1,x);%exact solution
plot(x,uexa,x,u,'+-r');grid on;
```

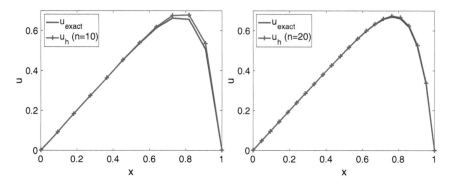

Fig. 5.7 Approximation of the convection–diffusion problem ($P1$ FEM), $\varepsilon = 0.1$, $\lambda = 1$, and $f = 1$ for $n = 10$ (left) and $n = 20$ (right).

```
en=strcat('u_h (n=',num2str(n));
legend('u_exact',strcat(en,')'))
```

For the tested values of λ and ε, we get a good approximation. This script is available in file FEM_M_ConvecDiffscript1.m.

5. The following script is available in FEM_M_ConvecDiffscript2.m.

```
lambda=1;eps=.1;
f = inline('ones(size(x))');
error=[ ];N=[ ];
for n=10:10:100
    A=FEM_F_ConvecDiffAP1(eps,lambda,n);
    b=FEM_F_ConvecDiffbP1(n,f);
    u=A\b;
    u=[0;u;0];
    x=(0:n+1)'/(n+1);
    uexa=FEM_F_ConvecDiffSolExa(eps,lambda,1,x);
    N=[N;n];error=[error; norm(uexa-u)/sqrt(n)];
end
plot(log(N),log(error),'o--r','LineWidth',2,'MarkerSize',10);
set(gca,'FontSize',20);grid on;
% determination of the slope
p=polyfit(log(N),log(error),1);
disp(p(1))
```

We get a straight line with slope approximately -2 (according to polyfit, the slope is -1.97).

Solution of Exercise 5.4.

See the script FEM_M_ConvecDiffscript3.m.

```
eps=0.01;lambda=1;
f = inline('ones(size(x))');
yes=1;
while yes
    n=input('enter n: ');
    A=FEM_F_ConvecDiffAP1(eps,lambda,n);
    b=FEM_F_ConvecDiffbP1(n,f);
    u=A\b;
    u=[0;u;0];
    x=(0:n+1)/(n+1);
    uexa=FEM_F_ConvecDiffSolExa(eps,lambda,1,x);
    plot(x,uexa,x,u,'+-r')
    drawnow
    Peclet=abs(lambda)/2/eps/(n+1);
    fprintf('Peclet number = %e \n',Peclet);
    yes=input('more ? yes=1, no=0 ')
end
```

We observe that the approximation is good for a Peclet number $Pe < 1$.

Solution of Exercise 5.5.

In the $P1$ method, the matrices $B_h^{(1)}$ and $C_h^{(1)}$ are tridiagonal. In the $P2$ method, the supports of the basis functions are larger and the matrices $B_h^{(2)}$ and $C_h^{(2)}$ are pentadiagonal. As in the $P1$ case, we can prove that matrix $B_h^{(2)}$ is symmetric, positive definite; matrix $C_h^{(2)}$ is skew-symmetric; and matrix $A_h^{(2)}$ is invertible.

Solution of Exercise 5.6.

The derivatives of the basis functions are

$$(\varphi_{h,2k+1}^{(2)})'(x) = \begin{cases} -8(x - x_{2k+1}^{(2)})/h^2 \text{ for } x \in I_k, \\ 0 \qquad\qquad\qquad \text{otherwise,} \end{cases}$$

and

$$(\varphi_{h,2k}^{(2)})'(x) = \begin{cases} 4(x - x_{2k+3/2}^{(2)})/h^2 \text{ for } x \in I_k, \\ 4(x - x_{2k-1/2}^{(2)})/h^2 \text{ for } x \in I_{k-1}, \\ 0 \qquad\qquad\qquad \text{otherwise.} \end{cases}$$

1.(a) Computation of $B_h^{(2)}$. Since the matrix is symmetric, only its upper triangular part is computed.

- Rows with odd indices: $(B_h^{(2)})_{2k+1,2k+1} = \int_0^1 [\varphi_{h,2k+1}^{(2)}{}'(x)]^2 dx =$
$\int_{x_{2k}^{(2)}}^{x_{2k+2}^{(2)}} [\frac{8}{h^2}(x - x_{2k+1}^{(2)})]^2 dx = \frac{16}{3}\frac{1}{h}$, $(B_h^{(2)})_{2k+1,2k+2} = -\frac{8}{3}\frac{1}{h}$, and
$(B_h^{(2)})_{2k+1,m} = 0$ for $m \geq 2k + 3$.
- Rows with even indices: $(B_h^{(2)})_{2k,2k} = \frac{14}{3}\frac{1}{h}$, $(B_h^{(2)})_{2k,2k+1} = -\frac{8}{3}\frac{1}{h}$,
$(B_h^{(2)})_{2k,2k+2} = \frac{1}{3}\frac{1}{h}$, and $(B_h^{(2)})_{2k,m} = 0$ for $m \geq 2k + 3$.

Thus, the upper triangular part of the matrix $B_h^{(2)}$ is

$$\frac{1}{3h}\begin{pmatrix}
16 & -8 & 0 & 0 & 0 & 0 & 0 \\
 & 14 & -8 & 1 & 0 & 0 & 0 \\
 & & 16 & -8 & 0 & 0 & 0 \\
 & & & 14 & -8 & 1 & 0 \\
 & & & & \ddots & \ddots & \ddots \\
 & & & & & \ddots & \ddots \\
 & & & & & & \ddots
\end{pmatrix}.$$

(b) Computation of $C_h^{(2)}$. The matrix being skew-symmetric, only its upper triangular part is computed.

- Rows with odd indices: $(C_h^{(2)})_{2k+1,2k+1} = 0$, $(C_h^{(2)})_{2k+1,2k+2} = \frac{2}{3}$,
and $(C_h^{(2)})_{2k+1,m} = 0$ for $m \geq 2k + 3$.
- Rows with even indices: $(C_h^{(2)})_{2k,2k} = 0$, $(C_h^{(2)})_{2k,2k+1} = \frac{2}{3}$,
$(C_h^{(2)})_{2k,2k+2} = -\frac{1}{6}$, and $(C_h^{(2)})_{2k,m} = 0$ for $m \geq 2k + 3$.

Thus, the upper triangular part of the matrix $C_h^{(2)}$ is

$$\frac{1}{6}\begin{pmatrix}
0 & 4 & 0 & 0 & 0 & 0 \\
 & 0 & 4 & -1 & 0 & 0 & 0 \\
 & & 0 & 4 & 0 & 0 & 0 \\
 & & & 0 & 4 & -1 & 0 \\
 & & & & \ddots & \ddots & \ddots \\
 & & & & & \ddots & \ddots \\
 & & & & & & \ddots
\end{pmatrix}.$$

2. See function FEM_F_ConvecDiffAP2.

3. $(b_h^{(2)})_{2k+1} = \int_0^1 f\varphi_{h,2k+1}^{(2)} dx = \int_{x_{2k}^{(2)}}^{x_{2k+1}^{(2)}} f\varphi_{h,2k+1}^{(2)} dx + \int_{x_{2k+1}^{(2)}}^{x_{2k+2}^{(2)}} f\varphi_{h,2k+1}^{(2)} dx.$

Using Simpson's formula on each interval, we obtain

$$\left(b_h^{(2)}\right)_{2k+1} \approx \frac{2}{3}hf\left(x_{2k+1}^{(2)}\right).$$

In the same way, $\left(b_h^{(2)}\right)_{2k} \approx \frac{1}{3}hf\left(x_{2k}^{(2)}\right)$. See function FEM_F_ConvecDiffbP2.

4. As for Exercise 5.3, we plot the logarithm of the error versus the logarithm of n. The curve is a straight line with slope approximately -4 (according to polyfit, the slope is -3.9). The error decreases faster than in the $P1$ case. See the script in FEM_M_ConvecDiffscriptVP2.m. Note the *natural* energy norm for this problem is

$$\int_0^1 |u_h'(x) - u'(x)|^2 dx = \sum_{\text{intervals } I_k} \int_{I_k} |u_h'(x) - u'(x)|^2 dx.$$

See references at the end of this chapter.

Solution of Exercise 5.7.

1. Let $x \in \mathbb{R}^n$ be a nonnull vector such that $(A - BD^{-1}C)x = 0$. The nonnull vector $y = (x^T, -(D^{-1}Cx)^T)^T \in \mathbb{R}^{2n+1}$ is such that $\widehat{A}_h^{(2)}y = 0$. However, the matrix $\widehat{A}^{(2)}$ is invertible. This leads to a contradiction. Hence, the square matrix $A - BD^{-1}C$ is injective and consequently invertible.
2.

```
A=FEM_F_ConvecDiffAP2(eps,lambda,n);
a=A(2:2:2*n+1,2:2:2*n+1);b=A(2:2:2*n+1,1:2:2*n+1);
c=A(1:2:2*n+1,2:2:2*n+1);d=A(1:2:2*n+1,1:2:2*n+1);
```

3. The following script is written in the file FEM_M_ConvecDiffscript4.m:

```
n=10;eps=0.01;lambda=1;
f = inline(ones(size(x))');
sm=FEM_ConvecDiffbP2(n,f);
nsm=sm(2:2:2*n+1)-b*inv(d)*sm(1:2:2*n+1);
u=(a-b*inv(d)*c)\nsm;          %computation of v
x=linspace(0,1,100);
uexa=FEM_F_ConvecDiffSolExa(eps,lambda,1,x);
plot(x,uexa);hold on
plot((1:n)/(n+1),u,'+');hold off;
```

Results for $n = 10$ and $n = 20$ are displayed in Fig. 5.8.
4. Fig. 5.9 displays the stabilized solutions for $n = 20$ and the $P1$ solutions for $n \in \{20, 40, 60, 80\}$. We observe that the $P1$ method with $n = 60$ (and more) gives better results than the $P2$ method with $n = 20$.

Solution of Exercise 5.8.

1. As observed in Fig. 5.10 (left), there is no boundary layer for nonzero integer values of a: the source term f transfers its oscillations to the solution u. This is no longer true for $a = 1.5$. In this case, we observe

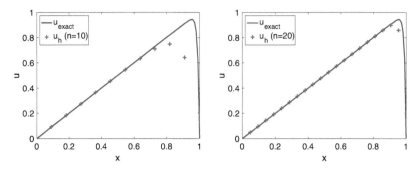

Fig. 5.8 Stabilized solution of the convection–diffusion problem for $\lambda = 1$, $\varepsilon = 0.01$, and $n = 10$ (left), $n = 20$ (right).

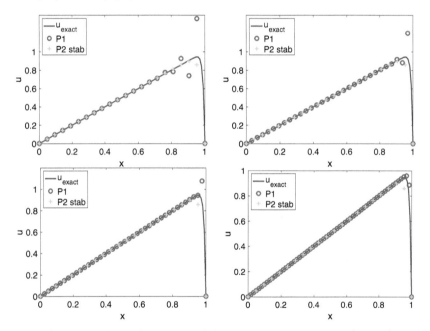

Fig. 5.9 Stabilized solution for $n = 20$ and the $P1$ solutions with $n = 20$ (top left), $n = 30$ (top right), $n = 40$ (bottom left), and $n = 60$ (bottom right).

a boundary layer. This observation is discussed in the next question. Here is the script that produces the figures. This script is available in file FEM_M_ConvecDiffscript5.m.:

```
n=100;lambda=1;eps=0.01;
x=(0:(n+1))'/(n+1);
A=FEM_F_ConvecDiffAP1(eps,lambda,n);
X=[ ];Y=[ ];
h=1/(n+1);tab=(1:n)'*h;
for af=1:5
```

```
        b=h*cos(af*pi*tab);
        y=A\b;y=[0; y; 0];
         X=[X x];Y=[Y y];
     end;
     plot(X,Y);
```

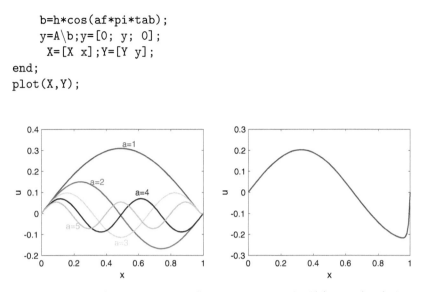

Fig. 5.10 Solution of the convection–diffusion equation. Left: $f(x) = \cos(a\pi x)$, for $a \in \{1, 2, 3, 4, 5\}$; right: $f(x) = \cos(\frac{3}{2}\pi x)$.

2. Computation of the exact solution.

   (a) It is easy to check that $\alpha + \beta e^{\theta x} - \frac{1}{\varepsilon}F_\theta$ is a solution of the differential equation (5.1).
   (b) The boundary conditions $u(0) = 0$ and $u(1) = 0$ allow the determination of α and β,

   $$\alpha = -\beta = \frac{1}{\varepsilon}\frac{F_\theta(1)}{1 - e^\theta},$$

   and the solution is

   $$u(x) = \frac{1}{\varepsilon(1 - e^\theta)}[(1 - e^{\theta x})F_\theta(1) - (1 - e^\theta)F_\theta(x)]. \qquad (5.14)$$

   (c) Let $I = \int_0^z \cos(a\pi y)e^{-\theta y}dy$ and $J = \int_0^z \sin(a\pi y)e^{-\theta y}dy$. Computing the real part of $I + iJ$, we get

   $$\int_0^z f(y)e^{-\theta y}dy = \frac{a\pi e^{-\theta z}\sin(a\pi z) + \theta - \theta e^{-\theta z}\cos(a\pi z)}{\theta^2 + \pi^2 a^2}$$

   and deduce the identity

   $$(\theta^2 + \pi^2 a^2)F_\theta(x) = e^{\theta x} - \cos(a\pi x) - \frac{\theta \sin(a\pi x)}{a\pi}.$$

   The solution of the problem (5.1) is the sum of the two terms $-\frac{1}{\varepsilon}F_\theta(x)$ and $\alpha + \beta e^{\theta x}$.

- The first one can be split into three parts:
  - $\dfrac{1}{\varepsilon}\dfrac{\theta}{(\theta^2+\pi^2a^2)}\dfrac{\sin(a\pi x)}{a\pi}$, whose limit is $\dfrac{\lambda}{a\pi}\sin(a\pi x)$ for ε going to 0,
  - $\dfrac{1}{\varepsilon}\dfrac{\cos(a\pi x)}{(\theta^2+\pi^2a^2)}$, whose limit is 0,
  - and $\gamma_1 = -\dfrac{1}{\varepsilon}\dfrac{e^{\theta x}}{\theta^2+\pi^2a^2}$.
- The only term in $\alpha + \beta e^{\theta x}$ that does not go to 0 is
$$\gamma_2 = \frac{1}{\varepsilon}\frac{e^{\theta}}{\theta^2+\pi^2a^2}\frac{1-e^{\theta x}}{1-e^{\theta}}.$$

Since the sum $\gamma_1 + \gamma_2$ goes to 0, we deduce that

$$\lim_{\varepsilon\to 0^+} u(x) = \frac{1}{\lambda a\pi}\sin(a\pi x).$$

3. The function u is continuous and satisfies the boundary condition $u(1)=0$; hence

$$\lim_{\varepsilon\to 0^+}\lim_{x\to 1} u(x) = \lim_{\varepsilon\to 0^+} u(1) = 0.$$

In addition, $\lim_{x\to 1}\lim_{\varepsilon\to 0^+} u(x) = \frac{1}{\lambda a\pi}\sin(a\pi)$. Consequently, for integer a, there is no boundary layer in the vicinity of $x=1$ since $\lim_{x\to 1}\lim_{\varepsilon\to 0^+} u(x) = \lim_{\varepsilon\to 0^+}\lim_{x\to 1} u(x) = 0$. On the contrary, for noninteger values of a, there exists a boundary layer since

$$\lim_{x\to 1}\lim_{\varepsilon\to 0^+} u(x) \neq \lim_{\varepsilon\to 0^+}\lim_{x\to 1} u(x).$$

Chapter References

K. Atkinson, W. Han, *Theoretical Numerical Analysis* (Springer, New York, 2001)

C. Bernardi, Y. Maday, F. Rapetti, *Discrétisations variationnelles de problèmes aux limites elliptiques, collection S.M.A.I. Mathématiques et Applications*, vol. 45 (Springer, Paris, 2004)

F. Brezzi, A. Russo, Choosing bubbles for advection-diffusion problems. Math. Model. Methods Appl. Sci. **4**(4) (1994)

I. Danaila, F. Hecht, O. Pironneau, *Simulation numérique en C++* (Dunod, Paris, 2003)

P. Joly, *Mise en œuvre de la méthode des éléments finis, collection S.M.A.I. Mathématiques et Applications* (Ellipses, Paris, 1990)

H. Le Dret, B. Lucquin, *Partial Differential Equations: Modeling, Analysis and Numerical Approximation* (Birkhäuser/Springer, Cham, 2016)

B. Lucquin, O. Pironneau, *Introduction to Scientific Computing* (Willey, Chichester, 1998)

Chapter 6
Solving a Differential Equation by a Legendre Spectral Method

Project Summary

Level of difficulty:	2
Keywords:	Spectral method, polynomial approximation, Gauss quadrature, orthogonal polynomials, Legendre polynomials, variational formulation
Application fields:	Computation of smooth solutions of PDEs with high accuracy

Introduction

Spectral methods are approximation techniques for the computation of solutions to ordinary or partial differential equations. They are based on a polynomial expansion of the solution. The precision of these methods is limited only by the regularity of the solution, in contrast to the finite difference and finite element methods. The approximation is based primarily on the weak formulation of the continuous problem. Test functions are polynomials and the integrals involved in the formulation are computed by suitable quadrature formulas. In this project we propose to implement a spectral method to solve the following boundary value problem defined on the interval $\Omega =]-1, 1[$:

$$\begin{cases} -u'' + cu = f, \\ u(-1) = 0, \\ u(1) = 0, \end{cases} \qquad (6.1)$$

with $f \in L^2(\Omega)$ and c a positive real number.

In the first part of the project we introduce some properties of the Legendre polynomials. These polynomials will be used to build a basis of the

© The Author(s), under exclusive license to Springer Nature Switzerland AG 2023   129
I. Danaila et al., *An Introduction to Scientific Computing*,
https://doi.org/10.1007/978-3-031-35032-0_6

approximation space. In the second part, we define the Legendre expansion of a function and compute the truncated Legendre expansion. To this end, we present a method to compute the integrals accurately, namely the Gauss quadrature formula. Finally, in the third part of the project, we implement the approximation of the differential equation (6.1) by a spectral method.

6.1 Some Properties of the Legendre Polynomials

Let $\mathbb{P}_n$ denote the set of all polynomials with degree less than or equal to a positive integer n and $(L_n)_{n \geq 0}$ the family of Legendre polynomials[1]

$$L_n(x) = \frac{1}{2^n n!} \frac{d^n}{dx^n} \left[(x^2 - 1)^n \right].$$
(6.2)

Note that these polynomials form an orthogonal basis on $]-1, 1[$ since for all integers n and m,

$$\int_{-1}^{1} L_n(x) L_m(x) dx = \frac{2}{2n + 1} \delta_{n,m},$$
(6.3)

where $\delta_{n,m}$ is the Kronecker symbol. The Legendre polynomials are solutions of the differential equation

$$[(1 - x^2) L_n'(x)]' + n(n + 1) L_n(x) = 0, \qquad n \geq 0,$$
(6.4)

and they satisfy the following three-term recurrence formula

$$\begin{cases} L_0(x) = 1, \\ L_1(x) = x, \\ (n + 1) L_{n+1}(x) = (2n + 1) x L_n(x) - n L_{n-1}(x), \quad \text{for } n \geq 1, \end{cases}$$
(6.5)

from which we deduce the special values $L_n(\pm 1) = (\pm 1)^n$.

Exercise 6.1 1. Write a function y=SPE_F_LegLinComb(x,c) that plots a linear combination of the Legendre polynomials of the form

$$y(x) = \sum_{k=1}^{p} c_k L_{k-1}(x).$$
(6.6)

The inputs of the program are

- an array (a vector) c that contains the coefficients c_k,
- an array x that contains the points of the grid.

---

[1] Note that Legendre polynomials are sometimes denoted as P_n instead of L_n, mostly to avoid confusion with Laguerre polynomials.

2. Write a script using this function with x corresponding to a fine discretization of the interval $[-1, 1]$ and c corresponding to the combination $L_0 - 2L_1 + 3L_5$. Plot y as a function of x, as in Fig. 6.1.
A solution of this exercise is proposed in Sect. 6.6. at page 139.

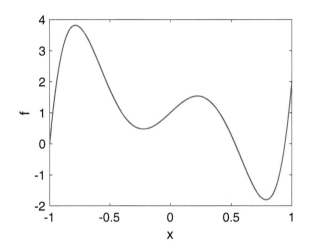

Fig. 6.1 Linear combination of Legendre polynomials $f = L_0 - 2L_1 + 3L_5$.

6.2 Gauss–Legendre Quadrature

Numerical quadratures (or rules) are efficient tools for computing an approximation of an integral (see Krommer and Ueberhuber (1994)). In the general case, no primitive function of the integrand is available, but the values of the integrand itself can be easily computed based on the following result, holding for smooth functions φ:

$$\int_{-1}^{1} \varphi(x)dx = \sum_{i=1}^{s} \varphi(x_i)w_i + R_s(\varphi), \tag{6.7}$$

where

1. the points x_i (called the *nodes* of the formula) are the zeros of the Legendre polynomial L_s,
2. the real numbers w_i (called the *weights* of the formula) are given by

$$w_i = \frac{2}{(1 - x_i^2)[L_s'(x_i)]^2}. \tag{6.8}$$

3. The remainder is $R_s(\varphi) = \dfrac{2^{2s+1}(s!)^4}{(2s+1)[(2s)!]^3}\varphi^{(2s)}(\xi)$,   for $\xi \in]-1, 1[$.

The Gauss–Legendre quadrature of order s is the approximation

$$\int_{-1}^{1} \varphi(x)dx \approx \sum_{i=1}^{s} \varphi(x_i)w_i. \tag{6.9}$$

This formula is exact for $\varphi \in \mathbb{P}_{2s-1}$, since in such a case the remainder $R_s(\varphi)$ is null.

To use the Gauss quadrature formula, we explain an efficient way to compute its weights and nodes. For all x, recurrence relations (6.5) for $j = 0, \ldots, s$ can be written in a compact matrix form,

$$Mu = xu + v, \tag{6.10}$$

where

$$M = \begin{pmatrix} 0 & 1 & & & \\ a_1 & 0 & b_1 & & \\ & \ddots & \ddots & \ddots & \\ & & a_{s-2} & 0 & b_{s-2} \\ & & & a_{s-1} & 0 \end{pmatrix}, \tag{6.11}$$

with $a_j = j/(2j+1)$, $b_j = (j+1)/(2j+1)$,

$$v = b_{s-1}L_s(x)\begin{pmatrix} 0 \\ \vdots \\ 0 \\ 1 \end{pmatrix}, \quad \text{and} \quad u = \begin{pmatrix} L_0(x) \\ \vdots \\ L_{s-1}(x) \end{pmatrix}. \tag{6.12}$$

Now let x be a zero of L_s; then $v = 0$ and the linear system (6.10) becomes

$$Mu = xu, \tag{6.13}$$

which means that *the zeros of L_s are the eigenvalues of the $s \times s$ tridiagonal matrix M.* To compute the weight w_i with formula (6.8), the recurrence formula (6.5) can be combined with another property of Legendre polynomials:

$$(1 - x^2)L'_s(x) = -sxL_s(x) + sL_{s-1}(x), \quad s \geq 1. \tag{6.14}$$

Since $L_s(x_i) = 0$, we get finally

$$w_i = \frac{2(1 - x_i^2)}{(sL_{s-1}(x_i))^2},$$

where $L_{s-1}(x_i)$ is computed by the recurrence formula (6.5).

Exercise 6.2

1. Write a function SPE_F_xwGauss that computes the weights and nodes of a Gauss–Legendre quadrature formula. Compare your results for $s = 8$ with the table below.

x_i	w_i
± 0.18343464249565	0.36268378337836
± 0.52553240991633	0.31370664587789
± 0.79666647741363	0.22238103445337
± 0.96028985649754	0.10122853629036

2. Write a script to validate this function: test the quadrature formula on various integrals with known exact values. In particular, check the accuracy of the formula for polynomials in $\mathbb{P}_{2s-1}$ and compare the exact and approximate values of the integral of e^x on $]-1, 1[$.

A solution of this exercise is proposed in Sect. 6.6. at page 140.

6.3 Legendre series expansions

We now associate to a function $f \in L^2(]-1, 1[)$ its Legendre expansion

$$\mathcal{L}(f) = \sum_{j=0}^{\infty} \hat{f}_j L_j,$$

with the Legendre coefficients $\hat{f}_j$ defined by

$$\hat{f}_j = \frac{2j+1}{2} \int_{-1}^{1} f(x) L_j(x) dx. \tag{6.15}$$

We also define the truncated expansion

$$\mathcal{L}_p(f) = \sum_{j=0}^{p} \hat{f}_j L_j.$$

This is an approximation of the function f, which is exact for polynomials in $\mathbb{P}_p$ since $(L_j)_{j=0}^{p}$ is an orthogonal basis of $\mathbb{P}_p$. The calculation of the coefficients $\hat{f}_j$ is done by a quadrature formula. The error induced by this approximation of $\hat{f}_j$ must be of the same order or negligible compared to the total error of the spectral method. Thus, it is necessary to use a high-order quadrature, and for this reason, the Gauss–Legendre quadrature is suitable.

Exercise 6.3

1. Write a script that computes the truncated Legendre expansion $\mathcal{L}_p(f)$ of a function f. The script includes the following steps:

- Compute the nodes and weights of the Gauss–Legendre quadrature for a given s.
- Compute the Legendre coefficients $(\hat{f}_k)_{k=0}^{P}$ (6.15) by the quadrature formula (6.9).
- Plot the function and its truncated Legendre expansion $\mathcal{L}_p(f)$ on $[-1, 1]$.

2. Compare with the plot in Fig. 6.2.
3. Justify the choice of the Gauss quadrature parameter s as a function of the degree p of the truncated series.
4. Test the script with less-regular functions, namely, the functions `abs(x)` and `sign(x)`.
5. Plot the error computed in the supremum norm $E_\infty(p) = \|f - \mathcal{L}_p(f)\|_\infty$ as a function of the truncation parameter p.

A solution of this exercise is proposed in Sect. 6.6. at page 140.

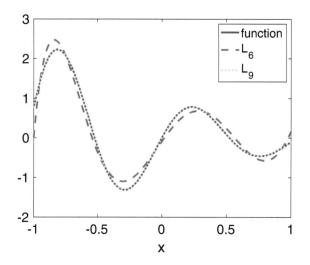

Fig. 6.2 Function $f(x) = \sin(6x)\exp(-x)$ and approximations $\mathcal{L}_6(f)$, and $\mathcal{L}_9(f)$.

The choice of the number s of Gauss nodes is related to the degree p of the truncated series expansion. The quadrature formula (6.9) with s nodes is exact on $\mathbb{P}_{2s-1}$. In addition, every polynomial $f \in \mathbb{P}_p$ is its own Legendre series $f = \mathcal{L}_p(f)$. In order to make the computation of the Legendre coefficients exact for $f \in \mathbb{P}_p$, it is necessary to take s such that $2s - 1 \geq 2p$, that is, $s \geq p + 1$. The reader will verify that for a very smooth function, let us say $f \in C^\infty(]-1, 1[)$, the error $E_\infty(p)$ decreases to zero as p goes to infinity. The convergence is faster than any power of $1/p$, as shown in the left panel of Fig. 6.3. In this figure, the error stops decreasing from $p \approx 20$, since the computer accuracy is then reached. In contrast, running the same script for the nonsmooth function $|x|$ exhibits a very slow convergence—roughly

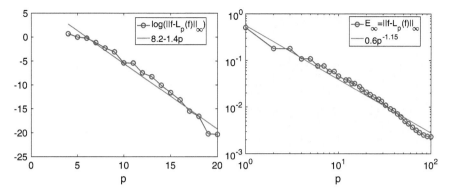

Fig. 6.3 Left panel: log of error $E_\infty(p) = \|f - \mathcal{L}_p(f)\|_\infty$ for $f(x) = \sin(6x)\exp(-x)$. The red line displays the best exponential fit $E_\infty(p) = \exp(8.2 - 1.4p)$. Right panel: error $E_\infty(p)$ for $f(x) = |x|$. The red line displays a roughly linear behavior $E_\infty(p)) = 0.6p^{-1.15}$.

linear—displayed in the right panel of Fig. 6.3. For the discontinuous function **sign**, the expansion does not converge to f in the supremum norm, although it does converge in the L^2 norm. The truncated series $\mathcal{L}_p(f)$ has oscillations. As p increases, the size of the oscillations decreases slowly, except near the discontinuity $x = 0$, where the oscillations remain. This problem is known as the Gibbs phenomenon. The results for functions $|x|$ and sign(x) are displayed in Fig. 6.4.

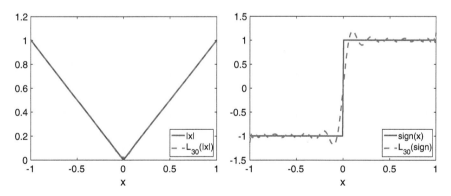

Fig. 6.4 Comparison of f with $\mathcal{L}_{30}(f)$. Left panel: $f(x) = |x|$; right panel: $f(x) =$ sign(x).

6.4 A Spectral Discretization

We consider a *weak formulation* of the problem (6.1) (see Chap. 5):

$$\begin{cases} \text{Find } u \in H_0^1(]-1,1[) \text{ such that for all } v \in H_0^1(]-1,1[) : \\ \int_{-1}^1 u'(x)v'(x)dx + c\int_{-1}^1 u(x)v(x)dx = \int_{-1}^1 f(x)v(x)dx. \end{cases} \quad (6.16)$$

Every regular solution of (6.16) is a solution to the problem (6.1). The space of test functions for this spectral method is the subset of $\mathbb{P}_m$ defined by

$$\mathbb{P}_m^0(\Omega) = \{p \in \mathbb{P}_m, p(-1) = p(1) = 0\}.$$

This linear space has dimension $m - 1$ and can be rewritten as

$$\mathbb{P}_m^0(\Omega) = \{p = (1 - x^2)q, \quad q \in \mathbb{P}_{m-2}\}.$$

Since $\mathbb{P}_m^0(\Omega)$ is included in $H_0^1(\Omega)$, we can easily define the variational approximation called the *spectral Galerkin method*:

$$\begin{cases} \text{Find } u_m \in \mathbb{P}_m^0(\Omega) \text{ such that for all } v_m \in \mathbb{P}_m^0(\Omega) : \\ \int_{-1}^1 u_m'(x)v_m'(x)dx + c\int_{-1}^1 u_m(x)v_m(x)dx = \int_{-1}^1 f(x)v_m(x)dx. \end{cases} \quad (6.17)$$

Functions $F_i = (1 - x^2)L_i'$ for $i = 1, \ldots, m - 1$, form a basis of $\mathbb{P}_m^0(\Omega)$. We denote this basis by $\mathcal{F}_m$. By linearity, the problem (6.17) is equivalent to the problem

$$\begin{cases} \text{Find } u_m \in \mathbb{P}_m^0(\Omega) \text{ such that for all } F_i \in \mathcal{F}_m : \\ \int_{-1}^1 u_m'(x)F_i'(x)dx + c\int_{-1}^1 u_m(x)F_i(x)dx = \int_{-1}^1 f(x)F_i(x)dx. \end{cases} \quad (6.18)$$

Let $\bar{u}_m = (u_{m,1}, \ldots, u_{m,m-1})^T$ be the vector whose entries are the coefficients of u_m in the basis $\mathcal{F}_m$:

$$\bar{u}_m = \sum_{i=1}^{m-1} u_{m,i}F_i. \quad (6.19)$$

By plugging this expansion into (6.18), we get a linear system for $\bar{u}_m$, which we write in matrix form as

$$A_m\bar{u}_m = b_m, \quad (6.20)$$

where the $(m-1) \times (m-1)$ matrix A_m and the vector $b_m \in \mathbb{R}^{m-1}$ are defined by

$$(A_m)_{i,j} = \frac{[i(i+1)]^2}{i+1/2}\delta_{i,j}$$

$$+ c\frac{i(i+1)\,j(j+1)}{(2i+1)\,(2j+1)}\left(\frac{4(2i+1)\delta_{i,j}}{(2i-1)(2i+3)} - \frac{2\delta_{i,j-2}}{2i+3} - \frac{2\delta_{i,j+2}}{2i-1}\right),$$

$$(b_m)_i = \int_{-1}^{1} f(x)F_i(x)dx = \frac{2i(i+1)}{2i+1}\left(\frac{1}{2i-1}\hat{f}_{i-1} - \frac{1}{2i+3}\hat{f}_{i+1}\right).$$

The terms $\hat{f}_i$ in the previous definition are the Legendre coefficients of the right-hand-side function f defined by (6.15).

Once the linear system is solved, (*i.e.*, the coefficients of u_m in the basis $\mathcal{F}_m$ are known), the coefficients of u_m in the Legendre basis $(L_j)_{j=0}^m$ are computed using the identities

$$(1 - x^2)L_i' = \frac{i(i+1)}{2i+1}(L_{i-1} - L_{i+1}). \tag{6.21}$$

Exercise 6.4

1. Write a program including the following steps:

   - Compute the matrix A_m and the vector b_m. This step includes the computation of the Legendre coefficients $\hat{f}_k$ of the source term of the differential equation.
   - Solve the linear system (6.20).
   - Compute the Legendre coefficients $\hat{u}_{m_k}$ of the numerical solution u_m using (6.21).
   - Plot on the same figure the function u and the numerical approximation u_m on $[-1,1]$.

2. Construction of an exact solution: take any reasonable function $u_e(x)$ such that $u_e(-1) = u_e(1) = 0$ and compute $f(x)$ in such a way that u_e solves the problem (6.1) with a constant c. Program the two functions $u_e(x)$ and $f(x)$. The function u_e will be used as a benchmark to compare the spectral solution and the exact solution.

3. Advantages of the method: compare the spectral solution to a solution computed by a second-order finite difference scheme leading to a linear system of identical dimension (see Section 10.2). Do you obtain the same precision? Compute the error between the exact solution and the spectral one. Increase the number of points in the finite difference discretization to get the same precision. Draw some conclusion on the respective advantages of the two methods.

A solution of this exercise is proposed in Sect. 6.6. at page 141.

6.5 Extensions to other orthogonal polynomials

The paragraph on the quadrature rules has several possible extensions. One can use a similar method to compute the nodes and weights of Gauss quadrature corresponding to other families of orthogonal polynomials. For example, the analogous formula to (6.7) for integrals on the real line $\mathbb{R}$ is

$$\int_{-\infty}^{+\infty} f(x)e^{-x^2}\,dx = \sum_{i=1}^{n} w_i f(x_i) + R_n(f). \tag{6.22}$$

Here the nodes x_i are the zeros of the Hermite polynomials (see below), the weights w_i are given by (Davis, 1975, Szegő, 1975)

$$w_i = \frac{2^{n-1}n!\sqrt{\pi}}{(nH_{n-1}(x_i))^2},$$

and the remainder is

$$R_n(f) = \frac{n!\sqrt{\pi}}{2^n(2n)!}f^{(2n)}(\xi).$$

The Hermite polynomials are defined by

$$\begin{cases} H_0(x) = 1, \\ H_1(x) = 2x, \\ 2xH_n(x) = H_{n+1}(x) + 2nH_{n-1}(x). \end{cases}$$

They are orthogonal for the inner product

$$\langle f, g \rangle = \int_{\mathbb{R}} f(x)g(x)e^{-x^2}\,dx.$$

Concerning the application of spectral methods to PDEs, one can think to generalize the example studied here in higher dimensions. Up to dimension two or three, the main feature of spectral methods is still the approximation by high-degree polynomials, using this time tensor products of polynomial bases. These objects have a high precision and turn out to be very efficient for simple geometries: rectangular prism or cylinder, for instance. The treatment of the Laplace equation in a square or cubic domain is thoroughly detailed by Bernardi et al. (1999) and Bernardi et al. (2004) (in French), with several possible types of boundary conditions (Dirichlet, Neumann, and Fourier). For complicated geometries, a domain decomposition technique is required (see for instance Wohlmuth (2001)). Detailed problems are also proposed in Bernardi and Maday (1997) and Bernardi et al. (2004), which can be handled starting from the case treated in this project, such as for instance the spectral discretization of the Dirichlet problem in an axisymmetric domain or the heat equation in one dimension.

6.6 Solutions and Programs

Solution of Exercise 6.1

The computation of the linear combination (6.6) is performed by the function SPE_F_LegLinComb. To compute the values of the Legendre polynomial of degree p at points $x_1, \ldots, x_n$, there is no need to store all the values of the polynomials of degree less than p. Only the values corresponding to degrees $p - 1$ and $p - 2$, which come into play in the recurrence relation (6.5), must be stored in two arrays pol1 and pol2, along with the current values that are stored in pol. The values of the linear combination are stored in an array y to which the terms $c_i p_i(x)$ are added as they are computed.

The graphical display of $L_0 - 2L_1 + 3L_5$ on the interval $[-1, 1]$ (see Fig. 6.1) is done in the script SPE_M_PlotLegPol.m. The function SPE_F_LegLinComb receives in its input argument the array $[0; -2; 0; 0; 0; 3]$ containing the coefficient values of the linear combination along with the points $x_i = -1 + (i - 1)/250$ for $i = 1, \ldots, 501$ at which this function must be displayed.

The MATLAB built-in function L=legendre(n,x) computes an array L with $n+1$ rows, whose $(m+1)$th row holds the values of the Legendre function L_n^m defined by

$$L_n^m(x) = (-1)^m (1 - x^2)^{m/2} \frac{d^m}{dx^m} L_n(x), \tag{6.23}$$

at points specified by the vector x. Therefore it is possible to compute the values of the Legendre polynomial with this function using only the first row of the computed array. Another method to compute the linear combination is to call the function legendre for all degrees from 0 up to p, to extract for each degree the first row of the output array, and to multiply it by the corresponding coefficient.

The script SPE_M_PlotLegPol.m compares the computing times required by the two methods, using the MATLAB stopwatch timers tic and toc before and after calling the function SPE_F_LegLinComb. The value returned by toc contains the computing time in seconds. The same thing is done again before and after the group of commands for the method using the legendre function. In order to get meaningful computing time estimates, it is best to increase the number of computing points to 500 points evenly spaced over the interval $[-1, 1]$ and to increase simultaneously the degree of the linear combination to 50. The ratio between the computing times is then higher than 100, which is unquestionably in favor of the script SPE_F_LegLinComb.m. Using the function legendre in this context implies a great number of redundant or useless computations.

Solution of Exercise 6.2

The computation of Gauss abscissas and integration weights is done by the function SPE_F_xwGauss.m. The abscissas are the eigenvalues of the matrix M defined by (6.11); the function therefore starts by building the matrix and using the MATLAB built-in function eig to compute its spectrum.

Once the abscissas and weights are computed and stored in two column vectors x and w, the quadrature formula (6.9) is encoded with a single MATLAB command. It consists in computing the scalar product of the vector w with the vector holding the values of the function at the integration abscissas x: I=w'*f(x);

The script SPE_M_TestIntGauss.m tests the quadrature formula on a smooth function, here the function e^x, and also compares this integration method with the method proposed by MATLAB, which is programmed in the built-in function quad.

The calling syntax is q = quad(@fun,a,b). This command returns the value of the integral of the function defined in fun.m between the bounds a and b to a default precision of 10^{-6}. The algorithm used in quad is the adaptive version of Simpson's rule (see (5.5) in Chap. 5), and one can specify the required precision by adding a fourth input argument q = quad(@fun,a,b,preci). It is also possible to get in the output the number of calls to the integrand that were performed: [q,nb] = quad(@fun,a,b,preci).

In order to compare the computing time of the function quad with the Gauss method, we use for the latter a number of points large enough to obtain the value of the integral with six significant digits. For instance, four points are sufficient in the case of the function e^x. The respective computing times are estimated using functions tic and toc as in the previous exercise. Simpson's method being less accurate than the Gauss quadrature on four points, it requires more evaluations of the integrand and is therefore slower. Hence, in the context of our project where a great number of integrals must be performed, using Gauss quadrature is more efficient.

Solution of Exercise 6.3

The comparison of the function with its truncated Legendre expansion is performed in the script SPE_M_AppLegExp.m. It produces Fig. 6.2, where the function $f(x) = \sin(6x)\exp(-x)$ is displayed along with $\mathcal{L}_6(f)$ and $\mathcal{L}_9(f)$. Slight modification produces Fig. 6.4 (a), where the function $f(x) = |x|$ is compared with its truncated series $\mathcal{L}_p(f)$ to order $p = 30$, and Fig. 6.4 (b), for $f(x) = \text{sign}(x)$.

This script calls the function SPE_F_CalcLegExp, which has the following input arguments:

- s: the degree of the Gauss quadrature to be used to compute the coefficients of the expansion (6.15),
- P: the degree of the truncated expansion,
- npt: the number of points in the interval $[-1, 1]$ where the expansion is computed,
- test: the name of the function whose expansion is computed, which should be defined either by an inline function or in a file.

The function returns:

- x: the npt abscissas,
- y: the values of the expansion at abscissas x,
- err: the error in the supremum between the function and its expansion, estimated on the values at abscissas x.

This function is also used in the script SPE_M_LegExpLoop.m to answer question 5 of the exercise, as illustrated in Fig. 6.3. The Legendre expansion of a test function and the error with the function itself is computed for different degrees, varying here between 2 and 30 with an increment of 2. The error in the supremum norm is then displayed as a function of the degree of the truncated expansion. For a smooth function, we expect exponential decrease for the error, which is actually what we obtain numerically for the function $f(x) = \sin(6x)\exp(-x)$. For less-smooth functions, for instance $f(x) = \text{abs}(x)$, the error decreases proportionally with $1/p$. Eventually, for a discontinuous function, such as $f(x) = \text{sign}(x)$, the error does not go to 0 when the degree of the expansion increases, due to the Gibbs phenomenon (see Fig. 6.4).

Solution of Exercise 6.4

We select here as a test case the function $u(x) = \sin(\pi x)\cos(10x)$, which satisfies the homogeneous Dirichlet boundary conditions $u(-1) = u(1) = 0$, and we program it in SPE_F_special.m. By setting the right-hand side to

$$f(x) = (\pi^2 + 130)\sin(\pi x)\cos(10x) + 20\pi\cos(\pi x)\sin(10x)$$

and the constant $c = 30$, the function u is a solution of problem (6.1). The right-hand side is programmed in the file SPE_F_fbe.m, using the second derivative of the solution function, which is programmed in SPE_F_specsec.m. The script SPE_M_SpecMeth.m computes the numerical solution using the spectral Galerkin method. It then compares it with the solution computed using the finite difference method. Figure 6.5 displays the exact solution, the spectral solution in $\mathbb{P}^0_{21}$, and the finite difference solution on 20 points. It is obtained by setting p=21 and mdf=21 in the script. It is clear that in this

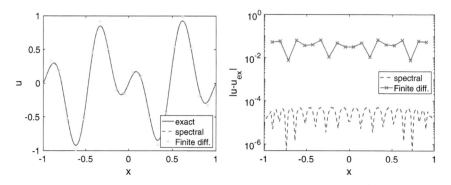

Fig. 6.5 Comparison of the exact, spectral, and finite difference solutions for $p = 21$ (left panel). Error (in log scale) between numerical and exact solutions (right panel).

example, the spectral method is much more accurate than the finite difference one, since the approximate solution in $\mathbb{P}_{21}^0$ cannot be distinguished from the exact one. The error in the supremum norm is $5 \cdot 10^{-5}$ when it is equal to $6 \cdot 10^{-2}$ for the second-order finite difference solution. Furthermore, a finite difference computation using about 800 points would be necessary to obtain the same order of error as with the spectral method.

On the other hand, as soon as the solution of the continuous problem is not regular enough, the performance of the spectral method in terms of accuracy drops and becomes comparable, and even in some cases worse than the performance of the finite difference method. The reader can easily verify this fact by building another test case, where the right-hand side f of equation (6.1) corresponds to a solution whose second derivative is a step function.

Chapter References

C. Bernardi, M. Dauge, Y. Maday, *Spectral Methods for Axisymmetric Domains. Numerical Algorithms and Tests due to Mejdi Azaiez, Series in Applied Mathematics*, vol. 3. (Gauthier-Villars, North-Holland, Amsterdam, 1999)

C. Bernardi, Y. Maday, *Spectral Methods in Handbook of Numerical Analysis*, vol. V (North-Holland, Amsterdam, 1997)

C. Bernardi, Y. Maday, F. Rapetti, *Discrétisations variationnelles de problèmes aux limites elliptiques, Mathématiques & Applications*, vol. 45 (Springer-Verlag, Mai, 2004)

P.J. Davis, *Interpolation and Approximation* (Dover Publications Inc, New York, 1975)

A.R. Krommer, C.W. Ueberhuber, *Numerical Integration on Advanced Computer Systems*, Lecture Notes in Computer Science, vol. 848. (Springer-Verlag, Berlin, 1994)

G. Szegő, *Orthogonal Polynomials*, vol. XXIII, 4th edn. (American Mathematical Society, Colloquium Publications, Providence, R.I., 1975)

B.I. Wohlmuth, *Discretization Methods and Iterative Solvers Based on Domain Decomposition*, Lecture Notes in Computational Science and Engineering, vol. 17. (Springer-Verlag, Berlin, 2001)

Chapter 7
High-order finite difference methods

Project Summary

Level of difficulty:	2
Keywords:	Explicit, implicit, compact finite difference schemes; second, fourth and sixth order schemes
Application fields:	Approximation of derivatives, solvers for PDEs

Finite difference (FD) methods are very popular for solving partial differential equations (PDEs) because of their simplicity. A single mathematical tool, namely the Taylor series expansion, is necessary to derive FD schemes to approximate derivatives. Such approximations are built locally at a point of the computational grid using an *approximation stencil* made of neighboring grid points. This stencil can be centered or decentered and thus easily adapted to the configuration of the computational domain or to the type of boundary conditions. The grid can be evenly spaced or stretched. Final resulting linear systems have generally a matrix with a band structure (i.e. non-zero coefficients are confined to a diagonal band) for which efficient solving algorithms can be used. All these advantages explain why the finite difference method is presented in almost all generic textbooks on numerical analysis or more specialized books (e.g. Hirsch 1988; Strikwerda 1989; LeVeque 2007). First- and second-order FD schemes are generally sufficient to build efficient numerical algorithms to solve linear or nonlinear PDEs. We use such schemes in several projects of this book: in the introductory Chap. 1 dealing with linear model PDEs, in Chap. 9 concerned with linear elasticity PDEs, in Chap. 10 illustrating the Schwarz domain decomposition method, and, finally, in Chaps. 12 and 15 dealing with highly nonlinear system of PDEs for fluid flows (Euler or Navier–Stokes equations).

Second-order FD schemes generally offer a good trade-off between accuracy and computational cost. However, using higher-order approximations is sometimes mandatory to enable the numerical system to capture particular details of the simulated phenomena (e.g. vortices in fluids or super-fluids, turbulence scaling laws, etc.). Well designed, high-order FD schemes do not add substantial computational cost to the overall numerical system. In the case of linear differential equations, as shown below, this additional cost is negligible. We present in this theoretical project a general method to derive high-order FD schemes with arbitrary approximation stencils. In addition to classical *explicit* schemes (the approximation is calculated explicitly using known values), we also introduce *implicit* schemes for which the approximation of derivatives involves the resolution of a linear system. The concept of implicit schemes is simple and elegant and enables high-accuracy approximations with a reduced stencil of grid points. Among implicit FD schemes, the so-called *compact* schemes are very popular because they are easy to implement and their accuracy is close to that of spectral methods presented in Chaps. 3 and 6. As an application of this project, we compare explicit/implicit high-order FD schemes for solving the 1D linear heat equation with Dirichlet or periodic boundary conditions.

7.1 Taylor series expansion and finite difference grid

We recall the main theorem needed for this project.

Theorem 7.1 (Taylor's theorem) *Let* $f : [a, b] \rightarrow \mathbb{R}, x \mapsto f(x)$ *be a real function and* $f \in C^{n+1}(]a, b[)$ *(i.e. its* $(n+1)$*st derivative is continuous), with* $n \geq 1$ *an integer. If* $x \neq x_a$ *are two points in the interval* $]a, b[$*, then there exists a value* ξ *in between* x *and* x_a *such that*

$$f(x) = f(x_a) + \sum_{k=1}^{n} \frac{(x - x_a)^k}{k!} f^{(k)}(x_a) + \underbrace{\frac{(x - x_a)^{n+1}}{(n + 1)!} f^{(n+1)}(\xi)}_{R_{n+1}}. \qquad (7.1)$$

We refer to (7.1) as the Taylor series expansion of $f(x)$ at point x_a. The remainder R_{n+1} was written in the differential (or Lagrange) form and, given the assumed regularity of the function f, we infer that

$$|R_{n+1}| \leq \frac{|x - x_a|^{n+1}}{(n + 1)!} \sup_{\xi \in [a,b]} |f^{(n+1)}(\xi)| = \frac{|x - x_a|^{n+1}}{(n + 1)!} \|f^{(n+1)}\|, \qquad (7.2)$$

where $\|.\|$ denotes the supremum (or uniform) norm. We shall use in the following Landau's notation (see also Chap. 1) and replace the remainder by $\mathcal{O}(|x - x_a|^{n+1})$, meaning that $|R_{n+1}| \leq C|x - x_a|^{n+1}$, with C a constant. In practice, it is convenient to consider points x and x_a separated by the (small) distance $h > 0$. We obtain the following Taylor series expansions:

$$f(x + h) = f(x) + \sum_{k=1}^{n} \frac{h^k}{k!} f^{(k)}(x) + \mathcal{O}(h^{n+1}), \tag{7.3}$$

$$f(x - h) = f(x) + \sum_{k=1}^{n} \frac{(-h)^k}{k!} f^{(k)}(x) + \mathcal{O}(h^{n+1}). \tag{7.4}$$

We apply now Taylor's theorem to find local approximations of the function f on the finite difference grid displayed in Fig. 7.1. We consider equidistant points $x_i = a + (i-1)h$, $i = 1, \ldots, (N+1)$, $h = (b-a)/N$ and use the short-hand notation: $f_i := f(x_i)$, $f_i' := f'(x_i)$, $f_i'' := f''(x_i)$.

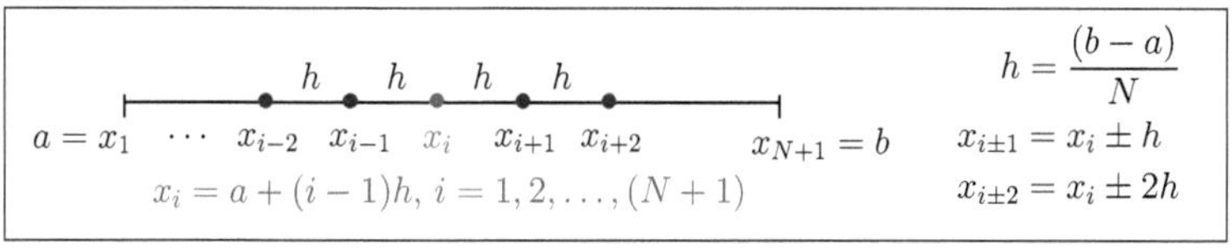

Fig. 7.1 Finite difference uniform grid.

Direct application of (7.3)-(7.4) with $x = x_i$ results in

$$f_{i+1} = f_i + h f_i' + \frac{h^2}{2} f_i'' + \frac{h^3}{6} f_i^{(3)} + \frac{h^4}{24} f_i^{(4)} + \frac{h^5}{120} f_i^{(5)} + \mathcal{O}(h^6), \tag{7.5}$$

$$f_{i-1} = f_i - h f_i' + \frac{h^2}{2} f_i'' - \frac{h^3}{6} f_i^{(3)} + \frac{h^4}{24} f_i^{(4)} - \frac{h^5}{120} f_i^{(5)} + \mathcal{O}(h^6). \tag{7.6}$$

In practice, we stop these expansions at the necessary order $n \geq 1$ and replace the remaining terms with $\mathcal{O}(h^{n+1})$.

In the next sections, we consider that values f_i of the function are known and we build local approximations of the derivatives $(\overline{f}_i' \approx f_i', \overline{f}_i'' \approx f_i'')$ using the so-called *finite difference schemes*.

7.2 Explicit finite difference schemes

Expansions (7.5) and (7.6) can be directly used to obtain FD schemes used in Chap. 1 for the first derivative:

$$f_i' = \frac{f_{i+1} - f_i}{h} + \mathcal{O}(h), \quad f_i' = \frac{f_i - f_{i-1}}{h} + \mathcal{O}(h). \tag{7.7}$$

In practice, the remainder $\mathcal{O}(h)$ is ignored: it thus defines the approximation error of the FD scheme. Following definitions in Chap. 1, these FD schemes are consistent (since the approximation error tends to zero if $h \to 0$) and they are of first-order accuracy (since the approximation error converges to zero as h^p, with $p = 1$). A second-order FD scheme for the first derivative is obtained by simply subtracting Eq. (7.6) from (7.5):

$$f_i' = \frac{f_{i+1} - f_{i-1}}{2h} + \mathcal{O}(h^2). \tag{7.8}$$

The sum of (7.6) and (7.5) simply gives the well-known centered second-order scheme for the second derivative:

$$f_i'' = \frac{f_{i+1} - 2f_i + f_{i-1}}{h^2} + \mathcal{O}(h^2). \tag{7.9}$$

These simple algebraic manipulations of Taylor series expansions suggest a more general method to derive FD schemes. This is the so-called *method of undetermined coefficients* and is summarized in the following algorithm.

Algorithm 7.1 (Approximation of $f_i^{(n)}$ by an explicit FD scheme)
Start by choosing the approximation stencil containing grid points x_m, close to the approximation point x_i (see Fig. 7.1). For example,
$m \in M = \{i+1, i+2, \ldots\}$ for a decentered forward scheme,
$m \in M = \{i-1, i-2, \ldots\}$ for a decentered backward scheme,
$m \in M = \{i \pm 1, i \pm 2, \ldots\}$ for a centered scheme, etc..

1. Write Taylor series expansions similar to (7.5)-(7.6) to approximate each value f_m, $m \in M$, <u>at the same grid point</u> x_i.
2. Insert these expansions in the following linear combination[1]

$$A f_i^{(n)} + \frac{1}{h^n} \left(a_0 f_i + \sum_{k \in M} a_k f_k \right) = R(h^p, f^{(n+1)}, f^{(n+2)}, \ldots) \tag{7.10}$$

and determine coefficients $\{A, a_0, (a_k)_{k \in M}\}$ such as
– the remainder R does not depend on values of f or its lower derivatives $f^{(m)}$ with $m \leq n$;
– the order of accuracy p is the highest possible (for the given approximation stencil).
3. Extract from (7.10) the FD scheme at order p to compute the approximation $\overline{f}_i^{(n)}$ of $f_i^{(n)}$:

$$\overline{f}_i^{(n)} = -\frac{1}{Ah^n} \left(a_0 f_i + \sum_{k \in M} a_k f_k \right). \tag{7.11}$$

As the best way to learn is from examples, let us take the case of a centered scheme for the first derivative. We consider a 5-point stencil (including the

---

[1] Since A is a constant, we divided the linear combination involving values of f by h^n to ensure the dimensional homogeneity of the expression.

point x_i), with $M = \{i \pm 1, i \pm 2\}$ (see Fig. 7.1) and the following linear combination (we use below more suggestive names for its coefficients than in the general algorithm):

$$Af_i' + \frac{a_+}{h} f_{i+1} + \frac{a_-}{h} f_{i-1} + \frac{a_0}{h} f_i + \frac{b_+}{h} f_{i+2} + \frac{b_-}{h} f_{i-2} = R. \qquad (7.12)$$

We first write the necessary Taylor series expansions and then multiply them by the constants of the linear combination (7.12):

$$\frac{a_+}{h} \times \left(f_{i+1} = f_i + hf_i' + \frac{h^2}{2} f_i'' + \frac{h^3}{6} f_i^{(3)} + \frac{h^4}{24} f_i^{(4)} + \frac{h^5}{120} f_i^{(5)} + \mathcal{O}(h^6) \right),$$

$$\frac{a_-}{h} \times \left(f_{i-1} = f_i - hf_i' + \frac{h^2}{2} f_i'' - \frac{h^3}{6} f_i^{(3)} + \frac{h^4}{24} f_i^{(4)} - \frac{h^5}{120} f_i^{(5)} + \mathcal{O}(h^6) \right),$$

$$\frac{b_+}{h} \times \left(f_{i+2} = f_i + (2h)f_i' + \frac{(2h)^2}{2} f_i'' + \frac{(2h)^3}{6} f_i^{(3)} + \frac{(2h)^4}{24} f_i^{(4)} + \frac{(2h)^5}{120} f_i^{(5)} + \mathcal{O}(h^6) \right),$$

$$\frac{b_-}{h} \times \left(f_{i-2} = f_i - (2h)f_i' + \frac{(2h)^2}{2} f_i'' - \frac{(2h)^3}{6} f_i^{(3)} + \frac{(2h)^4}{24} f_i^{(4)} - \frac{(2h)^5}{120} f_i^{(5)} + \mathcal{O}(h^6) \right).$$

After gathering all terms in (7.12), we obtain the following expression for the remainder:

$$\begin{aligned} R = \frac{1}{h} & [(a_+ + a_-) + a_0 + (b_+ + b_-)] f_i + \qquad (7.13) \\ & [(a_+ - a_-) + 2(b_+ - b_-) + A] f_i' + \\ & \frac{h}{2} [(a_+ + a_-) + 2(b_+ + b_-)] f_i'' + \\ & \frac{h^2}{6} [(a_+ - a_-) + 2^3(b_+ - b_-)] f_i^{(3)} + \\ & \frac{h^3}{24} [(a_+ + a_-) + 2^4(b_+ + b_-)] f_i^{(4)} + \\ & \frac{h^4}{120} [(a_+ - a_-) + 2^5(b_+ - b_-)] f_i^{(5)} + \mathcal{O}(h^5). \end{aligned}$$

The idea now is to eliminate (in order) a maximum of terms in the remainder to obtain the maximum accuracy possible. Since we introduced 6 coefficients in the linear combination, it is reasonable to try to cancel the first 6 terms in the remainder. We will see below that this approach is quite optimistic and finally it is wiser to deal with an undetermined linear system (less equations than unknowns) to ensure the existence of a non-trivial solution. We write however the complete linear system and exploit its particular structure as follows:

$$\begin{cases} a_+ + a_- + a_0 + b_+ + b_- & = 0 \Longrightarrow a_0 = -(a_+ + a_- + b_+ + b_-), \\ (a_+ - a_-) + 2(b_+ - b_-) + A = 0 \Longrightarrow A = -(a_+ - a_-) - 2(b_+ - b_-), \\ (a_+ + a_-) + 2^2(b_+ + b_-) & = 0 \Longrightarrow a_+ + a_- = -4(b_+ + b_-), \\ (a_+ - a_-) + 2^3(b_+ - b_-) & = 0, \\ (a_+ + a_-) + 2^4(b_+ + b_-) & = 0 \Longrightarrow a_+ + a_- = -16(b_+ + b_-), \\ (a_+ - a_-) + 2^5(b_+ - b_-) & = 0. \end{cases} \tag{7.14}$$

We obtain that necessarily $a_+ + a_- = 0$ and $b_+ + b_- = 0$. As a consequence, we have to restrain the system (7.14) to its first 5 equations. Using b_+ as parameter, we obtain the following solution:

$$\begin{cases} a_+ = -a_-, \\ b_- = -b_+, \\ a_+ - a_- = -8(b_+ - b_-) & \Longrightarrow a_+ = -a_- = -8b_+, \\ A = -(a_+ - a_-) - 2(b_+ - b_-) \Longrightarrow A = -2a_+ - 4b_+ = 12b_+, \\ a_0 = -(a_+ + a_- + b_+ + b_-) & \Longrightarrow a_0 = 0. \end{cases} \tag{7.15}$$

We go back now to the linear combination (7.12) and replace the obtained coefficients. To show the canceled terms, we present the linear combination as follows:

$$(12b_+)f_i' + (-8b_+)\frac{f_{i+1} - f_{i-1}}{h} + b_+ \frac{f_{i+2} - f_{i-2}}{h} = \tag{7.16}$$

$$\frac{1}{h}[(a_+ + a_-) + a_0 + (b_+ + b_-)] \; f_i \; +$$

$$[(a_+ - a_-) + 2(b_+ - b_-) + A] \; f_i' \; +$$

$$\frac{h}{2}[(a_+ + a_-) + 2(b_+ + b_-)] \; f_i'' \; +$$

$$\frac{h^2}{6}[(a_+ - a_-) + 2^3(b_+ - b_-)] \; f_i^{(3)} \; +$$

$$\frac{h^3}{24}[(a_+ + a_-) + 2^4(b_+ + b_-)] \; f_i^{(4)} \; +$$

$$\frac{h^4}{120}[(a_+ - a_-) + 2^5(b_+ - b_-)] \; f_i^{(5)} \; +$$

$$\frac{h^5}{720}[(a_+ + a_-) + 2^6(b_+ + b_-)] \; f_i^{(6)} \; +\mathcal{O}(h^6).$$

It is interesting to note that the term of order h^5 in the remainder also cancels. This was not requested when building the linear system (7.14), but it is a consequence of the form of solution (7.15). We finally obtain

$$f_i' = \frac{8(f_{i+1} - f_{i-1}) - (f_{i+2} - f_{i-2})}{12h} + \frac{h^4}{30}f_i^{(5)} + \mathcal{O}(h^6), \tag{7.17}$$

leading to the fourth-order centered FD scheme:

$$f_i' = \frac{4}{3}\frac{f_{i+1} - f_{i-1}}{2h} - \frac{1}{3}\frac{f_{i+2} - f_{i-2}}{4h} + \mathcal{O}(h^4). \tag{7.18}$$

Note that this is the maximum accuracy that could be obtained using a centered 5-point stencil. If we rewind all these calculations using a 3-point stencil, we recover the second-order FD scheme:

$$f_i' = \frac{f_{i+1} - f_{i-1}}{2h} - \frac{h^2}{6} f_i^{(3)} + \mathcal{O}(h^4). \tag{7.19}$$

Exercise 7.1 Derive decentered explicit FD schemes for the first derivative using the stencil $M = \{i+1, i+2\}$ (forward) and $M = \{i-1, i-2\}$ (backward). Show that

$$f_i' = \frac{-3f_i + 4f_{i+1} - f_{i+2}}{2h} + \frac{h^2}{3} f_i^{(3)} + \mathcal{O}(h^3), \tag{7.20}$$

$$f_i' = \frac{3f_i - 4f_{i-1} + f_{i-2}}{2h} - \frac{h^2}{3} f_i^{(3)} + \mathcal{O}(h^3). \tag{7.21}$$

Exercise 7.2 Derive decentered explicit FD schemes for the second derivative using the stencil $M = \{i+1, i+2, i+3\}$ (forward) and $M = \{i-1, i-2, i-3\}$ (backward). Show that

$$f_i'' = \frac{2f_i - 5f_{i+1} + 4f_{i+2} - f_{i+3}}{h^2} + \frac{11h^2}{12} f_i^{(4)} + \mathcal{O}(h^3), \tag{7.22}$$

$$f_i'' = \frac{2f_i - 5f_{i-1} + 4f_{i-2} - f_{i-3}}{h^2} + \frac{11h^2}{12} f_i^{(4)} + \mathcal{O}(h^3). \tag{7.23}$$

Exercise 7.3 Derive a centered explicit FD scheme for the second derivative using the stencil $M = \{i \pm 1, i \pm 2\}$. Show that

$$f_i'' = \frac{16(f_{i+1} + f_{i-1}) - 30f_i - (f_{i+2} + f_{i-2})}{12h^2} + \frac{h^4}{90} f_i^{(6)} + \mathcal{O}(h^6). \tag{7.24}$$

Hints: the expression of the remainder is similar to (7.13), with the difference that the coefficient A moves from the second to the third term. This changes the symmetries of the scheme (for this scheme $b_+ = b_-$ and $a_+ = a_-$).

7.3 Implicit finite difference schemes

We concluded from the previous section that the only way to increase the accuracy of an explicit scheme is to enlarge the size of the approximation stencil. This brings the drawback (besides the more complicated formulae) that such schemes are difficult to use near the borders of a domain. Implicit schemes circumvent this disadvantage and enable high accuracy with a reduced stencil. The idea behind is simple and ingenious: in Algorithm 7.1 extend the linear combination by including also the values of the derivative at points x_m of the stencil:

$$\left(A_0 f_i^{(n)} + \sum_{k \in M} A_k f_k^{(n)} \right) + \frac{1}{h^n} \left(a_0 f_i + \sum_{k \in M} a_k f_k \right) = R(h^p, f^{(n+1)}, f^{(n+2)}, \ldots).$$

$$(7.25)$$

The other steps of the algorithm remain the same. Once the coefficients of the linear combination are determined, we obtain a linear equation that connects the values of the approximation of the derivative at neighboring points of the grid:

$$A_0 \overline{f}_i^{(n)} + \sum_{k \in M} A_k \overline{f}_k^{(n)} = -\frac{1}{h^n} \left(a_0 f_i + \sum_{k \in M} a_k f_k \right), \qquad (7.26)$$

where $\overline{f}_i^{(n)}$ are now the approximations of $f_i^{(n)}$ since the remainder in (7.25) was neglected (as inevitably in practice). We note that the implicit scheme will not provide directly the value of $\overline{f}_i^{(n)}$ (as in explicit schemes), but it connects unknown values of the derivative at points neighboring the approximation point x_i. This connection is similar to that assumed in spectral methods (see Chaps. 3 and 6). Some implicit schemes (see below the compact schemes) are therefore known as schemes with spectral-like accuracy. As a final step of the algorithm for implicit schemes, we have to write Eq. (7.26) for all points i of the grid and solve the resulting linear system to get the values of the derivative approximation. This step doesn't exist for explicit schemes and implies an extra computational cost, that can be reduced if the structure of the system is simple. In practice, linear systems with *banded matrices* (i.e. sparse matrices with non-zero coefficients confined to a diagonal band) are solved with efficient algorithms. In the following, we consider 3-point stencils to obtain implicit schemes, since linear systems with tridiagonal matrices are very easy to solve.

Let us illustrate these features by writing a fourth-order implicit scheme for the first derivative. We consider the 3-point stencil with $M = \{i+1, i-1\}$ and the linear combination:

$$A_+ f'_{i+1} + A_0 f'_i + A_- f'_{i-1} + \frac{a_+}{h} f_{i+1} + \frac{a_0}{h} f_i + \frac{a_-}{h} f_{i-1} = R. \qquad (7.27)$$

The necessary Taylor series expansions are now (note the arrangement of terms involving the same derivative in columns to facilitate the assembly of terms in the linear combination):

$$\frac{a_+}{h} \times \left(f_{i+1} = f_i + h f_i' + \frac{h^2}{2} f_i'' + \frac{h^3}{6} f_i^{(3)} + \frac{h^4}{24} f_i^{(4)} + \frac{h^5}{120} f_i^{(5)} + \frac{h^6}{720} f_i^{(6)} + \mathcal{O}(h^7) \right),$$

$$\frac{a_-}{h} \times \left(f_{i-1} = f_i - h f_i' + \frac{h^2}{2} f_i'' - \frac{h^3}{6} f_i^{(3)} + \frac{h^4}{24} f_i^{(4)} - \frac{h^5}{120} f_i^{(5)} + \frac{h^6}{720} f_i^{(6)} + \mathcal{O}(h^7) \right),$$

$$A_+ \times \left(f_{i+1}' = f_i' + h f_i'' + \frac{h^2}{2} f_i^{(3)} + \frac{h^3}{6} f_i^{(4)} + \frac{h^4}{24} f_i^{(5)} + \frac{h^5}{120} f_i^{(6)} + \mathcal{O}(h^6) \right),$$

$$A_- \times \left(f_{i-1}' = f_i' - h f_i'' + \frac{h^2}{2} f_i^{(3)} - \frac{h^3}{6} f_i^{(4)} + \frac{h^4}{24} f_i^{(5)} - \frac{h^5}{120} f_i^{(6)} + \mathcal{O}(h^6) \right).$$

After collecting terms, the remainder of the linear combination takes the form:

$$R = \frac{1}{h} (a_+ + a_- + a_0) f_i + \tag{7.28}$$

$$[(a_+ - a_-) + (A_+ + A_0 + A_-)] f_i' +$$

$$\frac{h}{2} [(a_+ + a_-) + 2(A_+ - A_-)] f_i'' +$$

$$\frac{h^2}{3} [(a_+ - a_-) + 3(A_+ + A_-)] f_i^{(3)} +$$

$$\frac{h^3}{24} [(a_+ + a_-) + 4(A_+ - A_-)] f_i^{(4)} +$$

$$\frac{h^4}{120} [(a_+ - a_-) + 5(A_+ + A_-)] f_i^{(5)} + \mathcal{O}(h^5).$$

The analysis of the linear system resulting from canceling successive terms in the remainder leads to

$$\begin{cases} a_+ + a_- + a_0 = 0, \\ (a_+ - a_-) + (A_+ + A_0 + A_-) = 0, \\ (a_+ + a_-) + 2(A_+ - A_-) = 0 & \Longrightarrow a_+ + a_- = -2(A_+ - A_-), \\ (a_+ - a_-) + 3(A_+ + A_-) = 0, \\ (a_+ + a_-) + 4(A_+ - A_-) = 0 & \Longrightarrow a_+ + a_- = -4(A_+ - A_-), \\ (a_+ - a_-) + 5(A_+ + A_-) = 0. \end{cases} \tag{7.29}$$

We conclude that necessarily $a_+ + a_- = 0$ and $A_+ - A_- = 0$. We choose then A_+ as parameter and obtain the following solution:

$$\begin{cases} a_+ = -a_-, \\ A_- = A_+, \\ (a_+ - a_-) + 3(A_+ + A_-) = 0 & \Longrightarrow a_+ = -a_- = -3A_+, \\ (a_+ - a_-) + (A_+ + A_0 + A_-) = 0 \Longrightarrow A_0 = -2A_+ - 2a_+ = 4A_+, \\ a_+ + a_- + a_0 = 0 & \Longrightarrow a_0 = 0. \end{cases} \tag{7.30}$$

After replacing these coefficients in the linear combination (7.27), we obtain (the canceled terms in the remainder are again shown):

$$A_+ f'_{i+1} + 4A_+ f'_i + A_+ f'_{i-1} - 3A_+ \frac{f_{i+1} - f_{i-1}}{h} =$$

$$\frac{1}{h} (a_+ + a_- + a_0) \; f_i \; +$$

$$\frac{[(a_+ - a_-) + (A_+ + A_0 + A_-)]}{h} \; f'_i \; +$$

$$\frac{h}{2} [(a_+ + a_-) + 2(A_+ - A_-)] \; f''_i \; +$$

$$\frac{h^2}{3} [(a_+ - a_-) + 3(A_+ + A_-)] \; f^{(3)}_i \; +$$

$$\frac{h^3}{24} [(a_+ + a_-) + 4(A_+ - A_-)] \; f^{(4)}_i \; +$$

$$\frac{h^4}{120} [(a_+ - a_-) + 5(A_+ + A_-)] \; f^{(5)}_i \; +$$

$$\frac{h^5}{720} [(a_+ + a_-) + 6(A_+ - A_-)] \; f^{(6)}_i \; + \mathcal{O}(h^6).$$

Note the cancelation of term of order h^5, due to the symmetry of solution (7.30). We finally obtain

$$f'_{i+1} + 4f'_i + f'_{i-1} = \frac{3}{h} (f_{i+1} - f_{i-1}) + \frac{h^4}{30} f^{(5)}_i + \mathcal{O}(h^6), \qquad (7.31)$$

leading to the fourth-order implicit FD scheme:

$$f'_{i+1} + 4f'_i + f'_{i-1} = \frac{3}{h} (f_{i+1} - f_{i-1}) + \mathcal{O}(h^4). \qquad (7.32)$$

The major achievement is that the fourth order is reached with only a 3-point stencil. We recall that the same fourth order requested a 5-point stencil for an explicit scheme. To finally compute approximate values of the first derivative on the grid represented in Fig. 7.1, we write equations (7.32) for all inner points $i = 2, \ldots, N$ (this is another advantage of using only 3 points in the stencil). For boundary points $i = 1$ and $i = N+1$, we have to use boundary conditions prescribed by the formulation of the PDE to be solved. The final system with tridiagonal matrix could be solved using efficient algorithms (one of them, the Thomas algorithm is described in great detail in Chap. 15).

Exercise 7.4 Derive a centered implicit FD scheme for the second derivative using a 3-point stencil with $M = \{i+1, i-1\}$. Show that

$$f''_{i+1} + 10f''_i + f''_{i-1} = \frac{12}{h^2} (f_{i+1} - 2f_i + f_{i-1}) + \frac{h^4}{20} f^{(6)}_i + \mathcal{O}(h^6). \qquad (7.33)$$

7.4 Compact finite difference schemes

Computing derivatives with FD schemes can be formally reduced to solving the linear system

$$A\overline{F}_d = BF, \tag{7.34}$$

with $\overline{F}_d \in \mathbb{R}^{N+1}$ the vector of unknowns (values of the approximations of the derivative) and $F \in \mathbb{R}^{N+1}$ the vector of known values $f_i, i = 1, \ldots, N+1$. Matrices $A, B \in \mathbb{R}^{(N+1) \times (N+1)}$ depend on the FD scheme. For explicit schemes, $A = I$, where I is the identity matrix. If we ignore the first and last equations of the linear system (depending on boundary conditions), all previous FD schemes use sparse banded matrices: $A = I$ and B is penta-diagonal for explicit schemes (7.18) and (7.24); A and B are tridiagonal for implicit schemes (7.32) and (7.33). We have seen that increasing the order of accuracy requires larger approximation stencils and, consequently, larger bands in matrices A and/or B. Implicit schemes have the advantage to use stencils with a low number of points and thus deserve to be further explored. Fourth-order accuracy was obtained with a minimum stencil of 3 points. A legitimate question is how to further increase the accuracy of implicit schemes? Using a 5-point stencil could be a solution, but it brings the drawback of using penta-diagonal matrices for A and B. A much clever solution comes from the observation that in the algorithm for implicit schemes, we imposed the same tridiagonal structure for matrices A and B. Why not keeping the tridiagonal structure for A and enlarging the band of B? This ingenious idea was used by Lele (1992) to derive the so-called *compact FD schemes* using the same method of undetermined coefficients. Since the matrix A is the one involved in the inversion of the linear system, keeping a tridiagonal structure does not add supplementary computational effort. Matrix B affects only the right-hand side of the system and the cost of the multiplication BF is almost the same if B is tridiagonal or penta-diagonal.

Lele (1992) also astutely reformulated the linear combination used in the previous section by imposing several symmetries and thus reducing the complexity of the calculus. The method of undetermined coefficients for the first derivative became:[2]

$$f_i' + \alpha(f_{i+1}' + f_{i-1}') = a\frac{f_{i+1} - f_{i-1}}{2h} + b\frac{f_{i+2} - f_{i-2}}{4h} + R. \tag{7.35}$$

Note that the imposed symmetries are inspired from the formulation of explicit second-order schemes. The scheme implies tridiagonal and penta-diagonal matrices for A and B, respectively.[3] To determine coefficients $\{\alpha, a, b, c\}$, we proceed as in the previous section and use Taylor series expansions to gather:

---

[2] Compact schemes could be also obtained by using local rational approximations (or Padé approximations) (see, for instance, Brio et al 2010).

[3] A more general scheme was studied by Lele (1992)

$$a \times \left(\frac{f_{i+1} - f_{i-1}}{2h} = f_i' + \frac{h^2}{3!} f_i^{(3)} + \frac{h^4}{5!} f_i^{(5)} + \frac{h^6}{7!} f_i^{(7)} + \mathcal{O}(h^8) \right),$$

$$b \times \left(\frac{f_{i+2} - f_{i-2}}{2(2h)} = f_i' + \frac{(2h)^2}{3!} f_i^{(3)} + \frac{(2h)^4}{5!} f_i^{(5)} + \frac{(2h)^6}{7!} f_i^{(7)} + \mathcal{O}(h^8) \right),$$

$$\alpha \times \left(f_{i+1}' + f_{i-1}' = 2f_i' + 2\frac{h^2}{2!} f_i^{(3)} + 2\frac{h^4}{4!} f_i^{(5)} + 2\frac{h^6}{6!} f_i^{(7)} + \mathcal{O}(h^8) \right).$$

The linear combination (7.35) becomes

$$f_i' + \alpha(f_{i+1}' + f_{i-1}') - a\frac{f_{i+1} - f_{i-1}}{2h} - b\frac{f_{i+2} - f_{i-2}}{4h} =$$

$$(1 + 2\alpha - a - b) f_i' + \frac{h^2}{3!} \left(2\frac{3!}{2!}\alpha - a - 2^2 b \right) f_i^{(3)} +$$

$$\frac{h^4}{5!} \left(2\frac{5!}{4!}\alpha - a - 2^4 b \right) f_i^{(5)} + \frac{h^6}{7!} \left(2\frac{7!}{6!}\alpha - a - 2^6 b \right) f_i^{(7)} + \mathcal{O}(h^8).$$

Canceling the first three terms of the remainder results in the following linear system and solution:

$$\begin{cases} 1 + 2\alpha - a - b = 0, \\ 6\alpha - a - 4b = 0, \\ 10\alpha - a - 16b = 0 \end{cases} \implies \alpha = \frac{1}{3}, \quad a = \frac{14}{9}, \quad b = \frac{1}{9}. \qquad (7.36)$$

The sixth-order compact scheme for the first derivative is then

$$\frac{1}{3} f_{i-1}' + f_i' + \frac{1}{3} f_{i+1}' = \frac{14}{9} \frac{f_{i+1} - f_{i-1}}{2h} + \frac{1}{9} \frac{f_{i+2} - f_{i-2}}{4h} - \frac{4}{7!} h^6 f_i^{(7)} + \mathcal{O}(h^8).$$
$$(7.37)$$

Note that if we take $b = 0$ in the linear system (7.36) and consider only the first two equations, we obtain a fourth-order compact scheme with $\alpha = \frac{1}{4}$ and $a = \frac{3}{2}$; this is exactly the implicit fourth-order FD scheme (7.32), but presented as (Lele 1992):

$$\frac{1}{4} f_{i+1}' + f_i' + \frac{1}{4} f_{i-1}' = \frac{3}{2} \frac{f_{i+1} - f_{i-1}}{2h} + \mathcal{O}(h^4). \qquad (7.38)$$

Note that the implicit fourth-order scheme (7.32) was obtained from a linear combination without any symmetries imposed. The fact that a symmetric scheme was finally obtained is an a posteriori justification of the symmetries used in the construction of compact schemes.

$$f_i' + \alpha(f_{i+1}' + f_{i-1}') + \beta(f_{i+2}' + f_{i-2}') = a\frac{f_{i+1} - f_{i-1}}{2h} + b\frac{f_{i+2} - f_{i-2}}{4h} + c\frac{f_{i+3} - f_{i-3}}{6h} + R.$$

Compact schemes up to tenth order were obtained. We focus our presentation on the case $\beta = c = 0$, which gives the most popular compact schemes used in practice (sixth order).

For the second derivative, we use the linear combination:[4]

$$f_i'' + \alpha(f_{i+1}'' + f_{i-1}'') = a\frac{f_{i+1} - 2f_i + f_{i-1}}{h^2} + b\frac{f_{i+2} - 2f_i + f_{i-2}}{4h^2} + R, \quad (7.39)$$

and obtain the following linear system and solution for its coefficients:

$$\begin{cases} 1 + 2\alpha - a - b = 0, \\ 12\alpha - a - 4b = 0, \\ 30\alpha - a - 16b = 0 \end{cases} \implies \alpha = \frac{2}{11}, \quad a = \frac{12}{11}, \quad b = \frac{3}{11}. \quad (7.40)$$

The sixth-order compact scheme for the second derivative is then

$$\frac{2}{11}f_{i-1}'' + f_i'' + \frac{2}{11}f_{i+1}'' = \frac{12}{11}\frac{f_{i+1} - 2f_i + f_{i-1}}{h^2} + \frac{3}{11}\frac{f_{i+2} - 2f_i + f_{i-2}}{4h^2}$$
$$- \frac{23}{11 \cdot 7!}h^6 f_i^{(8)} + \mathcal{O}(h^8). \quad (7.41)$$

If $b = 0$ in the linear system (7.40) and consider only the first two equations, we obtain a fourth-order compact scheme with $\alpha = \frac{1}{10}$ and $a = \frac{12}{10}$; this is exactly the implicit FD scheme (7.33), but presented as (Lele 1992):

$$\frac{1}{10}f_{i+1}'' + f_i'' + \frac{1}{10}f_{i-1}'' = \frac{12}{10}\frac{f_{i+1} - 2f_i + f_{i-1}}{h^2} + \mathcal{O}(h^4). \quad (7.42)$$

7.5 Boundary conditions

Previous high-order schemes were obtained for inner grid points, far enough from boundaries to accommodate with the width of the approximation stencil. For example, the fourth-order schemes (7.32) and (7.33) can be applied for points $i = 2, \ldots N$ (see Fig. 7.1). Sixth-order schemes (7.37) and (7.41) have to avoid the first two and the last two grid points because of using $f_{i\pm2}$. We distinguish below between the periodic case and the non-periodic one and present in detail the final linear system for compact schemes.

7.5.1 Periodic Boundary Conditions

If the function f is periodic ($f_1 = f_{N+1}$), all schemes are applied only for grid points $i = 1, \ldots N$ (see Fig. 7.1). Square matrices A, B are of dimension $N_p = N$ and built using the periodic boundary conditions as follows. We can imagine *ghost* grid points such that $f_0 = f_{N_p}, f_{-1} = f_{N_p-1}$ and $f_{Np+1} = f_1$,

---

[4] Lele (1992) considered an extended linear combination by also including the terms $\beta\left(f_{i+2}'' + f_{i-2}''\right)$ and $c\frac{f_{i+3} - 2f_i + f_{i-3}}{9h^2}$.

$f_{N_p+2} = f_2$. The fourth-order FD scheme (7.32) for the first derivative leads to the following system (recall that $\overline{f}'_i$ is the approximation of the exact value of the derivative f'_i):

$$(i = 1) \qquad 4\overline{f}'_1 + \overline{f}'_2 + \overline{f}'_{N_p} = \frac{3}{h}f_2 - \frac{3}{h}f_{N_p},$$

$$(i = 2 \ldots N_p - 1) \qquad \overline{f}'_{i-1} + 4\overline{f}'_i + \overline{f}'_{i+1} = -\frac{3}{h}f_{i-1} + \frac{3}{h}f_{i+1},$$

$$(i = N_p) \qquad \overline{f}'_1 + \overline{f}'_{N_p-1} + 4\overline{f}'_{N_p} = \frac{3}{h}f_1 - \frac{3}{h}f_{N_p-1}, \qquad (7.43)$$

and corresponding matrices of size $N_p \times N_p$:

$$
A_4^{d1p} = \begin{pmatrix}
4 & 1 & \cdots & & 0 & \boxed{1} \\
1 & 4 & 1 & \cdots & & 0 \\
\vdots & \ddots & \ddots & \ddots & \ddots & \vdots \\
\vdots & & \ddots & \ddots & \ddots & \vdots \\
0 & & \cdots & 1 & 4 & 1 \\
\boxed{1} & 0 & & \cdots & 1 & 4
\end{pmatrix},
\quad
B_4^{d1p} = \frac{1}{h}\begin{pmatrix}
0 & 3 & \cdots & & 0 & \boxed{-3} \\
-3 & 0 & 3 & \cdots & & 0 \\
\vdots & \ddots & \ddots & \ddots & \ddots & \vdots \\
\vdots & & \ddots & \ddots & \ddots & \vdots \\
0 & & \cdots & -3 & 0 & 3 \\
\boxed{3} & 0 & & \cdots & -3 & 0
\end{pmatrix}. \qquad (7.44)
$$

Note that we preferred to write matrices with integer coefficients, obtained by multiplying the initial schemes by an integer constant. Similarly, for the fourth-order FD scheme (7.33) for the second derivative we obtain the system:

$$(i = 1) \qquad 10\overline{f}''_1 + \overline{f}''_2 + \overline{f}''_{N_p} = -\frac{24}{h^2}f_1 + \frac{12}{h^2}f_2 + \frac{12}{h^2}f_{N_p}, \qquad (7.45)$$

$$(i = 2 \ldots N_p - 1) \qquad \overline{f}''_{i-1} + 10\overline{f}''_i + \overline{f}''_{i+1} = \frac{12}{h^2}f_{i-1} - \frac{24}{h^2}f_i + \frac{12}{h^2}f_{i+1},$$

$$(i = N_p) \qquad \overline{f}''_1 + \overline{f}''_{N_p-1} + 10\overline{f}''_{N_p} = \frac{12}{h^2}f_1 + \frac{12}{h^2}f_{N_p-1} - \frac{24}{h^2}f_{N_p},$$

and corresponding matrices

$$
A_4^{d2p} = \begin{pmatrix}
10 & 1 & \cdots & & 0 & \boxed{1} \\
1 & 10 & 1 & \cdots & & 0 \\
\vdots & \ddots & \ddots & \ddots & \ddots & \vdots \\
\vdots & & \ddots & \ddots & \ddots & \vdots \\
0 & & \cdots & 1 & 10 & 1 \\
\boxed{1} & 0 & & \cdots & 1 & 4
\end{pmatrix},
\quad
B_4^{d2p} = \frac{1}{h^2}\begin{pmatrix}
-24 & 12 & \cdots & & 0 & \boxed{12} \\
12 & -24 & 12 & \cdots & & 0 \\
\vdots & \ddots & \ddots & \ddots & \ddots & \vdots \\
\vdots & & \ddots & \ddots & \ddots & \vdots \\
0 & & \cdots & 12 & -24 & 12 \\
\boxed{12} & 0 & & \cdots & 12 & -24
\end{pmatrix}. \qquad (7.46)
$$

The sixth-order scheme (7.37) for the first derivative uses a penta-diagonal B matrix and the periodicity has to be carefully taken into account:

$$(i = 1) \qquad 36\bar{f}'_1 + 12\bar{f}'_2 + 12\bar{f}'_{Np} = \frac{28}{h}f_2 + \frac{1}{h}f_3 - \frac{1}{h}f_{Np-1} - \frac{28}{h}f_{Np},$$

$$(i = 2) \qquad 12\bar{f}'_1 + 36\bar{f}'_2 + 12\bar{f}'_3 = -\frac{28}{h}f_1 + \frac{28}{h}f_3 + \frac{1}{h}f_4 - \frac{1}{h}f_{Np},$$

$$(i = 3\ldots N_p - 2) \qquad 12\bar{f}'_{i-1} + 36\bar{f}'_i + 12\bar{f}'_{i+1} = -\frac{1}{h}f_{i-2} - \frac{28}{h}f_{i-1} + \frac{28}{h}f_{i+1} + \frac{1}{h}f_{i+2},$$

$$(i = N_p - 1) \qquad 12\bar{f}'_{Np-2} + 36\bar{f}'_{Np-1} + 12\bar{f}'_{Np} = \frac{1}{h}f_1 - \frac{1}{h}f_{Np-3} - \frac{28}{h}f_{Np-2} + \frac{28}{h}f_{Np},$$

$$(i = N_p) \qquad 12\bar{f}'_1 + 12\bar{f}'_{Np-1} + 36\bar{f}'_{Np} = \frac{28}{h}f_1 + \frac{1}{h}f_2 - \frac{1}{h}f_{Np-2} - \frac{28}{h}f_{Np-1}.$$

$$(7.47)$$

Corresponding matrices are

$$A_6^{d1p} = \begin{pmatrix} 36 & 12 & \ldots & & 0 & \boxed{12} \\ 12 & 36 & 12 & \ldots & & 0 \\ \vdots & & \ddots & \ddots & \ddots & \vdots \\ \vdots & & & \ddots & \ddots & \vdots \\ 0 & & \ldots & 12 & 36 & 12 \\ \boxed{12} & 0 & & \ldots & 12 & 36 \end{pmatrix}, \quad B_6^{d1p} = \frac{1}{h}\begin{pmatrix} 0 & 28 & 1 & \ldots & & \boxed{-1} & \boxed{-28} \\ -28 & 0 & 28 & 1 & \ldots & 0 & \boxed{-1} \\ -1 & -28 & 0 & 28 & 1 & \ldots & 0 \\ \vdots & & \ddots & \ddots & \ddots & \ddots & \vdots \\ 0 & & \ldots & -1 & -28 & 0 & 28 & 1 \\ \boxed{1} & & & \ldots & -1 & -28 & 0 & 28 \\ \boxed{28} & \boxed{1} & & & \ldots & -1 & -28 & 0 \end{pmatrix}.$$

$$(7.48)$$

Similarly, for the sixth-order FD scheme (7.41) for the second derivative we obtain the system:

$$(i = 1) \qquad 44\bar{f}''_1 + 8\bar{f}''_2 + 8\bar{f}''_{Np} = -\frac{102}{h^2}f_1 + \frac{48}{h^2}f_2 + \frac{3}{h^2}f_3 + \frac{3}{h^2}f_{Np-1} + \frac{48}{h^2}f_{Np},$$

$$(i = 2) \qquad 8\bar{f}''_1 + 44\bar{f}''_2 + 8\bar{f}''_3 = \frac{48}{h^2}f_1 - \frac{102}{h^2}f_2 + \frac{48}{h^2}f_3 + \frac{3}{h^2}f_4 + \frac{3}{h^2}f_{Np},$$

$$(i = 3\ldots N_p - 2) \qquad 8\bar{f}''_{i-1} + 44\bar{f}''_i + 8\bar{f}''_{i+1} = \frac{3}{h^2}f_{i-2} + \frac{48}{h^2}f_{i-1} - \frac{102}{h^2}f_i + \frac{48}{h^2}f_{i+1} + \frac{3}{h^2}f_{i+2},$$

$$(i = N_p - 1) \qquad 8\bar{f}''_{Np-2} + 44\bar{f}''_{Np-1} + 8\bar{f}''_{Np} = \frac{3}{h^2}f_1 + \frac{3}{h^2}f_{Np-3} + \frac{48}{h}f_{Np-2} + \frac{102}{h}f_{Np-1} + \frac{48}{h^2}f_{Np},$$

$$(i = N_p) \qquad 8\bar{f}''_1 + 8\bar{f}''_{Np-1} + 44\bar{f}''_{Np} = \frac{48}{h^2}f_1 + \frac{3}{h^2}f_2 + \frac{3}{h^2}f_{Np-2} + \frac{48}{h^2}f_{Np-1} - \frac{102}{h^2}f_{Np}.$$

$$(7.49)$$

Corresponding matrices are

$$A_6^{d2p} = \begin{pmatrix} 44 & 8 & \ldots & & 0 & \boxed{8} \\ 8 & 44 & 8 & \ldots & & 0 \\ \vdots & & \ddots & \ddots & \ddots & \vdots \\ \vdots & & & \ddots & \ddots & \vdots \\ 0 & & \ldots & 8 & 44 & 8 \\ \boxed{8} & 0 & & \ldots & 8 & 44 \end{pmatrix}, \quad B_6^{d2p} = \frac{1}{h^2}\begin{pmatrix} -102 & 48 & 3 & \ldots & & \boxed{3} & \boxed{48} \\ 48 & -102 & 48 & 3 & \ldots & 0 & \boxed{3} \\ 3 & 48 & -102 & 48 & 3 & \ldots & 0 \\ \vdots & & \ddots & \ddots & \ddots & \ddots & \vdots \\ 0 & & \ldots & 3 & 48 & -102 & 48 & 3 \\ \boxed{3} & & & \ldots & 3 & 48 & -102 & 48 \\ \boxed{48} & \boxed{3} & & & \ldots & 3 & 48 & -102 \end{pmatrix}.$$

$$(7.50)$$

7.5.2 Non-Periodic Boundary Conditions

In this case we have to use all grid points $i = 1, 2, \ldots, (N+1)$ discretizing the space domain (see Fig. 7.1). Square matrices A, B for all schemes will be of dimension $N_p = N + 1$. Implicit schemes near boundaries must not increase the bandwidth of the matrix A of the final linear system. Consequently, the following linear combination is considered for the first derivative at the left boundary (Lele 1992):

$$f_i' + \alpha_l f_{i+1}' = \frac{1}{h}(a_l f_i + b_l f_{i+1} + c_l f_{i+2} + d_l f_{i+3}) + R. \tag{7.51}$$

Applying the same method to compute the coefficients $\{\alpha_l, a_l, b_l, c_l, d_l\}$, we obtain the following decentered (forward) FD implicit schemes:

$$\text{(3rd-order)} \quad f_i' + 2f_{i+1}' = \frac{1}{h}\left(-\frac{5}{2}f_i + 2f_{i+1} + \frac{1}{2}f_{i+2}\right) - \frac{h^3}{12}f_i^{(4)} + \mathcal{O}(h^4), \tag{7.52}$$

$$\text{(4th-order)} \quad f_i' + 3f_{i+1}' = \frac{1}{h}\left(-\frac{17}{6}f_i + \frac{3}{2}f_{i+1} + \frac{3}{2}f_{i+2} - \frac{1}{6}f_{i+3}\right) + \frac{h^4}{20}f_i^{(5)} + \mathcal{O}(h^5). \tag{7.53}$$

At the right boundary, the scheme is decentered toward the interior of the domain:

$$f_i' + \alpha_r f_{i-1}' = \frac{1}{h}(a_r f_i + b_r f_{i-1} + c_r f_{i-2} + d_r f_{i-3}) + R. \tag{7.54}$$

We obtain the following decentered (backward) FD implicit schemes:

$$\text{(3rd-order)} \quad f_i' + 2f_{i-1}' = \frac{1}{h}\left(\frac{5}{2}f_i - 2f_{i-1} - \frac{1}{2}f_{i-2}\right) + \frac{h^3}{12}f_i^{(4)} + \mathcal{O}\left(h^4\right), \tag{7.55}$$

$$\text{(4th-order)} \quad f_i' + 3f_{i-1}' = \frac{1}{h}\left(\frac{17}{6}f_i - \frac{3}{2}f_{i-1} - \frac{3}{2}f_{i-2} + \frac{1}{6}f_{i-3}\right) + \frac{h^4}{20}f_i^{(5)} + \mathcal{O}\left(h^5\right). \tag{7.56}$$

Note that schemes (7.55)-(7.56) for the right boundary are not identical to schemes (7.52)-(7.53) for the left boundary. The coefficients on the right-hand side of the schemes have opposite signs, which is typical for decentered FD schemes for odd derivatives. Applying the same scheme for the first derivative at both left and right boundaries is a common source of errors in programming compact schemes.

For the second derivative, we use the same type of scheme near the boundaries:

$$f_i'' + \alpha f_{i+1}'' = \frac{1}{h^2}(af_i + bf_{i+1} + cf_{i+2} + df_{i+3}) + R, \tag{7.57}$$

$$f_i'' + \alpha f_{i-1}'' = \frac{1}{h^2}(af_i + bf_{i-1} + cf_{i-2} + df_{i-3}) + R. \tag{7.58}$$

This time, we obtain the same coefficients for the schemes at left and right boundaries (we deal with a derivative of even order). The maximum accuracy possible is third order and it is achieved with the following schemes:

$$\text{(3rd-order)} \quad f_i'' + 11f_{i+1}'' = \frac{1}{h^2}(13f_i - 27f_{i+1} + 15f_{i+2} - f_{i+3})$$

$$+ \frac{h^3}{12}f_i^{(5)} + \mathcal{O}\left(h^4\right), \tag{7.59}$$

$$\text{(3rd-order)} \quad f_i'' + 11f_{i-1}'' = \frac{1}{h^2}(13f_i - 27f_{i-1} + 15f_{i-2} - f_{i-3})$$

$$- \frac{h^3}{12}f_i^{(5)} + \mathcal{O}\left(h^4\right). \tag{7.60}$$

In practice, the strategy of using compact sixth-order schemes for non-periodic cases is the following: for first points ($i = 1$ and $i = N_p = N + 1$ in Fig. 7.1) use the third-order decentered schemes; for second points ($i = 2$ and $i = N_p - 1$) use the fourth-order centered compact scheme; for inner points $3 \leq i \leq (N_p - 2)$ use the sixth-order compact centered scheme. More precisely, the linear system for the first derivative is built as

$$(i = 1) \qquad 2\overline{f}_1' + 4\overline{f}_2' = -\frac{5}{h}f_1 + \frac{4}{h}f_2 + \frac{1}{h}f_3,$$

$$(i = 2) \qquad \overline{f}_1' + 4\overline{f}_2' + \overline{f}_3' = -\frac{3}{h}f_1 + \frac{3}{h}f_3,$$

$$(i = 3 \dots N_p - 2) \quad 12\overline{f}_{i-1}' + 36\overline{f}_i' + 12\overline{f}_{i+1}' = -\frac{1}{h}f_{i-2} - \frac{28}{h}f_{i-1} + \frac{28}{h}f_{i+1} + \frac{1}{h}f_{i+2},$$

$$(i = N_p - 1) \quad \overline{f}_{N_p-2}' + 4\overline{f}_{N_p-1}' + \overline{f}_{N_p}' = -\frac{3}{h}f_{N_p-2} + \frac{3}{h}f_{N_p},$$

$$(i = N_p) \qquad 4\overline{f}_{N_p-1}' + 2\overline{f}_{N_p}' = -\frac{1}{h}f_{N_p-2} - \frac{4}{h}f_{N_p-1} + \frac{5}{h}f_{N_p}. \tag{7.61}$$

Corresponding matrices of size $N_p \times N_p$ are

$$A_6^{d1} = \begin{pmatrix} 2 & 4 & 0 & \dots & & 0 & 0 \\ 1 & 4 & 1 & \dots & & 0 & 0 \\ 0 & 12 & 36 & 12 & \dots & 0 & 0 \\ \vdots & \ddots & \ddots & \ddots & \ddots & & \vdots \\ 0 & 0 & \dots & 12 & 36 & 12 & 0 \\ 0 & 0 & \dots & 0 & 1 & 4 & 1 \\ 0 & 0 & & \dots & 0 & 4 & 2 \end{pmatrix}, \quad B_6^{d1} = \frac{1}{h}\begin{pmatrix} -5 & 4 & 1 & \dots & & 0 & 0 \\ -3 & 0 & 3 & 0 & \dots & 0 & 0 \\ -1 & -28 & 0 & 28 & 1 & \dots & 0 \\ \vdots & \ddots & \ddots & \ddots & \ddots & \ddots & \vdots \\ 0 & & \dots & -1 & -28 & 0 & 28 & 1 \\ 0 & & & \dots & 0 & -3 & 0 & 3 \\ 0 & 0 & & \dots & -1 & -4 & 5 \end{pmatrix}. \tag{7.62}$$

For the second derivative with sixth-order compact scheme, we obtain the following linear system

$$(i = 1) \qquad \overline{f}_1'' + 11\overline{f}_2'' = \frac{1}{h^2}\left(13f_1 - 27f_2 + 15f_3 - f_4\right), \qquad (7.63)$$

$$(i = 2) \qquad \overline{f}_1'' + 10\overline{f}_2'' + \overline{f}_3'' = \frac{1}{h^2}(12f_1 - 24f_2 + 12f_3),$$

$$(i = 3 \ldots N_p - 2) \quad 8\overline{f}_{i-1}'' + 44\overline{f}_i'' + 8\overline{f}_{i+1}'' = \frac{3}{h^2}f_{i-2} + \frac{48}{h^2}f_{i-1} - \frac{102}{h^2}f_i + \frac{48}{h^2}f_{i+1} + \frac{3}{h^2}f_{i+2},$$

$$(i = N_p - 1) \quad \overline{f}_{N_p-2}'' + 10\overline{f}_{N_p-1}'' + \overline{f}_{N_p}'' = \frac{1}{h^2}(12f_{N_p-2} - 24f_{N_p-1} + 12f_{N_p}),$$

$$(i = N_p) \qquad 11\overline{f}_{N_p-1}'' + \overline{f}_{N_p}'' = \frac{1}{h^2}\left(13f_{N_p} - 27f_{N_p-1} + 15f_{N_p-2} - f_{N_p-3}\right),$$

with corresponding matrices:

$$A_6^{d2} = \begin{pmatrix} 1 & 11 & 0 & \cdots & & & 0 & 0 \\ 1 & 10 & 1 & \cdots & & & 0 & 0 \\ 0 & 8 & 44 & 8 & \cdots & & & 0 \\ \vdots & \ddots & \ddots & \ddots & \ddots & \ddots & & \vdots \\ 0 & 0 & \cdots & 8 & 44 & 8 & 0 \\ 0 & 0 & \cdots & 0 & 1 & 10 & 1 \\ 0 & 0 & & \cdots & 0 & 11 & 1 \end{pmatrix}, \quad B_6^{d2} = \frac{1}{h^2} \begin{pmatrix} 13 & -27 & 15 & -1 & \cdots & & & 0 \\ 12 & -24 & 12 & 0 & \cdots & & 0 & 0 \\ 3 & 48 & -102 & 48 & 3 & & \cdots & 0 \\ \vdots & \ddots & & \ddots & \ddots & \ddots & & \vdots \\ 0 & \cdots & & 3 & 48 & -102 & 48 & 3 \\ 0 & & & \cdots & 0 & 12 & -24 & 12 \\ 0 & & & \cdots & -1 & 15 & -27 & 13 \end{pmatrix}.$$

$$(7.64)$$

Exercise 7.5 Write the expressions of matrices A_4^{d1}, B_4^{d1} (first derivative) and A_4^{d2}, B_4^{d2} (second derivative) for the fourth-order compact schemes and non-periodic boundary conditions. Hint: decentered (forward and backward) FD schemes are needed only for points $i = 1$ and $i = N_p$.

A solution of this exercise is proposed in Sect. 7.9 at page 170.

7.6 Summary of FD schemes

To have at hand all FD schemes derived in previous sections we summarized the main formulae in Table 7.1 for the first derivative and Table 7.2 for the second derivative.

7.7 Test of high-order FD schemes

The first test that must be addressed after the implementation of FD schemes is the assessment of their accuracy for the approximation of derivatives. We consider simple analytical functions and compare exact and numerical values for the first and second derivatives. This comparison consists of two steps:
(i) for a given grid (i.e. N is fixed) we apply different FD schemes and compute errors for each grid point $\varepsilon_i = |f_i'^{(num)} - f_i'^{(exact)}|$ (and similar for f''); when plotting $\varepsilon(x)$ we have to observe lower approximation errors when the order of the scheme increases;

Type	FD scheme	Order	First term of R
forward (explicit)	$f'_i = \dfrac{-3f_i + 4f_{i+1} - f_{i+2}}{2h} + R$	2	$\dfrac{h^2}{3} f_i^{(3)}$
(implicit)	$f'_i + 2f'_{i+1} = \dfrac{1}{h}\left(-\dfrac{5}{2}f_i + 2f_{i+1} + \dfrac{1}{2}f_{i+2}\right) + R$	3	$-\dfrac{h^3}{12} f_i^{(4)}$
(implicit)	$f'_i + 3f'_{i+1} = \dfrac{1}{h}\left(-\dfrac{17}{6}f_i + \dfrac{3}{2}f_{i+1} + \dfrac{3}{2}f_{i+2} - \dfrac{1}{6}f_{i+3}\right) + R$	4	$\dfrac{h^4}{20} f_i^{(5)}$
centered (explicit)	$f'_i = \dfrac{f_{i+1} - f_{i-1}}{2h} + R$	2	$-\dfrac{h^2}{6} f_i^{(3)}$
(explicit)	$f'_i = \dfrac{8(f_{i+1} - f_{i-1}) - (f_{i+2} - f_{i-2})}{12h} + R$	4	$\dfrac{h^4}{30} f_i^{(5)}$
(implicit)	$\dfrac{1}{4}f'_{i+1} + f'_i + \dfrac{1}{4}f'_{i-1} = \dfrac{3}{2}\dfrac{f_{i+1} - f_{i-1}}{2h} + R$	4	$\dfrac{h^4}{120} f_i^{(5)}$
(implicit)	$\dfrac{1}{3}f'_{i-1} + f'_i + \dfrac{1}{3}f'_{i+1} = \dfrac{14}{9}\dfrac{f_{i+1} - f_{i-1}}{2h} + \dfrac{1}{9}\dfrac{f_{i+2} - f_{i-2}}{4h} + R$	6	$-\dfrac{4}{7!}h^6 f_i^{(7)}$
backward (explicit)	$f'_i = \dfrac{3f_i - 4f_{i-1} + f_{i-2}}{2h} + R$	2	$\dfrac{h^2}{3} f_i^{(3)}$
(implicit)	$f'_i + 2f'_{i-1} = \dfrac{1}{h}\left(\dfrac{5}{2}f_i - 2f_{i-1} - \dfrac{1}{2}f_{i-2}\right) + R$	3	$\dfrac{h^3}{12} f_i^{(4)}$
(implicit)	$f'_i + 3f'_{i-1} = \dfrac{1}{h}\left(\dfrac{17}{6}f_i - \dfrac{3}{2}f_{i-1} - \dfrac{3}{2}f_{i-2} + \dfrac{1}{6}f_{i-3}\right) + R$	4	$\dfrac{h^4}{20} f_i^{(5)}$

Table 7.1 Commonly used explicit and implicit FD schemes for the first derivative.

(ii) to estimate numerically the accuracy p of the scheme and compare with its theoretical value, we run the codes written for point (i) for different values of N. For each run $m = 1, \ldots, N_r$, the value of N is doubled: $N = N_0 2^{m-1}$, with N_0 the starting value. This means that the grid step h is halved for each new run ($h = h_0/2^{m-1}$). The approximation error is computed for each run as the infinity norm $\epsilon = \max_{i=1,\ldots,N}(|f_i'^{(num)} - f_i'^{(exact)}|)$.

Type	FD scheme	Order	First term of R
forward (explicit)	$f_i'' = \dfrac{2f_i - 5f_{i+1} + 4f_{i+2} - f_{i+3}}{h^2} + R$	2	$\dfrac{11h^2}{12} f_i^{(4)}$
(implicit)	$f_i'' + 11f_{i+1}'' = \dfrac{1}{h^2}\left(13f_i - 27f_{i+1} + 15f_{i+2} - f_{i+3}\right) + R$	3	$\dfrac{h^3}{12} f_i^{(5)}$
centered (explicit)	$f_i'' = \dfrac{f_{i+1} - 2f_i + f_{i-1}}{h^2} + R$	2	$-\dfrac{h^2}{12} f_i^{(4)}$
(explicit)	$f_i'' = \dfrac{16(f_{i+1} + f_{i-1}) - 30f_i - (f_{i+2} + f_{i-2})}{12h^2} + R$	4	$\dfrac{h^4}{90} f_i^{(6)}$
(implicit)	$\dfrac{1}{10} f_{i+1}'' + f_i'' + \dfrac{1}{10} f_{i-1}'' = \dfrac{12}{10}\dfrac{f_{i+1} - 2f_i + f_{i-1}}{h^2} + R$	4	$\dfrac{h^4}{200} f_i^{(6)}$
(implicit)	$\dfrac{2}{11} f_{i-1}'' + f_i'' + \dfrac{2}{11} f_{i+1}'' = \dfrac{12}{11}\dfrac{f_{i+1} - 2f_i + f_{i-1}}{h^2}$ $\qquad\qquad + \dfrac{3}{11}\dfrac{f_{i+2} - 2f_i + f_{i-2}}{4h^2} + R$	6	$-\dfrac{23}{11\cdot 7!}h^6 f_i^{(8)}$
backward (explicit)	$f_i'' = \dfrac{2f_i - 5f_{i-1} + 4f_{i-2} - f_{i-3}}{h^2} + R$	2	$\dfrac{11h^2}{12} f_i^{(4)}$
(implicit)	$f_i'' + 11f_{i-1}'' = \dfrac{1}{h^2}\left(13f_i - 27f_{i-1} + 15f_{i-2} - f_{i-3}\right) + R$	3	$-\dfrac{h^3}{12} f_i^{(5)}$

Table 7.2 Commonly used explicit and implicit FD schemes for the second derivative.

Since theoretically $\epsilon \sim Ch^p$, the curve $\epsilon(h)$ using logarithmic axes will display as a line of slope p, i.e. $\log(\varepsilon) \sim \log(C) + p\,\log(h)$ (see also Chap. 1).

Periodic case

For the periodic case, we use the following test functions:

$$\begin{cases} f_1 : [0, 2\pi] \to \mathbb{R} \\ f_1(x) = \sin(x) \end{cases}, \qquad \begin{cases} f_2 : [0, 2\pi] \to \mathbb{R} \\ f_2(x) = \exp(\sin(x)) \end{cases}. \tag{7.65}$$

The domain $[0, 2\pi]$ is discretized with $N+1$ equidistant points (see Fig. 7.1), with $h = 2\pi/N$. The number of unknowns is $N_p = N$, since by periodicity

$f_1 = f_{N+1}$. We use FD schemes from Sect. 7.5.1 with matrices of size $N \times N$. A very simple way, although not the most efficient[5], to implement FD schemes is to use the general formulation (7.34), build matrices A and B and then use MATLAB libraries to solve the final linear system: $\overline{F}_d = A \backslash (B * F)$.

In the case of smooth periodic functions, as considered in our test, finite difference approximations are outperformed by spectral methods. Therefore, we implement, as a reference approximation on the same grid, the Fourier approach to compute derivatives. In Chap. 3, the decomposition of a periodic function $f : [0, 2\pi] \to \mathbb{R}$ using N Fourier modes was introduced as

$$f(x) = \sum_{k=0}^{N-1} \widehat{f}_k e^{ikx}, \tag{7.66}$$

where i is the imaginary unit ($i^2 = -1$), k the wave number and $\widehat{f}_k$ the spectral (Fourier) coefficients. The discrete version of (7.66), using equidistant grid points $x_j = (j-1)h$, $j = 1, \ldots, N$, with $h = 2\pi/N$ and N even, is then

$$f_j = \sum_{k=0}^{N-1} \widehat{f}_k e^{ik\frac{2\pi}{N}(j-1)} = \sum_{k=-N/2}^{N/2-1} \widehat{f}_k e^{ik\frac{2\pi}{N}(j-1)}. \tag{7.67}$$

Direct differentiation of (7.66) results in

$$f'(x) = \sum_{k=0}^{N-1} (ik\widehat{f}_k) e^{ikx}, \quad f''(x) = \sum_{k=0}^{N-1} (-k^2 \widehat{f}_k) e^{ikx}, \tag{7.68}$$

meaning that Fourier coefficients of the derivatives can be easily obtained from values of Fourier coefficients of the function itself. All we have to do is to use a Fast Fourier Transform (`fft` in MATLAB) to obtain coefficients $\widehat{f}_k$, compute coefficient for the derivative (for example $ik\widehat{f}_k$ for the first derivative), and finally apply an Inverse Fast Fourier Transform (`ifft`) to obtain the approximation of the derivative. The only delicate point in this simple approach is the setting of the vector containing wave numbers k. The arrangement of values $-N/2 \le k < N/2$ corresponding to decomposition (7.67) may

---

[5] The implementation of implicit FD schemes can be improved at several levels:
(i) since the matrix A is tridiagonal, special algorithms can be used, such as the Thomas algorithm in which the storage of the full matrix is avoided (only 3 vectors containing the diagonals are stored) and the LU algorithm is rewritten in an efficient way; this algorithm is described in detail in Chap. 15 for periodic or non-periodic tridiagonal matrices;
(ii) the computation of the right-hand side of the system can avoid the matrix-vector multiplication B*F and use instead explicit expressions of the right-hand side of the FD scheme;
(iii) when necessary to approximate derivatives of several functions on the same grid (or in 2D/3D settings), the matrix A is always the same; it is thus wise to factorize the matrix once (this is the most costly part of the LU algorithm) and solve each linear system (with different F) in two steps ($LY = BF$ and $U\overline{F}_d = Y$); this idea is also used in the Thomas algorithm applied to 2D problems in Chap. 15.

be different from one software to another. In MATLAB, the vector containing
the wave numbers has to be defined as (positive values first):

$$k = [0:N/2-1, -N/2:-1].$$ (7.69)

Exercise 7.6 Periodic case, assessment of approximation errors.

1. Write three MATLAB functions to compute approximations of the first
   derivative using the general scheme (7.34) and particular forms for matrices A and B corresponding to the following FD schemes: second-order
   explicit (7.8) (function dfdx=dfdx_o2(f,h)), fourth-order compact
   (7.43)-(7.44) (function dfdx=dfdx_o4(f,h)) and sixth-order compact
   (7.47)-(7.48) (function dfdx=dfdx_o6(f,h)).

2. Test FD approximations for periodic functions (7.65). In the main program, call each function computing the first derivative and superimpose
   in the same figure exact values and FD approximations using the three
   FD schemes. Compute errors $\varepsilon_i = |f_i'^{(num)} - f_i'^{(exact)}|$ at each grid point
   and plot in the same figure curves $\varepsilon(x)$ for each FD scheme.
   Numerical value: $N = 64$.

3. Repeat the same steps for the second derivative, using the following
   schemes: second-order explicit (7.9), fourth-order compact (7.45)-(7.46)
   and sixth-order compact (7.49)-(7.50).

4. Implement in function [dfdx,d2fdx2]=dfdx_spectral(f) the spectral
   approximation of derivatives based on (7.68) and (7.69). Compute the
   approximation error and compare with FD schemes.

A solution of this exercise is proposed in Sect. 7.9 at page 170.

Exercise 7.7 Periodic case, assessment of accuracy order.
Modify the main program written for Exercise 7.6 by introducing a loop
modifying the number of points following $N = N_0 \, 2^{m-1}$, with $m = 1, \ldots, N_r$.
Take $N_0 = 16$ and $N_r = 6$ (six runs). This practically amounts to including
the previous program in this loop. For each run, compute the infinity norm
$\epsilon = \max_{j=1,\ldots,N}(|f_j'^{(num)} - f_j'^{(exact)}|)$. Plot $\epsilon(h)$ in logarithmic coordinates
for each FD scheme (use loglog instead of plot). Plot in the same figure,
curves h^p, with $p = 2, 4, 6$, the theoretical accuracy order of each scheme. Try
to put into evidence that the numerically obtained curve $\epsilon(h)$ is parallel to
the theoretical curve h^p.

A solution of this exercise is proposed in Sect. 7.9 at page 174.

Non-periodic case

For the non-periodic case, we can still use test functions (7.65), but without assuming periodicity. We also add the following polynomial function, suggested by Laizet and Lamballais (2009) to test the accuracy of compact schemes:

$$
\begin{cases}
f_3 : [0,1] \to \mathbb{R}, \\[4pt]
f_3(x) = \dfrac{x^7}{7} - \dfrac{x^6}{2} + \dfrac{17x^5}{25} - \dfrac{9x^4}{20} + \dfrac{274x^3}{1875} - \dfrac{12x^2}{625}, \\[8pt]
f_3'(x) = x(x-1)\left(x - \dfrac{4}{5}\right)\left(x - \dfrac{3}{5}\right)\left(x - \dfrac{2}{5}\right)\left(x - \dfrac{1}{5}\right), \\[8pt]
f_3''(x) = 6x^5 - 15x^4 + \dfrac{68}{5}x^3 - \dfrac{27}{5}x^2 + \dfrac{548}{625}x - \dfrac{24}{625}.
\end{cases}
\tag{7.70}
$$

Recall that for this case the number of unknowns is $N_p = N+1$ (see Fig. 7.1). We use FD schemes from Sect. 7.5.2 with matrices of size $N_p \times N_p$.

Exercise 7.8 Non-periodic case, assessment of approximation errors.

1. In functions written for Exercise 7.6, introduce a flag for the periodicity (`iper=1` if the function is periodic, `iper=0` if not). Add the non-periodic case to all functions implementing FD schemes (for instance, `function dfdx=dfdx_o4(f,h,iper)`, `function d2fdx2=d2fdx2_o4(f,h,iper)`). Implement the new matrices for the non-periodic case:
   – second-order explicit (7.8) with (7.20)-(7.21) for the first and last grid points, fourth-order compact (see Exercise 7.5) and sixth-order compact (7.61)-(7.62) schemes for the first derivative,
   – second-order explicit (7.9) and (7.22)-(7.23) for the first and last grid points, fourth-order compact (see Exercise 7.5) and sixth-order compact (7.63)-(7.64) schemes for the second derivative.
2. Test FD approximations for functions (7.65) (without periodicity assumed) and function (7.70). Plot the approximation error $\varepsilon(x)$ for each FD scheme. Numerical value: $N = 64$.
3. Repeat the same steps for the second derivative.

A solution of this exercise is proposed in Sect. 7.9 at page 170.

Exercise 7.9 Non-periodic case, assessment of accuracy order.
Modify the program written for Exercise 7.8 to numerically assess the order of accuracy of FD schemes for the non-periodic case (follow the indications given for Exercise 7.7).

A solution of this exercise is proposed in Sect. 7.9 at page 174.

7.8 Solving boundary-value problems with high-order FD schemes

We show in this section how easy is to apply the FD formalism presented in previous sections to solve a linear boundary-value problem (BVP). The idea is to combine the linearity of the discretized differential operator with the linearity of the FD approximation (7.34). We consider the BVP

$$
\begin{cases}
-u''(x) + c(x)\, u(x) = g(x), & 0 < x < L, \\
u(0) = \alpha, \quad u(L) = \beta, & \text{(Dirichlet boundary conditions)},
\end{cases}
\tag{7.71}
$$

encountered in many fields of application: in elasticity, it models the deformation of a beam of stiffness coefficient $c(x)$ and submitted to a transverse charge $g(x)$ (see also Chap. 10); in heat transfer, it models the temperature distribution in a 1D fin, with $c(x)$ the coefficient of lateral heat losses by convection and $g(x)$ the distributed heat sources. It can be mathematically proved that problem (7.71) is well posed (i.e. the solution exists and is unique) if $c(x) > 0, \forall x \in]0, L[$. This is naturally the case for the physical phenomena described by this BVP!

To solve (7.71), we use the grid represented in Fig. 7.1 and write the discrete form of the BVP in the matrix form $U_d = CU - G$, with $U, U_d \in \mathbb{R}^{N_p}$ the vectors containing grid-point approximations of the solution and its second derivative, respectively. The diagonal matrix $C = diag((c_i)_{1 \le i \le N_p})$ and the vector $G = (g_i)_{1 \le i \le N_p}$ are given. Recall that $N_p = N + 1$ for this non-periodic problem. We shall discuss later the periodic case. The FD approximation (7.34) becomes in our case: $A^{2d}U_d = B^{2d}U$. Combining these two linear relations, we obtain the following linear system for the solution U:

$$
\underbrace{\left(A^{2d}C - B^{2d}\right)}_{M} U = \underbrace{A^{2d}G}_{S}.
\tag{7.72}
$$

Before solving the system (7.72), we have to reinforce the Dirichlet boundary conditions, by replacing the first equation of the system by $u_1 = \alpha$ and the last equation by $u_{N_p} = \beta$. This implies straightforward manipulations of the matrix M and right-hand side S. Since matrices A^{2d} and B^{2d} are already available for all FD schemes from previous exercises, solving the BVP (7.71) through (7.72) becomes an easy problem with MATLAB.

Exercise 7.10 Non-periodic case. Solve the BVP (7.71) with Dirichlet boundary conditions. Validate numerical results for the following *manufactured* solution: we set an *exact solution* u_{ex} that satisfies the boundary conditions and calculate the corresponding source function g:[6]

---

[6] The method of manufactured solutions is a general tool for the verification of calculations. It constructs an exact solution to a modified problem (a source term is added) related to

$$\begin{cases} u_{ex}(x) = (1-x)\,e^x, \\ \quad c(x) = x, \\ L = 1, \quad \alpha = 1, \quad \beta = 0, \end{cases} \implies g(x) = (1+2x-x^2)\,e^x. \qquad (7.73)$$

1. Write a main program that sets the discretization of problem (7.71) with expressions (7.73), using the grid in Fig. 7.1. Use previous programs to recover matrices A^{2d} and B^{2d} for different FD schemes (non-periodic case): second-order explicit (7.9) and (7.22)-(7.23) for the first and last grid points, fourth-order compact (see Exercise 7.5) and sixth-order compact (7.63)-(7.64) schemes. Use (7.72) to find the numerical solution vector U.
2. Plot in the same figure the exact solution and numerical solutions obtained with the three FD approximations. In a second figure, plot the approximation error $\varepsilon(x)$, with $\varepsilon_i = |(u_{ex})_i - u_i|$ calculated at each grid point.

A solution of this exercise is proposed in Sect. 7.9 at page 175.

Exercise 7.11 Periodic case. Consider the BVP (7.71) with periodic boundary conditions: $u(x) = u(x+L), \forall x \in \mathbb{R}$.

1. Repeat the steps of Exercise 7.10 for the periodic case, using the following *manufactured* solution obtained by superimposing simple waves:

$$\begin{cases} u_{ex}(x) = \sum_{i=1}^{2} A_i \sin\left(2\pi\lambda_i \dfrac{x}{L}\right), \\ c(x) = c, \end{cases} \implies \begin{cases} g(x) = c\,u_{ex}(x)+ \\ \sum_{i=1}^{2} A_i \left(\dfrac{2\pi\lambda_i}{L}\right)^2 \sin\left(2\pi\lambda_i \dfrac{x}{L}\right). \end{cases}$$

$$(7.74)$$

Numerical values: $N = 64$, $c = 1.2$, $A_1 = 1$, $A_2 = \frac{1}{4}$, $\lambda_i = 1$, $\lambda_2 = 10$. Test another case by adding waves to the manufactured solution (7.74).
2. Implement a spectral solver for this periodic case. Hint: use properties (7.68) to find that after applying the Fourier transform to the linear equation (7.71), we obtain that

$$\widehat{u}(k) = \frac{\widehat{g}}{c + k^2}, \qquad (7.75)$$

where k is the wave number defined by (7.69). Use the inverse Fourier transform to find the solution vector U.

A solution of this exercise is proposed in Sect. 7.9 at page 175.

Exercise 7.12 Following the example of Exercises 7.7 and 7.9, find numerically the accuracy order of the FD method used to solve the BVP (7.71) for both non-periodic (Exercise 7.10) and periodic (Exercise 7.11) cases. Vary the number of grid points as $N = N_0\, 2^{m-1}$, with $m = 1, \ldots, N_r$. Take $N_0 = 16$ and $N_r = 6$ (six runs).

A solution of this exercise is proposed in Sect. 7.9 at page 177.

---

the initial one. Even though in most cases these exact solutions are not physically realistic, this approach allows one to rigorously verify computations.

7.9 Solutions and programs

For Exercises 7.1, 7.2, 7.3, and 7.4, the reader is invited to apply the method of undetermined coefficients presented in Sects. 7.2 and 7.3 to obtain the requested FD schemes.

Solution of Exercise 7.5 (fourth-order compact scheme for non-periodic functions)

For the non-periodic case, since the fourth-order compact scheme uses tridiagonal matrices A and B, special treatment is needed for only points $i = 1$ and $i = N_p$. For these boundary points, we use third-order compact schemes, as we did for the sixth-order compact scheme.[7] We obtain the following matrices for the approximation of the first derivative:

$$
A_4^{d1} = \begin{pmatrix} 2 & 4 & 0 & \dots & & 0 & 0 \\ 1 & 4 & 1 & \dots & & 0 & 0 \\ 0 & 1 & 4 & 1 & \dots & 0 & 0 \\ \vdots & \ddots & \ddots & \ddots & \ddots & & \vdots \\ 0 & 0 & \dots & 1 & 4 & 1 & 0 \\ 0 & 0 & \dots & 0 & 1 & 4 & 1 \\ 0 & 0 & & \dots & 0 & 4 & 2 \end{pmatrix}, \quad
B_4^{d1} = \frac{1}{h} \begin{pmatrix} -5 & 4 & 1 & \dots & & 0 & 0 \\ -3 & 0 & 3 & 0 & \dots & 0 & 0 \\ 0 & -3 & 0 & 3 & 0 & \dots & 0 \\ \vdots & \ddots & \ddots & \ddots & \ddots & & \vdots \\ 0 & \dots & 0 & -3 & 0 & 3 & 0 \\ 0 & & \dots & 0 & -3 & 0 & 3 \\ 0 & 0 & & \dots & -1 & -4 & 5 \end{pmatrix} \quad (7.76)
$$

and for the second derivative:

$$
A_4^{d2} = \begin{pmatrix} 1 & 11 & 0 & \dots & & 0 & 0 \\ 1 & 10 & 1 & \dots & & 0 & 0 \\ 0 & 1 & 10 & 1 & \dots & & 0 \\ \vdots & \ddots & \ddots & \ddots & \ddots & & \vdots \\ 0 & 0 & \dots & 1 & 10 & 1 & 0 \\ 0 & 0 & \dots & 0 & 1 & 10 & 1 \\ 0 & 0 & & \dots & 0 & 11 & 1 \end{pmatrix}, \quad
B_4^{d2} = \frac{1}{h^2} \begin{pmatrix} 13 & -27 & 15 & -1 & \dots & & 0 \\ 12 & -24 & 12 & 0 & \dots & 0 & 0 \\ 0 & 12 & -24 & 12 & 0 & \dots & 0 \\ \vdots & \ddots & \ddots & \ddots & \ddots & & \vdots \\ 0 & \dots & 0 & 12 & -24 & 12 & 0 \\ 0 & & \dots & 0 & 12 & -24 & 12 \\ 0 & & \dots & -1 & 15 & -27 & 13 \end{pmatrix}. \quad (7.77)
$$

Solution of Exercises 7.6 and 7.8 (assessment of approximation errors for periodic or non-periodic cases)

Since periodic and non-periodic cases involve the same code structure, we provide a single main MATLAB program HFD_M_test_FD.m to test the accuracy of different FD schemes for the approximation of derivatives. Each test case is identified by specifying the name of the function implementing the

---

[7] We can also use the fourth-order compact scheme for boundary points if we can afford to enlarge the bandwidth of the matrix B.

analytical expression of the test function, the length L of the space domain and the flag iper (1 if periodic, 0 otherwise). We can set:

- name_fex='funcTRIG' calling HFD_F_funcTRIG.m implementing f_1 from (7.65); both flags iper=1 or iper=0 (the periodicity is not accounted for) are possible valid choices;
- name_fex='funcEXPSIN' calling HFD_F_funcEXPSIN.m implementing f_2 from (7.65); again, both flags iper=1 or iper=0 are possible;
- name_fex='funcPOLY' calling HFD_F_funcPOLY.m implementing the polynomial function f_3 from (7.70); only iper=0 is a valid choice.

For each test function, we compute FD approximations for the first (dfdx) and second derivative (d2fdx2) using the following functions:

- HFD_F_dfdx_o2.m and HFD_F_d2fdx2_o2.m for the second-order scheme;
- HFD_F_dfdx_o4.m and HFD_F_d2fdx2_o4.m for the fourth-order compact scheme;
- HFD_F_dfdx_o6.m and HFD_F_d2fdx2_o6.m for the sixth-order compact scheme.

For each of these six functions, we use similar headers of the type [dudx,A,B]= HFD_F_dfdx_o2(u,hx,iper) meaning that the function has to provide the approximation of the derivative and the corresponding matrices A and B. These matrices are not used here, but will be needed for solving the linear BVP (7.71). As commented before, this is not the most efficient way to implement the FD schemes, but it offers a quick way to code theoretical formulae.

The program then calls successively the functions computing second, fourth, and sixth-order FD approximations and computes approximation errors ε_i for each grid point. For the periodic case, the spectral approximation of first and second derivatives-based Eqs. (7.68) and (7.69) is implemented in function HFD_F_dfdx_spectral.m. To simplify the graphical display, we provide a unique function HFD_F_add_plot.m that is able to create a new plot or to add a curve on an existing figure. With the help of this function, we can easily superimpose different curves on the same graph.

The obtained results are summarized for the periodic case in Fig. 7.2 and in Figs. 7.3, 7.4 for the non-periodic case. We display the first and second derivative and approximation errors. For the periodic case (Fig. 7.2) we notice an uniform (for all grid points) decrease of the approximation error by several orders of magnitude from second-order FD scheme to the fourth-order scheme and the sixth-order scheme. For example, for the the first derivative of function f_1 and $N = 64$, we switch from a maximum error $\varepsilon_{\max} = 1.6 \cdot 10^{-3}$ (second order), to $5.1 \cdot 10^{-7}$ (fourth order) and $4.2 \cdot 10^{-10}$ (sixth order). This important gain in accuracy is obtained with the same computation effort. The spectral precision remains the reference result for the periodic case, with maximum error $\varepsilon_{\max} = 7.4 \cdot 10^{-15}$ for the same example. For the test function f_2 (more challenging case), the gain in accuracy is less spectacular, but still several orders of magnitude are gained in the precision of the approximation when using high-order compact schemes.

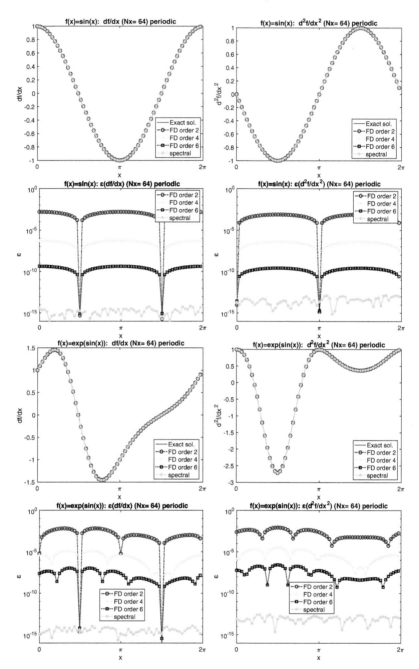

Fig. 7.2 Test of FD schemes for two periodic functions: $f_1(x) = \sin(x)$ and $f_2(x) = \exp(\sin(x))$ defined on the interval $[0, 2\pi]$. Second-order explicit (7.8), fourth-order compact (7.43), and sixth-order compact (7.47) schemes for the first derivative. Second-order explicit (7.9), fourth-order compact (7.45), and sixth-order compact (7.49) schemes for the second derivative. Comparison with the spectral method.

For the non-periodic case (Fig. 7.3), we notice that the reduction of approximation errors is not uniform over the space domain. The approximation error for grid points near boundaries is degraded for high-order schemes. This was expected, since lower order schemes were used for the first and second grid points near the boundaries. The fact that in the linear system for compact schemes all grid points are connected makes this low approximation *contaminate* several grid points near the boundaries. This is a well-known effect for compact schemes for non-periodic problems. However, the gain of accuracy observed for the periodic case is maintained for most of the points inside the space domain. For the polynomial function f_3 (Fig. 7.4), the accuracy of the sixth-order scheme reaches the machine precision (10^{-15}) for the second derivative and inner points. This is equivalent to the spectral accuracy, with the major difference that in this non-periodic case classical Fourier spectral methods are not possible to use. Due to their high accuracy, compact schemes are often referred to as *spectral-like* FD schemes.

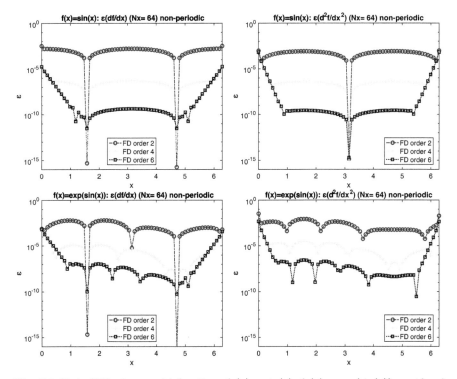

Fig. 7.3 Test of FD schemes with functions $f_1(x) = \sin(x)$, $f_2(x) = \exp(\sin(x))$ considered as non-periodic. Second-order explicit (7.8), fourth-order compact (7.76) and sixth-order compact (7.61) schemes for the first derivative. Second-order explicit (7.9), fourth-order compact (7.77) and sixth-order compact (7.63) schemes fors the second derivative.

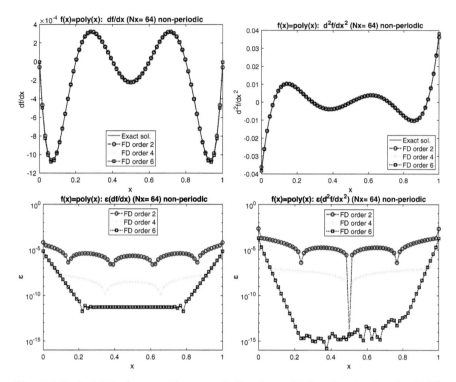

Fig. 7.4 Test of FD schemes with non-periodic polynomial function f_3, given by (7.70). Same FD schemes as in Fig. 7.3.

Solution of Exercises 7.7 and 7.9 (assessment of accuracy order)

Assessing the accuracy of FD schemes involves slight modifications of the previous program. Practically speaking, the program HFD_M_test_FD_accuracy.m simply embraces the old program in a loop varying the value of N. Each passage in the loop defines a new value of h and consequently the matrices for all schemes have to be recomputed. For each value of h, the maximum error is stored in a vector ϵ.

Figures 7.5 and 7.6 display, in logarithmic coordinates, the curves $\epsilon(h)$ and theoretical convergence curves h^p, with $p = 2, 4, 6$. For the periodic case (Fig. 7.5), the value of ϵ is computed using all grid points and expected accuracy orders are nicely recovered numerically. Note the typical behavior of accuracy curves for values of the error close to 10^{-12}: the error surprisingly increases because of round-off errors, which, in this range of very low values, dominate the approximation error of the FD scheme.

For the non-periodic case (Fig. 7.6), given the variation of the error displayed in Figs. 7.3 and 7.4, it is more delicate to obtain theoretical accuracy orders. If the value of ϵ is computed considering only inner points, theoretical

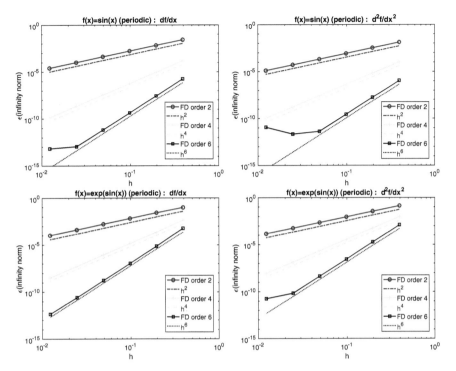

Fig. 7.5 Accuracy of FD schemes for the computation of the first and second derivatives. Periodic case, with test functions and FD schemes as in Fig. 7.2.

values for p are recovered. Otherwise, a lower order of accuracy is obviously obtained (to be tested). Note from Fig. 7.4 that approximation errors are very low for the sixth-order compact scheme applied to the second derivative, and thus it is not possible to compute an order of accuracy for this case (see the last panel in Fig. 7.6).

Solution of Exercises 7.10 and 7.11 (FD solver for a linear BVP)

Given the structure of previous programs for computing FD approxima- tions of derivatives, we treat the periodic and non-periodic case of the BVP (7.71) in the same main MATLAB program HFD_M_solve_BVP.m. The flag iper will distinguish between the periodic or non-periodic cases. Func- tion HFD_F_BVP_setting.m sets the test case, with the exact solution, the corresponding right-hand side vector G and diagonal matrix C. Matri- ces A and B are returned by the function implementing the FD scheme (e.g. [d2udx2,A,B]=HFD_F_d2fdx2_o2(uex,hx,iper);) and then used in the function HFD_F_BVP_system.m which computes the solution of the linear system (7.72). For the non-periodic case, the first and the last line of the

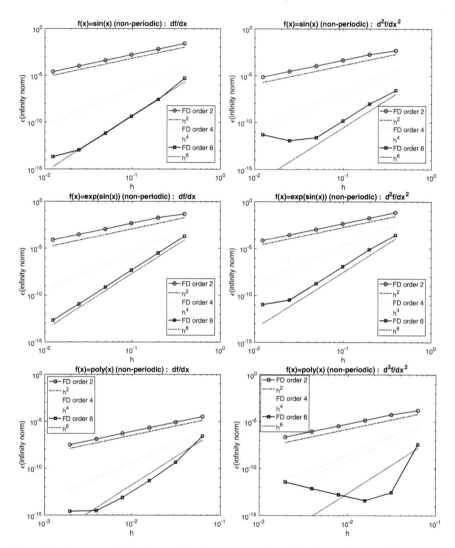

Fig. 7.6 Accuracy of FD schemes for the computation of the first and second derivatives. Non-periodic case, with test functions and FD schemes as in Fig. 7.3.

final system are modified to take into account the Dirichlet boundary conditions.

For each FD scheme, the numerical solution is computed and displayed in the first figure. Local approximation errors ε_i are computed at each grid points and then plotted in the second figure. For the periodic case, the spectral solver using (7.75) is implemented in function HFD_F_BVP_spectral.m.

Figure 7.7 shows the computed solution for the two cases. Approximation errors are considerably reduced when using high-order FD compact schemes instead of the classical second-order scheme. For the non-periodic case, spectral-like accuracy is obtained with the sixth-order scheme and $N = 64$.

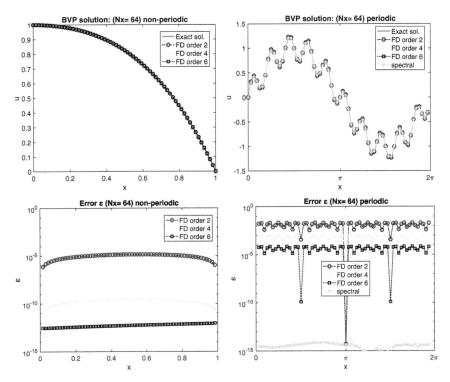

Fig. 7.7 Numerical solution u of the boundary-value problem (7.71) with (left) Dirichlet (non-periodic) and (right) periodic boundary conditions. Approximation errors ε for three FD schemes (second, fourth, and sixth order) and (for the periodic case) spectral method.

For the periodic case, the spectral methods offers, as expected, the best accuracy. It should be noted that the exact solution for this case is perfectly adapted to a Fourier decomposition, since it was built by superimposing two simple waves. On the other side, the finite difference discretization is not sufficiently refined to capture the fastest oscillations (for $N = 64$ we can see that only 10 grid points discretize the shortest wavelength).

Solution of Exercise 7.12 (accuracy of FD solvers for the BVP)

The accuracy of FD schemes used to solve the linear BVP (7.71) is assessed in the main MATLAB program HFD_M_solve_BVP_accuracy.m. We use again a loop to vary the grid-step h and store the infinity norm ϵ of the approximation errors for each grid resolution. The log-log plot of $\epsilon(h)$ is presented in Fig. 7.8

and shows the expected theoretical orders of accuracy (i.e. the slopes of the lines are 2, 4, and 6) for both periodic and non-periodic cases.

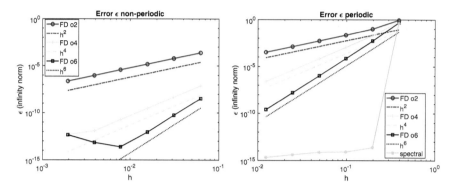

Fig. 7.8 Accuracy of FD schemes used to solve the linear BVP (7.71).

Chapter References

C. Hirsch, *Numerical Computation of Internal and External Flows*. (John Wiley & Sons, 1988)

S. Laizet, E. Lamballais, High-order compact schemes for incompressible flows: A simple and efficient method with quasi-spectral accuracy. J. Comput. Phys. **228**, 5989–6015 (2009)

S.K. Lele, Compact finite difference schemes with spectral-like resolution. J. Comput. Phys. **103**, 16–42 (1992)

R.J. LeVeque, *Finite Difference Methods for Ordinary and Partial Differential Equations*. (Society for Industrial and Applied Mathematics, 2007)

J.C. Strikwerda, *Finite Difference Schemes and Partial Differential Equations*. (Wadsworth and Brooks/Cole, 1989)

M. Brio, G. Webb, A. Zakharian, *Numerical Time-Dependent Partial Differential Equations for Scientists and Engineers*. (Academic Press, 2010)

Chapter 8
Signal Processing: Multiresolution Analysis

Project Summary

Level of difficulty:	1
Keywords:	Approximation, multiresolution analysis, wavelets
Application fields:	Signal processing, image processing

8.1 Introduction

This chapter is devoted to a short introduction to multiresolution analysis (MRA). This is a very promising field in mathematics, with numerous theoretical and practical developments in engineering applications. Over the past three decades, wavelet functions have proven to be a very efficient tool for dealing with problems arising from data compression, and signal and image processing. Famous examples of applications are the FBI fingerprint database, and the image coding standard MPEG3.

8.2 Approximation of a Function: Theoretical Aspect

8.2.1 Piecewise Constant Functions

In this section we introduce the basic ideas of multiresolution analysis. Let Ω be the interval $[0, 1[$, and consider a function $f \in L^1(\Omega)$. For any arbitrary fixed integer $j \geq 0$ we define the intervals $\Omega_j^k = [2^{-j}k, 2^{-j}(k+1)[$ for $k = 0, 1, \ldots, 2^j - 1$. We then approximate the function f by its projection $P_j f$

© The Author(s), under exclusive license to Springer Nature Switzerland AG 2023   179
I. Danaila et al., *An Introduction to Scientific Computing*,
https://doi.org/10.1007/978-3-031-35032-0_8

onto the family of functions which are constant on intervals Ω_j^k (see Fig. 8.2). The value of $P_j f$ on Ω_j^k is computed as

$$P_j^k f = 2^j \int_{\Omega_j^k} f(t)\, dt, \quad \text{for } k = 0, 1, \ldots, 2^j - 1.$$

We also introduce $\phi = \chi_{[0,1[}$, the characteristic function[1] of Ω, and note first that ϕ satisfies the following property, known as *the two-scale relation*:

$$\forall x \in \Omega, \quad \phi(x) = \phi(2x) + \phi(2x - 1). \tag{8.1}$$

This relation is "plotted" in Fig. 8.1.

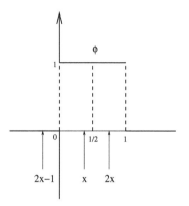

Fig. 8.1 The two-scale relation (Haar basis).

We remark that $x \mapsto \phi(2^j x - k)$ is the characteristic function of the interval $\Omega_j^k = [2^{-j}k, 2^{-j}(k+1)[$, and we redefine $P_j f$ as

$$\forall x \in \Omega, \quad P_j f(x) = \sum_{k=0}^{2^j - 1} P_j^k f \phi(2^j x - k).$$

Since Ω is a bounded domain, $f \in L^1(\Omega)$ whenever $f \in L^2(\Omega)$, and $P_j f$ is then an element of the vector space

$$V_j = \left\{ f \in L^2(\Omega),\ f_{|\Omega_j^k} \text{ is constant, for } k = 0, 1, \ldots, 2^j - 1 \right\}.$$

The space V_j has finite dimension $\dim V_j = 2^j$. For $k = 0, 1, \ldots, 2^j - 1$, we define the functions ϕ_j^k as

---

[1] In the following, $\chi_{[a,b[}$ is the characteristic function of the interval $[a, b[$.

$$\forall x \in \Omega, \quad \phi_j^k(x) = 2^{j/2}\phi(2^j x - k). \qquad (8.2)$$

The 2^j functions ϕ_j^k span V_j and are orthonormal relative to the L^2 scalar product: $\langle f, g \rangle = \int_\Omega f(t)g(t)\, dt$. Using this orthonormal basis, we can write

$$\forall x \in \Omega, \quad P_j f(x) = \sum_{k=0}^{2^j-1} \langle f, \phi_j^k \rangle \phi_j^k(x) = \sum_{k=0}^{2^j-1} c_j^k \phi_j^k(x),$$

where the coefficients c_j^k are the components of $P_j f$ in the $\{\phi_j^k\}_{k,j}$ basis; they are computed according to

$$c_j^k = \langle f, \phi_j^k \rangle = \int f(t)\phi_j^k(t)\, dt = 2^{j/2} \int_{\Omega_j^k} f(t)\, dt. \qquad (8.3)$$

The application P_j is then the orthogonal projection onto V_j relative to the $L^2(\Omega)$ scalar product. Consider now two arbitrary integers $j' > j \geq 0$, and define as previously two spaces $V_{j'}$ and V_j. The basis functions $\{\phi_{j'}^{k'}\}_{k',j'}$ of the space $V_{j'}$ are constant on intervals $\Omega_{j'}^{k'}$ of length $2^{-j'}$, while the V_j basis functions $\{\phi_j^k\}_{k,j}$ are constant on intervals Ω_j^k of length $2^{-j} > 2^{-j'}$. Because $V_{j'} \subset V_j$, the function $P_{j'}f$ is a more accurate approximation of f than $P_j f$, in the sense that $\|P_{j'}f - f\|_2 < \|P_j f - f\|_2$. It can be proven that this approximation $P_j f$ converges to f in $L^2(\Omega)$ as j goes to infinity (see Fig. 8.2). Furthermore, when $f \in C^0(\Omega)$, the approximation $P_j f$ converges to f according to the uniform norm: $\lim_{j \to +\infty} \|f - P_j f\|_\infty = 0$.

For an arbitrary fixed integer $j \geq 0$, we consider now the two spaces V_j and V_{j+1}. From (8.1) we may write, for any $f \in L^2(\Omega)$ and for $k = 0, 1, \ldots, 2^j - 1$,

$$\sqrt{2} \int f(t)\phi_j^k(t)\, dt = \int f(t)\phi_{j+1}^{2k}(t)\, dt + \int f(t)\phi_{j+1}^{2k+1}(t)\, dt.$$

This leads to a first relation connecting the coefficients c_j^k and $c_{j+1}^{k'}$:

$$c_j^k = (c_{j+1}^{2k} + c_{j+1}^{2k+1})/\sqrt{2}, \text{ for } k = 0, 1, \ldots, 2^j - 1. \qquad (8.4)$$

Remark 8.1 In the case of an unbounded domain ($\Omega = \mathbb{R}$ for example), we may change the definition of the space V_j to

$$V_j = \left\{ f \in L^2(\mathbb{R}), \ f_{|\Omega_j^k} \text{ is constant, } k \in \mathbb{Z} \right\}.$$

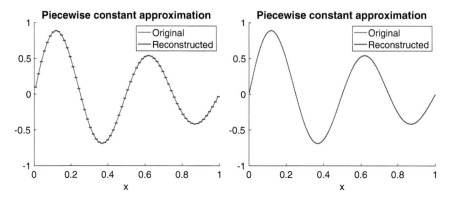

Fig. 8.2 Approximating function $f(x) = e^{-x} \sin(4\pi x)$ (in red) using mean values (in blue) on 2^j intervals: Left: $j = 6$; right: $j = 8$.

8.2.2 Decomposition of the Space V_J

Consider now an arbitrary fixed integer $J \geq 0$. Then for any integer j satisfying $J > j \geq 0$, we define successive functional spaces $V_j, V_{j+1}, \ldots, V_J$, that satisfy $V_j \subset V_{j+1} \subset \cdots \subset V_J$. For any given function $f \in L^2(\Omega)$, a standard way to write the orthogonal projection of f on the subspace V_{j+1} is to consider $P_{j+1}f$ as the orthogonal projection of f onto V_j along with a correction term:

$$P_{j+1}f = P_j f + (P_{j+1}f - P_j f) = P_j f + Q_j f. \qquad (8.5)$$

This relation introduces the new operator $Q_j = P_{j+1} - P_j$, which is actually the orthogonal projection operator onto W_j, the orthogonal complement of V_j in V_{j+1}: $V_{j+1} = V_j \oplus W_j$. It is easy to check that the function $\psi = \chi_{[0,1/2[} - \chi_{[1/2,1[}$ satisfies

$$\psi(x) = \phi(2x) - \phi(2x - 1). \qquad (8.6)$$

Now we consider the functions ψ_j^k defined by

$$\forall x \in \Omega, \quad \psi_j^k(x) = 2^{j/2}\psi(2^j x - k), \text{ for } k = 0, 1, \ldots, 2^j - 1. \qquad (8.7)$$

The 2^j functions ψ_j^k span W_j and are orthonormal relative to the L^2 scalar product. Then for an arbitrary function $f \in L^2(\Omega)$, we compute the coefficients d_j^k according to

$$d_j^k = \langle f, \psi_j^k \rangle = \int f(t)\psi_j^k(t)\, dt$$

$$= 2^{j/2} \int_{\Omega_{j+1}^{2k}} f(t)\, dt - 2^{j/2} \int_{\Omega_j^{2k+1}} f(t)\, dt. \tag{8.8}$$

The coefficient d_j^k is the *fluctuation* of f on the interval Ω_j^k. Using (8.3), we write first

$$d_j^k = (c_{j+1}^{2k} - c_{j+1}^{2k+1})/\sqrt{2}, \tag{8.9}$$

and then, by adding (8.9) to (8.4), we get

$$\sqrt{2}\, c_{j+1}^{2k} = c_j^k + d_j^k, \text{ for } k = 0, 1, \ldots, 2^j - 1, \tag{8.10}$$

while subtracting (8.9) from (8.4) leads to

$$\sqrt{2}\, c_{j+1}^{2k+1} = c_j^k - d_j^k, \text{ for } k = 0, 1, \ldots, 2^j - 1. \tag{8.11}$$

These relations are connected to the space decomposition $V_{j+1} = V_j \oplus W_j$. We gather these results in the useful relations (8.12) and (8.13), basic elements of the decomposition and reconstruction algorithms:

$$\begin{cases} c_j^k = (c_{j+1}^{2k} + c_{j+1}^{2k+1})/\sqrt{2}, \\ d_j^k = (c_{j+1}^{2k} - c_{j+1}^{2k+1})/\sqrt{2}, \text{ for } k = 0, 1, \ldots, 2^j - 1; \end{cases} \tag{8.12}$$

$$\begin{cases} c_{j+1}^{2k} = (c_j^k + d_j^k)/\sqrt{2}, \\ c_{j+1}^{2k+1} = (c_j^k - d_j^k)/\sqrt{2}, \text{ for } k = 0, 1, \ldots, 2^j - 1. \end{cases} \tag{8.13}$$

Before going any further, we remark that it is possible to iterate the space decomposition process according to

$$V_J = V_{J-1} \oplus W_{J-1} = V_{J-2} \oplus W_{J-2} \oplus W_{J-1} = \cdots$$

$$= V_0 \oplus W_0 \oplus \cdots \oplus W_{J-2} \oplus W_{J-1}. \tag{8.14}$$

Since the functions ϕ_j^k (respectively ψ_j^k) span an orthonormal basis of V_j (respectively W_j), we are now able to define many orthonormal bases of V_J. Among all possibilities, the emphasis is put on two particular bases: the *canonical basis*, generated by functions ϕ_J^k,

$$P_J f = \sum_{k=0}^{2^J - 1} c_J^k \phi_J^k, \tag{8.15}$$

and the so-called *Haar basis*, spanned by ϕ_0^0 and all functions ψ_j^k, for $j = 0, 1, 2, \ldots, J-1$ and $k = 0, 1, \ldots, 2^j - 1$:

$$P_J f = c_0^0 \phi_0^0 + \sum_{j=0}^{J-1} \sum_{k=0}^{2^j-1} d_j^k \psi_j^k. \tag{8.16}$$

Remark 8.2 Since $\phi_0^0 = \phi = \chi_{[0,1[}$ is the characteristic function of the whole domain Ω, the coefficient c_0^0 is simply the mean value of f on Ω: $c_0^0 = \int_\Omega f(t)\, dt$.

Remark 8.3 It is worth noting that the family of functions $\{\phi_J^k\}_{k=0}^{2^J-1}$, which form an orthonormal basis of the finite-dimensional space V_J, converges to an orthonormal basis of the infinite-dimensional space $L^2(\Omega)$ as J goes to infinity. So the coefficients c_J^k are also the components of f in the corresponding basis of $L^2(\Omega)$. Note also that the family of functions $\{\phi\} \cup \{\psi_j^k\}_{k=0, j=0}^{k=2^j-1, j=J-1}$, an orthonormal basis of the finite-dimensional space V_J, converges to an orthonormal basis of the infinite-dimensional space $L^2(\Omega)$ as J goes to infinity. The coefficients c_0^0 and d_j^k are the components of f in the corresponding basis of $L^2(\Omega)$.

8.2.3 Decomposition and Reconstruction Algorithms

In this section, we look at the standard operations required to switch from the expression of a function in the canonical basis of V_J to its expression in the Haar basis, and conversely.

A. Let f be a function in $L^2(\Omega)$ and J an arbitrary fixed integer. We compute the 2^J coefficients c_J^k, either exactly if we have an expression for f, or approximately using a sampling of the 2^J values of f on intervals Ω_k^J. Starting from these 2^J coefficients c_J^k, we successively compute the coefficients c_j^k and d_j^k according to the following algorithm:

$$
\begin{array}{l}
\textbf{for } j = J-1, \ldots, 1, 0 \textbf{ compute} \\
\quad \textbf{for } k = 0, 1, \ldots, 2^j - 1 \textbf{ compute} \\[4pt]
\quad\quad c_j^k = (c_{j+1}^{2k} + c_{j+1}^{2k+1})/\sqrt{2} \\[4pt]
\quad\quad d_j^k = (c_{j+1}^{2k} - c_{j+1}^{2k+1})/\sqrt{2} \\[4pt]
\quad \textbf{end} \\
\textbf{end}
\end{array} \tag{8.17}
$$

Decomposition algorithm.

This calculation is referred to as the *analysis* or *decomposition algorithm* of f. We may represent step j of this algorithm by the symbolic scheme

$$c_j^k$$
$$(0 \le k < 2^j)$$
$$\swarrow \quad \searrow$$
$$c_{j-1}^k \qquad\qquad d_{j-1}^k$$
$$(0 \le k < 2^{j-1}) \; (0 \le k < 2^{j-1})$$

Once computed, the 2^j coefficients d_j^k are not used in the next steps of the decomposition algorithm; only the 2^j values of the coefficients $\{c_j^k\}_k$ are required in order to compute the coefficients $\{c_{j-1}^k\}_k$ and $\{d_{j-1}^k\}_k$. The computational cost of step j in algorithm (8.17) is that of computing $2 \times 2^{j-1}$ coefficients, that is, exactly 2^{j+1} operations. The computational cost of the decomposition algorithm, required to obtain the values of the 2^J coefficients, is then

$$2^{J+1} + \cdots + 2^j + \cdots + 2^2 + 1 = 2^{J+2} = 4 \times 2^J \text{ operations.}$$

This may be considered as an optimal value, since we are computing 2^J outputs from 2^J inputs for a cost of $O(2^J)$ operations.

B. Conversely, assume that we know c_0^0, the mean value of f on Ω, and all other coefficients d_j^k for $j = 0, \ldots, J-1$. Then we retrieve all the coefficients c_j^k using the following algorithm:

> **for** $j = 0, \ldots, J-1$ **compute**
> > **for** $k = 0, 1, \ldots, 2^j - 1$ **compute**
> > $$c_{j+1}^{2k} = (c_j^k + d_j^k)/\sqrt{2}$$
> > $$c_{j+1}^{2k+1} = (c_j^k - d_j^k)/\sqrt{2}$$
> > **end**
> **end**

(8.18)

Reconstruction algorithm.

This calculation is referred to as the *synthesis* or the *reconstruction algorithm* of f. We represent step j of this algorithm by the symbolic scheme

$$c_{j-1}^k \qquad\qquad d_{j-1}^k$$
$$(0 \le k < 2^{j-1}) \; (0 \le k < 2^{j-1})$$
$$\searrow \quad \swarrow$$
$$c_j^k$$
$$(0 \le k < 2^j)$$

We remark again that the computational cost of step j of this algorithm is that of computing $2 \times 2^{j-1}$ coefficients, that is, 2^{j+1} operations. The total computational cost of the reconstruction algorithm is also $O(2^J)$ operations.

Both algorithms (8.17) and (8.18) are efficient tools for obtaining the components of a function in the *Haar basis* from its components in the *canonical basis*, and conversely. There exist many other orthonormal bases of the space V_J and as many corresponding algorithms; we shall see two further examples of such algorithms, which have strong similarity to (8.17) and (8.18). This calculation is referred to as a *multiscale analysis* or *multiresolution analysis*.

8.2.4 Importance of Multiresolution Analysis

We now look more closely at the *data compression* aspect included in the multiresolution analysis formulation. We assume first that the function f is constant ($f = C \neq 0$) on Ω, and compare two expressions for $P_J f$. The first is the representation of f in the canonical basis. According to (8.3), all the 2^J coefficients c_J^k are equal to C, and are thus different from zero. To represent $P_J f$ in this basis, an array of 2^J components is required. From another point of view, the expression for $P_J f$ in the Haar basis needs to compute coefficients $c_{J-1}^{k'}$ and $d_{J-1}^{k'}$ from the c_J^k's according to (8.17). We obtain immediately $c_{J-1}^{k'} = C$ and $d_{J-1}^{k'} = 0$ for $k' = 0, \ldots, 2^{J-1} - 1$. Going further in the computation, we see that $c_j^k = C$ and $d_j^k = 0$ for $j = J-1, \ldots, 0$. Finally, there is only one nonzero coefficient in the Haar basis: $c_0^0 = C$.

In information processing, attention is focused on the most condensed expression of a signal, in order to compute information, store it in memory, or send it through a network. Many algorithms are dedicated to the compression or decompression of data without any loss. We understand with the previous example how useful multiresolution may be in that case. In a more general case, when a function f is no longer constant, the coefficient c_j^k stands for the mean value of f on Ω_j^k, while the coefficient d_j^k represents the variation of f at the scale j. For any slowly varying function f, many coefficients d_j^k have a small value and may be neglected. Then the number of significant coefficients in the Haar basis representation is far smaller than the number of coefficients c_J^k present in the canonical basis. Conversely, a large coefficient d_j^k is associated with a fast variation of the function f within Ω_j^k. This property is of great interest when one wants to look automatically for the singularities of a function. As an illustration we just mention that special events in the sky are automatically detected by computers analyzing thousands of photographs of stars captured daily by telescopes.

Remark 8.4 Fourier analysis is known to be useful for dealing with oscillating signals. It is a very accurate way to capture the frequencies hidden in a signal, but its important drawback is the lack of spatial localization of these oscillations, due to the use of cosine or sine functions oscillating on the whole domain. This drawback is not present in multiresolution analysis, where the basis functions have supports limited to the Ω_k^j. Unfortunately, the frequency localization is then less accurate than with the Fourier basis.

8.3 Multiresolution Analysis: Practical Aspect

In this section, we deal with a very simple example in order to understand the practical efficiency of the multiresolution analysis theory. But before doing this, it remains to clarify how we will store the different coefficients arising from the previous algorithms. Let $\Omega \subset \mathbb{R}$ be a bounded interval, f a function defined on Ω, and $J > 0$ an arbitrary fixed integer. Using the formulas of algorithm (8.17), we compute for $j = J - 1, \ldots, 0$ coefficients c_j^k and d_j^k, for $k = 0, 1, \ldots, 2^j - 1$. These coefficients are stored in the following way: We first compute coefficients c_J^k according to (8.3) and store them in an array $[c_J]$ of 2^J components. Then we compute c_0^0 and $\{\{d_j^k\}_k\}_j$ using (8.17), and store them in an array $[d_J]$ of 2^J components. We begin at step J of algorithm (8.17) by imposing $[d_J] = [c_J]$. Then at step $J - 1$ the 2^{J-1} coefficients $c_{J-1}^{k'}$ are stored in the first half of array $[d_J]$ (components 1 up to 2^{J-1}), while the 2^{J-1} coefficients $d_{J-1}^{k'}$ are stored in the second half of array $[d_J]$ (components $2^{J-1} + 1$ up to 2^J). At step $J - 2$ the 2^{J-2} coefficients $c_{J-2}^{k''}$ are stored in the first quarter of the array (components 1 up to 2^{J-2}), thus erasing the now useless values of coefficients $c_{J-1}^{k'}$ for $k' = 1, 2, \ldots, 2^{J-2}$. Then the 2^{J-2} coefficients $d_{J-2}^{k''}$ are stored in the second quarter of the array (components $2^{J-2} + 1$ up to 2^{J-1}), thus erasing the remaining useless values of coefficients $c_{J-1}^{k'}$ for $k' = 2^{J-2} + 1, 2, \ldots, 2^{J-1}$. Note that during this operation the $[d_J]$ components from $2^{J-1} + 1$ up to 2^J, which correspond to coefficients $d_{J-1}^{k'}$, are not modified. Proceeding in this way until step $j = 0$, we finally get the following array $[d_J]$:

$$[d_J] = [c_0^0, d_0^0, d_1^0, d_1^1, \ldots, d_j^0, d_j^1, \ldots, d_j^{2^j - 1}, \ldots, d_{J-1}^0, \ldots, d_{J-1}^{2^{J-1} - 1}]. \quad (8.19)$$

This storage is related to the decomposition of the space V_J according to the scheme

$$V_J \to \begin{cases} W_{J-1} \\ \\ V_{J-1} \to \begin{cases} W_{J-2} \\ \\ V_{J-2} \to \begin{cases} \cdots \\ \\ \cdots \to \begin{cases} W_0 \\ \\ V_0 \end{cases} \end{cases} \end{cases} \end{cases}$$

8.4 Multiresolution Analysis: Implementation

Let f be the function defined on $\Omega = [0, 1]$ by $f(x) = \exp(-x)\sin(4\pi x)$. We choose $J = 10$ and compute the arrays $[c_J]$ and $[d_J]$ associated with $P_J f$.

Exercise 8.1 1. Write a program that computes all coefficients c_J^k according to (8.17). Store these coefficients in an array $[c_J]$ with 2^J components.
2. Using the decomposition algorithm (8.17), compute for $j = J - 1, \ldots, 0$ all coefficients c_j^k and d_j^k, for $k = 0, 1, \ldots, 2^j - 1$. Store these coefficients in an array $[d_J]$ with 2^J components, as detailed in the previous section.
3. Write a program that computes all coefficients c_J^k from the $[d_J]$ components, according to the reconstruction algorithm (8.18). Check the results.

A solution of this exercise is proposed in Sect. 8.7 at page 201. Using these direct and inverse transformation programs, one can perform some numerical experiments. We shall deal now with an example of a compression algorithm.

Exercise 8.2 1. Calculate the number of coefficients in an array $[d_J]$ whose absolute values are greater than $\varepsilon = 2^{-J/2} \times 10^{-3}$.
2. Copy array $[d_J]$ into a new array $[d_J^\varepsilon]$ and set to zero every component of $[d_J^\varepsilon]$ whose absolute value is less than ε ($d_j^{\varepsilon,k} = 0$ when $|d_j^k| < \varepsilon$).
3. Compute the array $[c_J^\varepsilon]$ from $[d_J^\varepsilon]$ using the reconstruction algorithm (8.18).
4. Visualize the resulting signal and compare both curves representing $P_J f$ and $P_J^\varepsilon f$.
5. Study the variations of $\|P_J^\varepsilon f - P_J f\|_2$ and the number of nonzero coefficients in $[d_J^\varepsilon]$ as ε varies.

A solution of this exercise is proposed in Sect. 8.7 at page 201. Table 8.1 displays results of this experiment (with $f(x) = \exp(-x)\sin(4\pi x)$ and $J = 10$). The number of nonzero coefficients is reported in front of the threshold value, with the corresponding relative error $\|P_J^\varepsilon f - P_J f\|_2 / \|P_J f\|_2$. Figure 8.3 plots two signals reconstructed after thresholding. We emphasize here the compression capability of the method: using only 352 coefficients instead of the 1024 sample values, we obtain 0.8% relative error.

Threshold	Coefficients	Relative error
0.1000	78	0.0402
0.0500	121	0.0258
0.0100	352	0.0080
0.0050	563	0.0042
0.0010	947	0.0003
0.0005	987	0.0001
0.0001	1015	0.00001

Table 8.1 Thresholding (Haar wavelet).

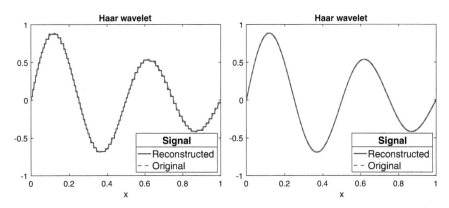

Fig. 8.3 Reconstruction after thresholding: $J = 10$. Left: $\varepsilon = 0.1$; right: $\varepsilon = 0.01$.

8.5 Introduction to Wavelet Theory

8.5.1 Scaling Functions and Wavelets

If you have correctly performed the previous numerical experiments, you deserve our warmest congratulations for your first steps in the fabulous world of wavelets! When Haar proposed (in 1910!) the construction of an orthonormal basis of the space $L^2(\Omega)$ like the one discussed previously, he was in fact far from imagining the practical importance of his discovery. Wavelet theory was founded in the sixties, arising from an idea of a petroleum engineer named Morlet, who was looking for algorithms well suited to seismic signal processing, and more accurate than Fourier analysis (see Meyer, 1990).

The Haar basis construction is the foundation of multiresolution analysis. In the associated terminology, functions ϕ_j^k are the called *scaling functions*, while functions ψ_j^k are *wavelet functions*, or *wavelets*, for short. What is the appearance of a wavelet? Have a look at the previous numerical experiments:

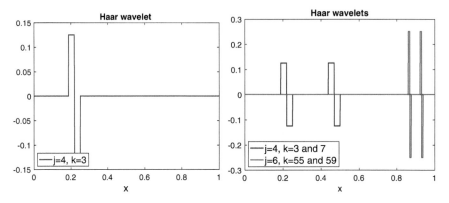

Fig. 8.4 Haar wavelets ψ_j^k for $(J = 10)$. Left: single wavelet; right: shifted wavelets.

for $J = 10$ we begin by setting to zero all components of the $[d_J]$ and give the value 1 to only one of the coefficients d_j^k;[2] then using the reconstruction algorithm (8.18), we obtain the associated wavelet ψ_j^k, as plotted in Fig. 8.4 (left). Note that the integers k and j have to satisfy the following conditions: $0 < j < J$ and $0 \leq k < 2^j$. Choose now another integer $k' \neq k$ such that $0 \leq k < 2^j$ and repeat the experiment. This leads to a wavelet $\psi_j^{k'}$, which appears (see Fig. 8.4 (right)) to be shifted from wavelet ψ_j^k by an offset of $2^{-j}(k' - k)$. Both wavelets belong to the subspace W_j and are hence called level-j wavelets. All level-j wavelets (they are altogether 2^j in level j) are obtained from any one of them by a $p\, 2^{-j}$ shift (p an integer).

Now, if we choose two integers j' and k'' such that $0 < j' \neq j < J$ and $0 \leq k'' < 2^{j'}$, the corresponding wavelet $\psi_{j'}^{k''}$ has features similar to the previous wavelets: same global shape but different sizes (amplitude has been multiplied by $2^{j'-j}$, while the length has been divided by the same factor (see Fig. 8.4 (right)).

The whole space $V_J \subset L^2(\Omega)$ is then generated by direct summation of the subspace V_0 and all orthogonal subspaces W_j, each of which is spanned by 2^j wavelet functions. When J goes to infinity, we retrieve the structure introduced by Haar (see Remark 8.2): $L^2(\Omega)$ is a direct summation of finite-dimensional orthogonal subspaces. Moreover, for any $f \in L^2(\Omega)$, the multiresolution analysis can be summarized in

---

[2] The level j coefficient d_j^k is stored in the $(2^j + k + 1)$th component of $[d_J]$ (see (8.19)).

$$P_J f = \langle f, \phi_0^0 \rangle \phi_0^0 + \sum_{j=0}^{j=J-1} \sum_{k=0}^{2^j-1} \langle f, \psi_j^k \rangle \psi_j^k,$$

$$f = \langle f, \phi_0^0 \rangle \phi_0^0 + \sum_{j \geq 0} \sum_{k=0}^{2^j-1} \langle f, \psi_j^k \rangle \psi_j^k, \qquad (8.20)$$

$$f = P_J f + \sum_{j \geq J} \sum_{k=0}^{2^j-1} \langle f, \psi_j^k \rangle \psi_j^k.$$

Hence the coefficients c_0^0 and d_j^k are the components of f in the corresponding basis of $L^2(\Omega)$. From that point of view, the wavelet theory appears to be a powerful tool for approximating functions. Since the relation $V_{j+1} = V_j \oplus W_j$ holds at any level $j < J$, we may consider the subspace W_j as a set of *detail functions*, that is, the functions we have to add to the functions of V_j, in order to retrieve all V_{j+1} functions.

By fixing an integer $J < +\infty$, we restrain the space description to the scale 2^{-J}, and consequently we are unable to capture the variation $|f(x) - f(x')|$ when $|x - x'| < 2^{-J}$. On the other hand, we may write $\|P_J f - f\|_2 < C\, 2^{-J}$ for any function f in $L^2(\Omega)$ with bounded variation. This means we know exactly the accuracy of an approximation $P_J f$ of a given f; moreover, we also know the price to pay to improve this result: we have to compute at least 2^J new coefficients.

In the previous example, all *scaling functions* are derived from the same function ϕ by the relation $\phi_j^k(x) = \phi(2^j x - k)$. Likewise, the relation $\psi_j^k(x) = \psi(2^j x - k)$ shows that all *wavelet functions* are derived from the same function ψ, sometimes called the *mother wavelet* function. In this first study, both ϕ and ψ are discontinuous functions; it follows that the approximating function $P_J f$ is also discontinuous, even when f is continuous. How is one to get a more regular approximation? Much work has been done to answer the question; there exist abstract necessary conditions on the pair (ϕ, ψ) in order to generate a general framework for multiresolution analysis. In short, it is possible to build continuous wavelet approximations for a given continuous function; however, these aspects of wavelet theory are beyond our scope. We refer to Cohen and Ryan (1995), Cohen (2000, 2003), Daubechies (1992), and Mallat (1997) for further details. In the following sections, we introduce two examples of continuous wavelets: the Schauder and the Daubechies wavelets.

For the sake of simplicity we shall limit our study to periodic continuous functions on Ω. Although wavelet theory is able to address the general case, it needs some technical modifications that we want to skip here.

8.5.2 The Schauder Wavelet

We follow here the outline of the previous section and introduce a new V_J space definition:

$$V_j = \left\{ f \in C^0(\Omega),\ f_{|\Omega_j^k}\ \text{is affine, for}\ k = 0, 1, \ldots, 2^j - 1 \right\} \subset L^2(\Omega).$$

We are now dealing with piecewise linear functions continuous on Ω. We consider first the function ϕ defined by

$$\phi(x) = \max\left(0, 1 - |x|\right). \tag{8.21}$$

The function ϕ satisfies the two-scale relation (see Fig. 8.5)

$$\phi(x) = \frac{1}{2}\phi(2x - 1) + \phi(2x) + \frac{1}{2}\phi(2x + 1). \tag{8.22}$$

Then we define the functions ϕ_j^k by scaling and shift:

$$\phi_j^k(x) = 2^{j/2}\phi(2^j x - k),\ \text{for}\ k = 0, 1, \ldots, 2^j - 1. \tag{8.23}$$

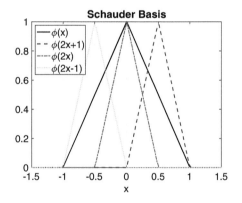

Fig. 8.5 The two-scale relation (Schauder basis).

The functions ϕ_j^k are known as *hat functions* in the finite element method; though they do not span an orthogonal basis of V_j, we shall admit that they provide the Schauder wavelets by the definition $\psi_j^k = \phi_{2j-1}^k$, for $k = 0, 1, \ldots, 2^j - 1$. It is not the only eligible choice in that case, but this is the simplest one (for more details see Mallat (1997)). Figure 8.6 displays a set of Schauder wavelets when $J = 3$. For any function f we consider again the two representations

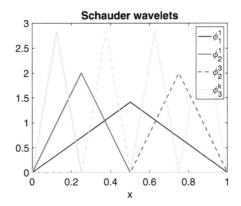

Fig. 8.6 Some Schauder wavelets ($J = 3$) : ϕ_1^1 (blue), ϕ_2^1 and ϕ_2^3 (red), ϕ_3^k for k=1,3,5,7 (green).

$$P_J f = \sum_{k=0}^{2^J - 1} c_J^k \phi_J^k \quad \text{and} \quad P_J f = c_0^0 \phi_0^0 + \sum_{j=0}^{J-1} \sum_{k=0}^{2^j - 1} d_j^k \psi_j^k. \qquad (8.24)$$

This definition leads to the decomposition formulas

$$\begin{cases} c_j^k = \sqrt{2}\, c_{j+1}^{2k}, \\ d_j^k = \sqrt{2}\, \left[c_{j+1}^{2k+1} - \dfrac{1}{2}(c_{j+1}^{2k} + c_{j+1}^{2k+2}) \right], \text{ for } k = 0, 1, \ldots, 2^j - 1, \end{cases} \qquad (8.25)$$

as well as the reconstruction formulas

$$\begin{cases} c_{j+1}^{2k} = \dfrac{\sqrt{2}}{2}\, c_j^k, \\ c_{j+1}^{2k+1} = \dfrac{\sqrt{2}}{2}\, \left[d_j^k + \dfrac{1}{2}(c_j^k + c_j^{k+1}) \right], \text{ for } k = 0, 1, \ldots, 2^j - 1. \end{cases} \qquad (8.26)$$

Any coefficient c_j^k stands here for the (normalized) *value* of f at the point $x_j^k = 2^j x - k$, and a usual interpretation of (8.25) is that the decomposition algorithm proceeds by elimination, keeping only point values of even indices when going from level $j+1$ to level j. Similarly, the coefficient d_j^k appears to be the difference between the (normalized) odd-index point value c_{j+1}^{2k+1} and the linear interpolation of the adjacent (normalized) even-index point values c_{j+1}^{2k} and c_{j+1}^{2k+2}; it is known as the *detail*, that is, the value to be added to the current level-j values c_j^k and c_j^{k+1}, in order to obtain the level-$(j+1)$ value c_{j+1}^{2k+1}, as can be seen in the reconstruction algorithm (8.26). This process is similar to the multiresolution idea introduced in (8.5) and (8.20). Figure 8.7 shows a graphical interpretation of this computation.

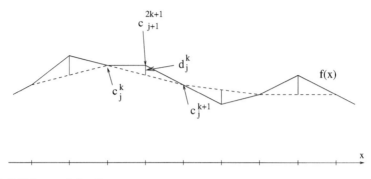

Fig. 8.7 Values and details.

A computer implementation of the relations (8.25)–(8.26) is easily obtained from the previous algorithms (8.17)–(8.18). This is a very attractive property of wavelet theory. All decomposition and reconstruction algorithms are similar to (8.17) and (8.18); switching from a particular wavelet basis to another one results from a slight change in the formulas. Moreover, this change arises only from the corresponding two-scale relations. Consequently, both $[c_J]$ and $[d_J]$ share the same structure of 2^J-component arrays. The general computation, common to all decomposition and reconstruction algorithms, is known as the *Mallat transform* (see Mallat, 1997).

8.5.3 Implementation of the Schauder Wavelet

Let f be the function defined on $\Omega = [0, 1]$ by $f(x) = \exp(-x)\sin(4\pi x)$. We choose $J = 10$ and compute the arrays $[c_J]$ and $[d_J]$ associated with $P_J f$.

Exercise 8.3 1. Write a program that computes all coefficients c_J^k according to (8.25). Store these coefficients in an array $[c_J]$ with 2^J components.
2. Using the decomposition algorithm (8.25), compute for $j = J - 1, \ldots, 0$ all coefficients c_j^k and d_j^k, for $k = 0, 1, \ldots, 2^j - 1$. Store these coefficients in an array $[d_J]$ with 2^J components, as detailed in the previous section.
3. Write a program that computes all coefficients c_J^k from the $[d_J]$ components, according to the reconstruction algorithm (8.26). Check the results.

Remark 8.5 The use of periodic functions requires a particular treatment at both edges of the domain. More precisely, formulas (8.25) and (8.26) use the relation $c_{j+1}^{2^j} = c_{j+1}^0$ for $j = 1, 2, \ldots, J - 1$.

A solution of this exercise is proposed in Sect. 8.7 at page 201. We deal again with an example of the compression algorithm.

Exercise 8.4 1. Calculate the number of coefficients in the array $[d_J]$ whose absolute values are greater than $\varepsilon = 2^{-J/2} \times 10^{-3}$.
2. Copy array $[d_J]$ in a new array $[d_J^\varepsilon]$ and set to zero each component of $[d_J^\varepsilon]$ whose absolute value is less than ε ($d_j^\varepsilon = 0$ when $|d_j| < \varepsilon$).
3. Compute the array $[c_J^\varepsilon]$ from $[d_J^\varepsilon]$ using the reconstruction algorithm (8.26).
4. Visualize the resulting signal and compare both curves representing $P_J f$ and $P_J^\varepsilon f$.
5. Study the variations of $\|P_J^\varepsilon f - P_J f\|_2$ and the number of nonzero coefficients in $[d_J^\varepsilon]$ as ε varies.

A solution of this exercise is proposed in Sect. 8.7 at page 201. Table 8.2 displays results of this experiment (with $f(x) = \exp(-x)\sin(4\pi x)$ and $J = 10$). The number of nonzero coefficients is reported in front of the threshold value, with the corresponding relative error $\|P_J^\varepsilon f - P_J f\|_2 / \|P_J f\|_2$. Figure 8.8 plots two signals reconstructed after thresholding. We emphasize the spectacular compression capacity of the method: using only 77 coefficients arising from 1024 values, we obtain a 0.2% relative error. We see that for a smaller number of significant components in array $[d_J]$, we get a better approximation with the Schauder wavelet than with the Haar wavelet. This is not a big surprise because $P_J f$ is now a continuous approximation of the same continuous function f. Note that there is a small increase of the computational cost, due to the use of more coefficients in formulas (8.25) and (8.26).

Threshold	Coefficients	Relative error
0.1000	25	0.0155
0.0500	41	0.0072
0.0100	77	0.0020
0.0050	102	0.0010
0.0010	205	0.0003
0.0005	246	0.0002
0.0001	459	0.00005

Table 8.2 Thresholding (Schauder wavelet).

8.5.4 The Daubechies Wavelet

Is it possible to improve these results? Cohen (2003) has proven that a multiresolution analysis is available as soon as there exists a generalized two-scale relation such as

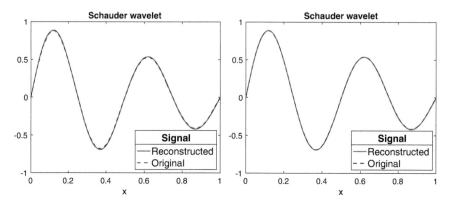

Fig. 8.8 Reconstruction after thresholding: $J = 10$. Left: $\varepsilon = 0.10$; right: $\varepsilon = 0.01$.

$$\phi(x) = \sum_{k \in \mathbb{Z}} h_k \phi(2x - k). \tag{8.27}$$

In signal-processing theory, the h_k's are the components of an array h, called a *filter*. Knowledge of a filter is a necessary and sufficient condition to build a multiresolution analysis. From (8.27) one may write the complementary relation

$$\psi(x) = \sum_{k \in \mathbb{Z}} \tilde{h}_k \psi(2x - k). \tag{8.28}$$

Both relations are related to decomposition and reconstruction algorithms, as previously established in (8.17)–(8.18) for the Haar wavelet, or (8.25)–(8.26) for the Schauder wavelet. Daubechies (1992) has proven that the mother wavelet regularity depends on the *filter length*, that is, the number of nonzero coefficients h_k used in relation (8.27). A general method for defining compact-support wavelets with arbitrary regularity has been proposed, introducing the *Daubechies wavelets* family. To put it in a nutshell, the more nonzero coefficients appear in the two-scale relation (8.27), the more accurate is the wavelet approximation. To end with this study, we shall deal now with the Daubechies wavelet D4, which is defined by the following formulas:

Decomposition:

$$\begin{cases} c_j^k = C_0\, c_{j+1}^{2k-1} + C_1\, c_{j+1}^{2k} + C_2\, c_{j+1}^{2k+1} + C_3\, c_{j+1}^{2k+2}, \\ d_j^k = C_3\, c_{j+1}^{2k-1} - C_2\, c_{j+1}^{2k} + C_1\, c_{j+1}^{2k+1} - C_0\, c_{j+1}^{2k+2}. \end{cases} \tag{8.29}$$

Reconstruction:

$$\begin{cases} c_{j+1}^{2k} = C_3\, c_j^{k-1} - C_0\, d_j^{k-1} + C_1\, c_j^k - C_2\, d_j^k, \\ c_{j+1}^{2k+1} = C_2\, c_j^k + C_1\, d_j^k + C_0\, c_j^{k+1} + C_3\, d_j^{k+1}. \end{cases} \tag{8.30}$$

According to Daubechies (1992), the values of C_k are respectively

$$\begin{cases} C_0 = \dfrac{1+\sqrt{3}}{4\sqrt{2}}, \qquad C_1 = \dfrac{3+\sqrt{3}}{4\sqrt{2}}, \\[3mm] C_2 = \dfrac{3-\sqrt{3}}{4\sqrt{2}}, \qquad C_3 = \dfrac{1-\sqrt{3}}{4\sqrt{2}}. \end{cases} \qquad (8.31)$$

8.5.5 Implementation of the Daubechies Wavelet D4

Let f be the function defined on $\Omega = [0,1]$ by $f(x) = \exp(-x)\sin(4\pi x)$. We choose $J = 10$ and compute the arrays $[c_J]$ and $[d_J]$ associated with $P_J f$.

Exercise 8.5 1. Write a program that computes all coefficients c_J^k according to (8.29). Store these coefficients in an array $[c_J]$ with 2^J components.
2. Using the decomposition algorithm (8.25), compute for $j = J - 1, \ldots, 0$ all coefficients c_j^k and d_j^k, for $k = 0, 1, \ldots, 2^j - 1$. Store these coefficients in an array $[d_J]$ with 2^J components, as detailed in the previous sections.
3. Write a program that computes all coefficients c_J^k from the $[d_J]$ components, according to the reconstruction algorithm (8.30). Check the results.
4. Visualize a Daubechies wavelet (see Fig. 8.9; what a surprise!).

Remark 8.6 As previously noticed, we consider here periodic functions, and special formulas are required to treat the domain edges.

A solution of this exercise is proposed in Sect. 8.7 at page 202. We deal again with an example of a compression algorithm.

Exercise 8.6 1. Calculate the number of coefficients in array $[d_J]$ whose absolute values are greater than $\varepsilon = 2^{-J/2} \times 10^{-3}$.
2. Copy array $[d_J]$ in a new array $[d_J^\varepsilon]$ and set to zero each component of $[d_J^\varepsilon]$ whose absolute value is less than ε ($d_j^\varepsilon = 0$ when $|d_j| < \varepsilon$).
3. Compute the array $[c_J^\varepsilon]$ from $[d_J^\varepsilon]$ using the reconstruction algorithm (8.26).
4. Visualize the resulting signal and compare both curves representing $P_J f$ and $P_J^\varepsilon f$.
5. Study the variations of $\|P_J^\varepsilon f - P_J f\|_2$ and the number of nonzero coefficients in $[d_J^\varepsilon]$ as ε varies.

A solution of this exercise is proposed in Sect. 8.7 at page 202. Table 8.3 displays results of this experiment (with $f(x) = \exp(-x)\sin(4\pi x)$ and $J = 10$). The number of nonzero coefficients is reported in front of the threshold value, with the corresponding relative error $\|P_J^\varepsilon f - P_J f\|_2 / \|P_J f\|_2$. Figure 8.10 plots two signals reconstructed after thresholding. We emphasize again the spectacular compression capacity of the method: by using only 78 coefficients arising from 1024 values, we obtain a 0.3% relative error.

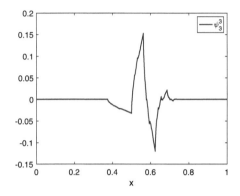

Fig. 8.9 The Daubechies wavelet D4.

Threshold	Coefficients	Relative error
0.1000	30	0.0200
0.0500	43	0.0108
0.0100	78	0.0031
0.0050	108	0.0015
0.0010	207	0.0003
0.0005	233	0.0002
0.0001	454	0.00007

Table 8.3 Thresholding (Daubechies wavelet D4).

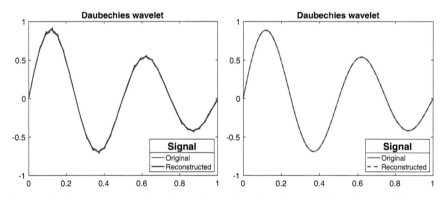

Fig. 8.10 Reconstruction after thresholding: $J = 10$. Left: $\varepsilon = 0.10$; right: $\varepsilon = 0.01$.

8.6 Generalization: Image Processing

Generalization of the previous results to image processing is straightforward. We might define a wavelet function of two variables $\psi_2(x, y)$; meanwhile, the tensor product is easier to deal with. We then consider a mother wavelet of the form $\psi_2(x, y) = \psi(x)\psi(y)$. This choice introduces a 2D Mallat transform, where the image to treat is a matrix $[c_J]$: we begin to proceed to a row-by-row decomposition. The resulting transformed rows are stored in a matrix $[\tilde{c}_J]$; we proceed then to a decomposition of the $[\tilde{c}_J]$ columns, and the final results are stored (column by column) in a matrix $[d_J]$. This matrix contains all the components of the image in the wavelet basis. It is then possible to compress this object using a thresholding algorithm, and the compressed data are stored in a matrix $[d_J^\varepsilon]$. The use of a column-by-column reconstruction algorithm followed by a row-by-row reconstruction algorithm provides a new image $[c_J^\varepsilon]$ from $[d_J^\varepsilon]$. From a practical point of view, some operations may be performed using the $[d_J]$ representation directly rather than the $[c_J]$ initial one:

- Two distinct images may be compared in the (compressed) wavelet format; this is very helpful for saving computing time. As an example of such utilization, we cite the research of suspected fingerprints in a criminal fingerprints database. As the information is more condensed into wavelet storage, comparisons go very fast.
- Storage in $[d_J]$ format is also useful to detect singularities because they are associated with large values of coefficients $[d_J]_{k,l}$. So the presence (or absence) of such coefficients may reveal special features of the original image $[c_J]$ very quickly.

8.6.1 Image Processing: Implementation

We assume here that F is an image defined as a 2D pixel array $[c_J]$.

Exercise 8.7 1. Write a procedure performing decomposition and reconstruction of a given image $[c_J]$ for all three wavelet functions described in the previous sections.
2. Check the thresholding compression algorithm.
3. Compare and visualize all results.

A solution of this exercise is proposed in Sect. 8.7 at page 202. Figure 8.11 and Fig. 8.12 display original and reconstructed images. For a threshold value $\varepsilon = 10^{-3} \times (2^J)^2$, the numbers of nonzero components in $[d_J^\varepsilon]$ are respectively $nbc_H^\varepsilon = 3231$ using the Haar wavelet, $nbc_S^\varepsilon = 10184$ using the Schauder wavelet, and $nbc_D^\varepsilon = 2702$ using the D4 Daubechies wavelet. The original document is a 256×256 pixel image, corresponding to a matrix $[c_J]$ with 65536 coefficients.

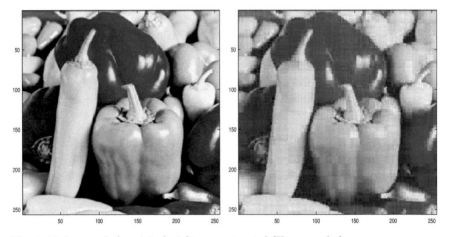

Fig. 8.11 Images. Left: original; right: reconstructed (Haar wavelet).

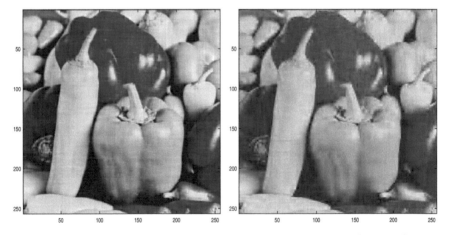

Fig. 8.12 Reconstructed images. Left: Schauder wavelet; right: Daubechies wavelet.

Multiresolution analysis is rich in theoretical and practical developments. Numerous projects are progressing all around the world, making it one of the most active areas of research in the mathematical sciences. Readers will find a large literature on this subject. Among many papers of interest, we cite Cohen (2000, 2003), Daubechies (1992), Mallat (1997), Meyer (1990), Mallat (2016), Flandrin (2018), and Mallat (2023).

8.7 Solutions and Programs

Solution of Exercise 8.1

The file MRA_F_haar.m provides the procedure related to decomposition and reconstruction algorithm (8.17) and (8.18). A flag parameter allows one to switch from decomposition ($[c_J] \longrightarrow [d_J]$) to reconstruction ($[d_J] \longrightarrow [c_J]$) formulas.
The file MRA_M_haar_ex1.m contains the script that generates a sampling of a function on the interval $[0, 1]$ (for this particular example, the function contained in the file MRA_F_function.m is defined by $f(x) = \exp(-x)\sin(4\pi x)$. This sampling is then used to define a piecewise constant function with the help of the procedure MRA_F_pwcte. Decomposition and reconstruction computations are then performed by the procedure MRA_F_haar.

Solution of Exercise 8.2

The file MRA_M_haar_ex2.m contains the script for performing the same computations as MRA_M_haar_ex1, with the difference that all coefficients whose absolute values are smaller than the threshold are set to zero. This procedure is used for the compression tests of Table 8.1 and Fig. 8.3.

Solution of Exercise 8.3

The file MRA_F_schauder.m provides the procedure related to decomposition and reconstruction algorithm (8.25) and (8.26). A flag parameter allows one to switch from decomposition ($[c_J] \longrightarrow [d_J]$) to reconstruction ($[d_J] \longrightarrow [c_J]$) formulas.
The file MRA_M_schauder_ex1.m contains the script that generates a sampling of a function on the interval $[0, 1]$ (for this particular example, the function contained in the file MRA_F_function.m is defined by $f(x) = \exp(-x)\sin(4\pi x)$. This sampling is then used to define a piecewise constant function with the help of the procedure MRA_F_pwcte. Decomposition and reconstruction computations are then performed by the procedure MRA_F_schauder.

Solution of Exercise 8.4

The file MRA_M_schauder_ex2.m contains the script for performing the same computations as MRA_M_schauder_ex1, with the difference that all coefficients whose absolute values are smaller than the threshold are set to zero. This procedure is used for the compression tests of Table 8.2 and Fig. 8.8.

Solution of Exercise 8.5

The file `MRA_F_daube4.m` provides the procedure related to decomposition
and reconstruction algorithm (8.29) and (8.30). A flag parameter allows one
to switch from decomposition ($[c_J] \longrightarrow [d_J]$) to reconstruction ($[d_J] \longrightarrow [c_J]$)
formulas.

The file `MRA_M_daube4_ex1.m` contains the script that generates a sampling of
a function on the interval $[0, 1]$ (for this particular example, the function contained in file `MRA_F_function.m` is defined by $f(x) = \exp(-x)\sin(4\pi x)$. This
sampling is then used to define a piecewise constant function with the help of
the procedure `MRA_F_pwcte`. Decomposition and reconstruction computations
are then performed by the procedure `MRA_F_daube4`.

Solution of Exercise 8.6

The file `MRA_M_daube4_ex2.m` contains the script for performing the same
computations as `MRA_M_daube4_ex1`, with the difference that all coefficients
whose absolute values are smaller than the threshold are set to zero. This
procedure is used for the compression tests of Table 8.3 and Fig. 8.10.

Solution of Exercise 8.7

The files `MRA_M_haar_ex3.m`, `MRA_M_schauder_ex3.m`, and `MRA_M_daube4_ex3.m`
contain the scripts that read an image from the file `pepper.jpg` and then perform
decomposition, compression, and reconstruction steps with the Haar wavelet
(respectively Schauder and Daubechies wavelets). Figures 8.11 and 8.12 were
obtained in this way. Note that in these procedures, decomposition and
reconstruction are performed by successive uses of 1D decomposition and the
reconstruction algorithm.

Chapter References

A. Cohen, Wavelet methods in numerical analysis, in *Handbook of Numerical
    Analysis*, ed. by P.G. Ciarlet, J.L. Lions, vol. VII (North-Holland, Amster-
    dam, 2000)
A. Cohen, *Numerical Analysis of Wavelet Methods* Studies in mathematics
    and its applications. (North-Holland, Amsterdam, 2003)
A. Cohen, R. Ryan, *Wavelets and Multiscale Signal Processing* (Chapman
    and Hall, London, 1995)
I. Daubechies, *Ten Lectures on Wavelets* (Society for Industrial and Applied
    Mathematics, Philadelphia, Pennsylvania, 1992)

S.G. Mallat, *A Wavelet Tour of Signal Processing* (Academic Press, New York, 1997)

Y. Meyer, *Ondelettes et Opérateurs. Tomes I à III* (Hermann, Paris, 1990)

H. Ammari, S. Mallat, I. Waldspurger, H. Wang, Wavelets methods for shape perception in electro-sensing, imaging, multi-scale and high contrast partial differential equations. AMS eBooks Collection, vol 660 (2016)

P. Flandrin, *Explorations in Time-Frequency Analysis* (Cambridge University Press, 2018)

S. Mallat, G. Rochette, S. Zhang, *Theoretical Physics Wavelets, Analysis, Genomics. An Indisciplinary Tribute to Alex Grossmann* (Springer, 2023)

Chapter 9
Elasticity: Elastic Deformation of a Thin Plate

Project Summary

Level of difficulty:	2
Keywords:	Finite difference method, Laplacian, bilaplacian
Application fields:	Linear elasticity: deformation of a membrane or plate

9.1 Introduction

We study in this chapter the deformation of a thin plate. In our example, the plate is part of a condenser microphone, such as one may find inside a telephone (or a cellular phone). When the user speaks, the plate (which is in fact a metalized plastic diaphragm) moves in response to changes in the acoustic pressure induced by sound waves. Since the plate is also the side of an electric capacitor, its dynamic deformations infer variations of the electric potential, which is amplified to generate a measurable signal. For the sake of simplicity, we shall consider here a thin rectangular plate in the device displayed in Fig. 9.1.

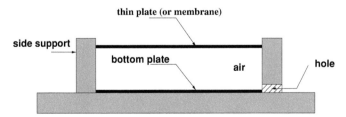

Fig. 9.1 Sketch of the pressure sensor (side-view).

© The Author(s), under exclusive license to Springer Nature Switzerland AG 2023   205
I. Danaila et al., *An Introduction to Scientific Computing*,
https://doi.org/10.1007/978-3-031-35032-0_9

9.2 Modeling Elastic Deformations (Linear Problem)

As a first stage of approximation, we shall neglect electrostatic forces in the device and consider that the plate bends exclusively because of the difference between the inside and outside values of the acoustic pressure (see Fig. 9.2). The pressure is assumed constant inside the device, and we take into account only variations of the outside acoustic pressure. There are two physical models relating the deformation f_a to the pressure value P_a:

- for a high-strained plate,

$$-c_1 \Delta f_a = P_a, \tag{9.1}$$

- and for a low-strained plate (the term "membrane" is then more appropriate than "plate"),

$$c_2 \Delta^2 f_a = P_a. \tag{9.2}$$

The coefficients c_1 and c_2 are physical constants depending on the material and defined as

$$c_1 = T \quad \text{and} \quad c_2 = \frac{Ee^3}{12(1 - \nu)},$$

where e is the thickness of the plate, T is the mechanical stress, E is Young's modulus, and ν is the Poisson coefficient.

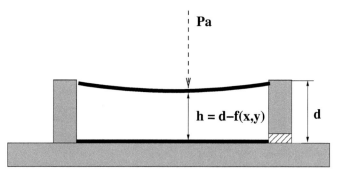

Fig. 9.2 Deformation of the plate.

In previous equations the symbol Δ denotes the Laplacian, a differential operator defined in two dimensions as

$$\Delta f_a = \frac{\partial^2 f_a}{\partial x^2} + \frac{\partial^2 f_a}{\partial y^2}.$$

The bilaplacian (or biharmonic operator) Δ^2 is defined accordingly as $\Delta^2 f_a = \Delta(\Delta f_a)$.

In order to have a general formulation of the problem, we shall consider in the following the "mixed" equation

$$c_2 \Delta^2 f_a - c_1 \Delta f_a = P_a. \tag{9.3}$$

This is a partial differential equation (PDE) of fourth order. For any physically acceptable value of P_a, there exists a solution f_a to equation (9.3) (for mathematical details see Ciarlet (1978, 2000)). In fact, the solution is not unique, since for any harmonic function f_h (i.e., a function such that $\Delta f_h = 0$), $f_a + f_h$ is also a solution. This is a direct consequence of the linearity of the Laplacian and bilaplacian.[1]

To ensure uniqueness (which is a crucial feature for the success of a numerical computation) we shall prescribe appropriate boundary conditions. We want the solution satisfying a realistic condition: the plate is assumed to be fastened along the four sides of the rectangle. This means that the deformation f_a is null all along the edge of the membrane:

$$f_a|_{\partial \Omega} = 0, \tag{9.4}$$

where $\partial \Omega$ denotes the boundary of the domain Ω covered by the plate. This is called a *Dirichlet homogeneous boundary condition* and is a sufficient condition to obtain the uniqueness of the solution of equation (9.1), because the Laplacian is a second-order differential operator (for the proof, see, for example, Ciarlet (2000)). When considering equation (9.2) or (9.3), a supplementary boundary condition is required, since the bilaplacian is a fourth-order differential operator. Denoting by $\mathbf{n}$ the outward normal vector[2] to $\partial \Omega$, this supplementary condition simulates the elastic "clamping" of the plate along all the boundaries:

$$\left. \frac{\partial f_a}{\partial n} \right|_{\partial \Omega} = 0. \tag{9.5}$$

The boundary condition (9.5) is referred to mathematically as a *Neumann condition*.

9.3 Modeling Electrostatic Forces (Nonlinear Problem)

Relax the pressure for a while, and have a new look at Fig. 9.1. The plate and the bottom of the cavity are both made of metallic material and form the two parts of a capacitor whose dielectric is the air within the cavity. A dielectric material is a substance that is poor conductor of electricity, but an

---

[1] $\Delta(f_a + f_h) = \Delta f_a + \Delta f_h$ and consequently $\Delta^2(f_a + f_h) = \Delta^2 f_a + \Delta^2 f_h$.

[2] The normal vector, often simply called the "normal", to a surface is a vector perpendicular to it.

efficient support of electrostatic fields. So, when both parts of the capacitor have different electric potential values, there exists a force bringing them closer.

We start with basic relationships expressing the electrostatic energy W and force F:

$$W = \frac{1}{2}CU^2 = \frac{\varepsilon SU^2}{2h}, \qquad F = -\frac{dW}{dh} = \frac{\varepsilon SU^2}{2h^2},$$

as functions of the capacitance C, the potential difference U, the air permittivity ε, the surface S of the plate, and the capacitor thickness h (i.e., the distance between the top and bottom plates; see Fig. 9.2). The electrostatic pressure acting on the plate is then obtained as

$$P_e = \frac{F}{S} = \frac{\varepsilon U^2}{2h^2}.$$

As with the acoustic pressure, the effect of the electrostatic pressure is to bend the plate. Consequently, the resulting deformation f_e is the solution of an equation similar to (9.3), with modified right-hand side P_e. The difference between the two cases is that the electrostatic pressure P_e is no longer a constant, like P_a, but depends on the position (x, y) since (see Fig. 9.2) $h = d - f_e(x, y)$.

In conclusion, the mathematical model taking into account the electrostatic forces consists of the following nonlinear PDE:

$$c_2\Delta^2 f_e - c_1\Delta f_e = P_e(f_e) = \frac{\varepsilon U^2}{2(d - f_e(x, y))^2}, \qquad (9.6)$$

with Dirichlet and Neumann boundary conditions (which make the solution unique)

$$f_e = 0 \quad \text{and} \quad \frac{\partial f_e}{\partial n} = 0 \quad \text{on} \quad \partial\Omega. \qquad (9.7)$$

9.4 Numerical Discretization of the Problem

In this section, we shall not worry about the right-hand side of the model equation (9.3) or (9.6) and discuss only the discretization of the differential operators (Laplacian and bilaplacian). We consider, for example, the equation (9.3) with boundary conditions (9.7).

Since it is not generally possible to obtain an exact (analytical) form of the solution f_a, we shall compute an approximate solution on a regular mesh representing the rectangular plate $\Omega = [0, L_x] \times [0, L_y]$ (see Fig. 9.3). We use the notation $M_{i,j}$ for the grid point of coordinates (x_i, y_j), with

$$x_i = i \cdot h_x, \quad i = 0, \ldots, mx + 1, \quad h_x = L_x/(mx + 1), \qquad (9.8)$$
$$y_j = j \cdot h_y, \quad j = 0, \ldots, my + 1, \quad h_y = L_y/(my + 1). \qquad (9.9)$$

Note that we have to compute only $mx \cdot my$ discrete values $f_{i,j}$ ($i = 1, \ldots, mx$, $j = 1, \ldots, my$), approximating the values $f_a(M_{i,j}) = f_a(x_i, y_j)$ of the exact solution, because the values on the boundaries are known (i.e., $f_{0,j} = f_{mx+1,j} = f_{i,0} = f_{i,my+1} = 0$). These values are obtained by approximating the differential operators in (9.3) using the *finite difference* method (see Chaps. 1 and 7, or for more details Strikwerda (1989)).

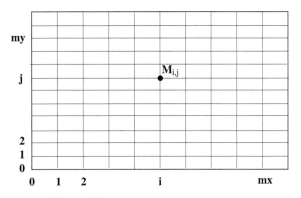

Fig. 9.3 A regular mesh (or grid).

To begin with, we address the case of the Laplacian. The easiest way to approximate second derivatives in this differential operator is to use centered differences, leading to the well-known 5-point (difference) scheme:

$$-(\Delta_5 f)_{i,j} = \frac{1}{h_x^2}(-f_{i+1,j} + 2f_{i,j} - f_{i-1,j})$$
$$+ \frac{1}{h_y^2}(-f_{i,j+1} + 2f_{i,j} - f_{i,j-1}). \qquad (9.10)$$

This scheme is second-order accurate at any point of the grid, that is,

$$-(\Delta_5 f)_{i,j} = -(\Delta f)_{i,j} + O(h_m^2), \quad \text{with} \quad h_m = \max(h_x, h_y).$$

We proceed in the same manner to discretize the bilaplacian Δ^2. We first substitute in equation (9.10) all $f_{i,j}$ by $-(\Delta_5 f)_{i,j}$:

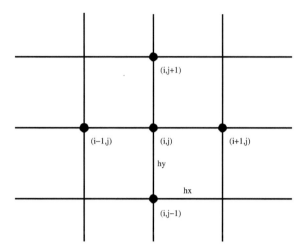

Fig. 9.4 Discretization of the 2D Laplacian with a 5-point scheme.

$$(\Delta_{13}^2 f)_{i,j} = \frac{1}{h_x^2}\left(-(\Delta_5 f)_{i+1,j} + 2(\Delta_5 f)_{i,j} - (\Delta_5 f)_{i-1,j}\right)$$

$$+ \frac{1}{h_y^2}\left(-(\Delta_5 f)_{i,j+1} + 2(\Delta_5 f)_{i,j} - (\Delta_5 f)_{i,j-1}\right).$$

Then, by inserting the expression of $-(\Delta_5 f)_{i,j}$ according to (9.10), we obtain the so-called 13-point scheme (see Fig. 9.5), which is also an approximation of second order:

$$(\Delta_{13}^2 f)_{i,j} = \frac{1}{h_y^4} f_{i,j-2}$$

$$+ \frac{2}{h_x^2 h_y^2} f_{i-1,j-1} - \left(\frac{4}{h_x^2 h_y^2} + \frac{4}{h_y^4}\right) f_{i,j-1} + \frac{2}{h_x^2 h_y^2} f_{i+1,j-1}$$

$$+ \frac{1}{h_x^4} f_{i-2,j} - \left(\frac{4}{h_x^2 h_y^2} + \frac{4}{h_x^4}\right) f_{i-1,j}$$

$$+ \left(\frac{8}{h_x^2 h_y^2} + \frac{6}{h_x^4} + \frac{6}{h_y^4}\right) f_{i,j} \tag{9.11}$$

$$- \left(\frac{4}{h_x^2 h_y^2} + \frac{4}{h_x^4}\right) f_{i+1,j} + \frac{1}{h_x^4} f_{i+2,j}$$

$$+ \frac{2}{h_x^2 h_y^2} f_{i-1,j+1} - \left(\frac{4}{h_x^2 h_y^2} + \frac{4}{h_y^4}\right) f_{i,j+1} + \frac{2}{h_x^2 h_y^2} f_{i+1,j+1}$$

$$+ \frac{1}{h_y^4} f_{i,j+2}.$$

Finally, the discrete form of our PDE reads

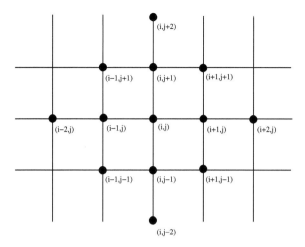

Fig. 9.5 Discretization of the 2D bilaplacian with a 13-point scheme.

$$c_2(\Delta_{13}^2 f)_{i,j} - c_1(\Delta_5 f)_{i,j} = P(M_{i,j}) = P_{i,j}, \qquad (9.12)$$

where P stands for either acoustic or electrostatic pressure. These equations, written for any grid point $M_{i,j}$, with $i = 1, 2, \ldots, mx$ and $j = 1, 2, \ldots, my$, form a linear system whose unknowns are the $mx \cdot my$ values $f_{i,j}$. It is not difficult to observe that the discretization (9.11) is not well posed for grid points near the boundaries, since it involves "ghost" points that do not exist (for example, the equation for $i = 1$ and any j requires the value of $f_{-1,j}$, which is not defined). We are fortunately rescued from this critical situation by the (Neumann) boundary condition on the normal derivative (9.5), which allows us to define such ghost points. Indeed, the derivative $\partial f / \partial n$ can be discretized by the first-order backward finite differences

$$(f_{i,j} - f_{i-1,j})/h_x = 0 \quad \text{and} \quad (f_{i,j} - f_{i,j-1})/h_y = 0,$$

for any point $M_{i,j}$ located on the boundary (i.e., $i = 0$ or $mx + 1$ and $j = 0$ or $my + 1$). Keeping in mind that $f_{i,j} = 0$ in any point $M_{i,j}$, following the (Dirichlet) boundary condition (9.5), we deduce that $f_{i,j} = 0$ for any ghost point.

Remark 9.1 We can use a simple programming trick when computing the $(mx \cdot my)^2$ matrix of the linear system (9.12). For the rows corresponding to $i = 1$ or $j = 1$ (and similarly, for $i = mx$ or $j = my$), we simply set the nonexistent (ghost) values to zero (i.e., if $i < 1$ or $j < 1$ and similarly for $i > mx$ or $j > my$). The right-hand side of the system is thus not affected.

Remark 9.2 Since one should expect from the discussion on the uniqueness of the solution of the continuous PDE (9.3), implementing discrete boundary conditions results in rendering the matrix of the linear system (9.12) invertible.

9.5 Programming Tips

9.5.1 Modular Programming

In the previous section, different logical steps have been pointed out when implementing the finite difference method:

1. definition of the coordinates (x_i, y_j) of any mesh point,
2. construction of the linear system (9.12),
3. introduction of the boundary conditions,
4. solving the linear system,
5. visualization of the results.

Any scientific package has to deal separately with each item of this list by associating a specialized computing procedure. These procedures are called *modules*. Results (outputs) of a given step module are data (inputs) for the next step module. Several modules may exist for the same logical step; in this case they all have to share similar formatted input and provide similar formatted output.

For the present study, we shall also proceed step by step, by setting up progressively the numerical operators. First, we neglect the effect of the electrostatic pressure in order to check the programs required to solve the linear problem (9.12). Then we shall deal with the nonlinear problem by iterating on successive linear problems.

9.5.2 Program Validation

Some questions can be asked when one uses a numerical approximation to solve a problem such as (9.3). Are we sure the good solution is computed with an effective procedure? How many points are required in order to get an accurate numerical solution? There is a simple way to answer these questions: it consists in solving a problem for which an exact solution is known, and then comparing the computed result to the exact one.

Let us consider, for example, the Laplace equation (9.1). We may choose a more or less complicated solution of the PDE, as for example

$$\tilde{f}_a(x, y) = 100 \sin(3.7\pi x) \sin(5.4\pi y) + (3.7x - 5.4y), \tag{9.13}$$

and calculate the corresponding right-hand side (considering $c_1 = 1$):

$$\tilde{P}_a(x, y) = -\Delta \tilde{f}_a(x, y) = 100(3.7^2 + 5.4^2)\pi^2 \sin(3.7\pi x) \sin(5.4\pi y). \tag{9.14}$$

A program solving the PDE $-\Delta f_a = P_a$, with boundary conditions $f_a|_{\partial\Omega} = g(x, y)$, needs two inputs: $P_a(x, y)$ and $g(x, y)$. Inserting in the program (as discrete input data) the expression (9.14) for $P_a(x, y)$ and (9.13) for $g(x, y)$, we should obtain numerical values $f_{i,j}$ close to the exact values $\tilde{f}_a(x_i, y_j)$ at the same grid points (x_i, y_j). We are now able to compare the two solutions (exact and numerical) qualitatively by plotting the results in the same graphical window and quantitatively by computing, for example, the following relative approximation error:

$$\text{Error} = \frac{\sum_{i,j} |\tilde{f}_a(x_i, y_j) - f_{i,j}|^{1/2}}{\sum_{i,j} |\tilde{f}_a(x_i, y_j)|^{1/2}}. \tag{9.15}$$

This error has to be "reasonably" small and to diminish when the number of grid points is increased. If this is not the case, the program must be checked.

The same validation procedure can be used for the PDEs (9.2) and (9.3).

9.6 Solving the Linear Problem

In order to solve the linear problem (9.12) we note that

1. $n = mx \cdot my$, is the total number of grid points,
2. Ah_5 is the matrix associated with the 5-point scheme (including boundary conditions),
3. Ah_{13} is the matrix associated with the 13-point scheme (including boundary conditions),
4. b_5, b_{13} are the corresponding right-hand sides.

Exercise 9.1 1. Write a program generating all the coordinates (x_i, y_j) of the grid points $M_{i,j}$, for $i = 1, 2, \ldots, mx$ and $j = 1, 2, \ldots, my$.
2. Write a program computing the matrix Ah_5 and the corresponding right-hand side b_5, in order to solve equation (9.1).
3. Write a program computing the matrix Ah_{13} and the corresponding right-hand side b_{13}, in order to solve equation (9.2).
4. Write a program computing the matrix Ah and the corresponding right-hand side b, in order to solve the complete problem (9.3), including the boundary conditions. Solve this problem for a given value of the pressure.
5. Visualize the results.

N.B. All the programs must be checked using the validation procedure described above.

A solution of this exercise and related procedures are described in Sect. 9.8 at page 216. We show here (see Fig. 9.6) a plot of the numerical solution obtained from validating the program solving the linear problem (9.3). The right-hand side of the PDE was calculated such that (9.13) becomes the exact solution. Even though a coarse mesh was used ($nx = 20$ and $ny = 30$), the numerical solution is very close to the exact one. The relative error (9.15) for this numerical experiment has the value Error $= 0.0375$. This error diminishes as the number of grid points increases (Error $= 0.0095$ for $nx = 40$ and $ny = 60$), but the computing time is considerably larger for this last run!

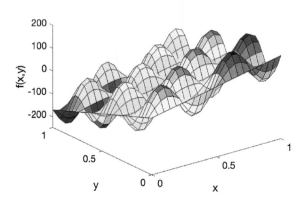

Fig. 9.6 Numerical solution obtained in validating the program solving the linear problem (9.3) by "imposing" the exact solution (9.13) on the boundary.

9.7 Solving the Nonlinear Problem

We now address the nonlinear problem, corresponding to a more realistic case when the plate is part of a microphone and subject to both acoustic and electrostatic pressures.

9.7.1 A Fixed-Point Algorithm

The resulting nonlinear problem is

$$c_2 \Delta^2 f - c_1 \Delta f = P_a + P_e(f), \tag{9.16}$$

with Dirichlet and Neumann boundary conditions

$$f = 0 \quad \text{and} \quad \frac{\partial f}{\partial n} = 0 \quad \text{on} \quad \partial \Omega. \tag{9.17}$$

To solve this problem we use a fixed-point algorithm. We start by solving the linear problem (9.3) corresponding to $P_e = 0$; we denote by f_0 this solution. We define then the sequence $\{f_k\}_{k \in N}$ of solutions of successive linear problems:

$$c_2 \Delta^2 f_{k+1} - c_1 \Delta f_{k+1} = P_a + P_e(f_k). \tag{9.18}$$

Since the fixed-point algorithm is an iterative method, we have to choose a stopping criterion to decide whether a solution f_k is accurate enough to be a good approximation of the exact solution. A classical criterion is based on the relative variation of the approximate solution f_k:

$$\max_{x,y} |f_{k+1}(x,y) - f_k(x,y)| < \varepsilon \max_{x,y} |f_k(x,y)|, \tag{9.19}$$

where ε is the convergence threshold.

9.7.2 Numerical Solution

Once again, we shall first consider a test problem before solving (9.16) in order to validate the procedures. We choose the same test solution f as previously (9.13) and compute the corresponding right-hand side. For this case, we also have to impose the expression of the nonlinear term $P_e(f)$, depending on the solution. For example, we can use the function defined by

$$P_e(f) = \frac{100}{(200 - f)^2}.$$

For this choice and for the same grid ($nx = 20$ and $ny = 30$), the solution converges after only two iterations of the fixed-point algorithm (the convergence threshold is fixed to $\varepsilon = 0.001$). We check that the plot of the solution is similar to that displayed in Fig. 9.6. More details on the solution procedures can be found in Sect. 9.8 at page 217.

We consider now a more realistic choice of the values of the physical parameters[3] appearing in problem (9.16). The atmospheric pressure value is set to 10^5 [Pa] or [N/m²], the acoustic pressure P_a then goes from 10^{-3} [Pa] to 10 [Pa]. It is established that the human ear can perceive pressure variations from 2×10^{-5} [Pa] up to 2 [Pa]. We may choose, without any damage to one's hearing or numerical procedures, the value of 1 [Pa] for the acoustic pressure variation.

---

[3] As physical units, we use [Pa] = Pascal, [N] = Newton, [m] = meter, [mm] = millimeter, [μm] = micrometer [F] = Faraday, [V] = Volt.

We assume that the plate is made of silicon; for such a material Young's modulus is $E = 1.3 \cdot 10^{11}$ [Pa] and the Poisson coefficient is $\nu = 0.25$. The device displayed in Fig. 9.1 has the following characteristic dimensions: length 1 [mm], width 1 [mm], and thickness $e = 1$ [μm]. The mechanical stress of the plate is $T = 100$ [N/m]. Concerning the capacitor, the thickness (without pressure variations) is $d = 5$ [μm]. The dry-air permittivity is $\varepsilon = 8.85 \cdot 10^{-12}$ [F/m], and the polarization potential is $V = 25$ [V]. The mesh of the plate, as displayed in Fig. 9.3, has $nx = 20$ and $ny = 30$ grid points, resulting in a total number of 600 discretization points, and the same number of unknowns.

Exercise 9.2 1. Modify the procedure used to solve Exercise 9.1 in order to use the above realistic data for the linear problem (9.3).
2. Write a program implementing the fixed-point algorithm.
3. Solve the nonlinear problem (9.16).
4. Visualize the results.

A solution of this exercise is proposed in Sect. 9.8 at page 217.

Hint: We first solve the acoustic problem (9.3) with boundary conditions (9.7) and obtain a deformation as plotted in the left panel of Fig. 9.7. The maximum deformation value is located at the center of the plate (max f_a = 0.080 [μm]).

In the next step, we consider the complete problem (9.16) including boundary conditions (9.17). The fixed-point algorithm will converge within three iterations when we use the stopping criterion (9.19) with $\varepsilon = 0.001$. The maximum deformation of the plate (see right panel of Fig. 9.7) is reached again at the center of the plate (max $f_e = 0.077$ [μm]).

Remark 9.3 It is important to note that relative values of the acoustic pressure P_a and the polarization potential V were chosen in order to respect the constraint max $f_a < d$ (see Fig. 9.2).

9.8 Solutions and Programs

9.8.1 Solution of Exercise 9.1

The script ELAS_M_plate_ex.m solves the problem (9.3) and computes the test solution defined in the function ELAS_F_solution.m. This script calls the functions ELAS_F_lap_matrix and ELAS_F_lap_rhs, which compute the linear system obtained from the discretization of the equation (9.1). Similarly, functions ELAS_F_bilap_matrix.m and ELAS_F_bilap_rhs.m compute the linear system corresponding to equation (9.2). The computed test solution is plotted in Fig. 9.6.

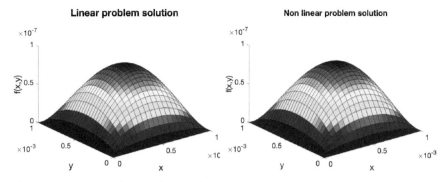

Fig. 9.7 Numerical solution of the realistic microphone problem. Left: linear; right: non-linear.

9.8.2 Solution of Exercise 9.2

The main program solving the nonlinear problem (9.6) with realistic coefficients is ELAS_M_microphone_ex. It also contains the fixed-point algorithm. Functions ELAS_F_lap_matrix and ELAS_F_lap_rhs (for the Laplacian part of the PDE) and in the files ELAS_F_bilap_matrix and ELAS_F_bilap_rhs (for the bilaplacian part of the PDE) are used as previously. The function ELAS_F_pressure defines the nonlinear term. The obtained numerical solutions are displayed in Fig. 9.7.

9.8.3 Further Comments

In this section, we address the important point of the construction of the matrices resulting from the approximation of the operators. The use of the rectangular mesh (displayed in Fig. 9.3), with a lexical ordering of the nodes,[4] added to the 5-point scheme approximation of the Laplacian (see Fig. 9.4), lead altogether to a very particular pattern of the matrix Ah_5, displayed in the left panel of Fig. 9.8. Such a matrix is called *banded*, of bandwidth $2mx + 1$, because its coefficients satisfy the relation $(Ah_5)_{ij} = 0$ if $|i - j| > mx$. Note that this matrix is *sparse* because it contains only $5mx \cdot my$ nonzero coefficients. For the same reasons, the matrix associated with the bilaplacian (see right panel of Fig. 9.8) is banded of bandwidth $4mx + 1$, and sparse with $13mx \cdot my$ nonzero coefficients. These properties are useful for reducing storage, because increasing the number of unknowns leads to

---

[4] We order the nodes starting from the bottom line, from left to right $1, 2, \ldots, mx$; then we continue with the line just above, from left to right $mx + 1, mx + 2, \ldots, 2mx$, and so on.

huge use of memory. Scientific programs have to deal carefully with these
properties; thankfully MATLAB is a user-friendly environment and provides
very simple ways to build such matrices. For example, the following procedure
(**ELAS_F_lap_matrix**) computes the matrix Ah_5:

```
n=nx*ny;
h2x=hx*hx;h2y=hy*hy;
Ah5=sparse(n,n);
Dx=toeplitz( [2.d0 -1.d0 zeros(1,nx-2) ] ) ;
Dx = Dx / h2x ;
Dy=eye(nx,nx) ;
Dy = - Dy / h2y ;
Dx = Dx - 2.d0 * Dy ;
for k=1:(ny-1)
    i=(k-1)*nx ; j=k*nx ;
    Ah5( (i+1) : (i+nx) , (i+1) : (i+nx) ) = Dx ;
    Ah5( (j+1) : (j+nx) , (i+1) : (i+nx) ) = Dy ;
    Ah5( (i+1) : (i+nx) , (j+1) : (j+nx) ) = Dy ;
end ;
i=(ny-1)*nx ;
Ah5( (i+1) : (i+nx) , (i+1) : (i+nx) ) = Dx ;
```

This program calls MATLAB built-in functions:

1. **sparse** is used to declare a low-storage sparse matrix;
2. **toeplitz** is used to define a Toeplitz matrix; here Dx is an $nx \times nx$ symmetric tridiagonal matrix whose entries are $(Dx)_{ii} = 2$ and $(Dx)_{ii-1} = -1$;
3. **eye** is used to define the $nx \times nx$ identity matrix;
4. then, the nonzero coefficients of Ah_5 are defined "block by block", using Dx to set diagonal blocks, respectively Dy for off-diagonal blocks.

Unfortunately, such a structure occurs in a very particular case, strongly
depending on the geometry: for a nonrectangular mesh or with a random
ordering of the nodes, the resulting matrix has a less-regular pattern (see for
instance Chap. 14). Nevertheless, it remains sparse because this property is
related only to the approximation scheme.

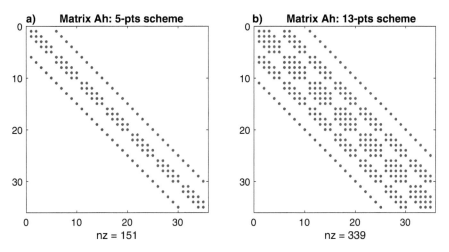

Fig. 9.8 Left: Matrix Ah_5, right: Matrix Ah_{13} for nx=7 and ny=9. nz is the number of nonzero elements.

Chapter References

P.G. Ciarlet, *The Finite Element Method for Elliptic Problems* (North Holland, Amsterdam, 1978)

P.G. Ciarlet, *Mathematical Elasticity*, vol. I, II, III. (North Holland, Amsterdam, 2000)

J.C. Strikwerda, *Finite Difference Schemes and Partial Differential Equations*. (Wadsworth and Brooks/Cole, 1989)

Chapter 10
Domain Decomposition Using a Schwarz Method

Project Summary

Level of difficulty:	2
Keywords:	Domain decomposition, Schwarz method with overlapping, Laplacian discretization, 1D and 2D finite difference
Application fields:	Thermal analysis, steady heat equation

10.1 Principle and Application Field of Domain Decomposition

Realistic modeling of physical problems often involves systems of partial differential equations (PDE), usually nonlinear, and is defined on domains that can have both a large size and a complex shape. In most cases, the selected numerical method requires that one discretize the domain, and the number of degrees of freedom can easily be more than what the available computer will handle. Modeling of the airflow around an aircraft with 3D finite elements requires, for instance, the discretization of the surrounding domain with a few million points, with several unknowns to determine at each point. The numerical scheme can furthermore be implicit and hence involve the resolution of linear systems with this impressive number of unknowns. If we do not have a supercomputer at hand, which only very specialized research centers do, the matrix of such a system cannot even fit within the memory of the computer.

A simple answer to this technological lock is to subdivide the problem into smaller ones, that is, to compute the solution piecewise, as the solution of problems defined on subdomains of the initial one. Eventually, the global solution is the union of all the solutions of partial problems.

This method can also simplify the resolution of a problem set originally on complexly shaped domains, by selecting a decomposition in which each subdomain has a simpler, elementary shape, making the local solution simpler to compute. Another possible extension is the coupling of equations in order to treat interactions between two different physical phenomena defined on neighboring domains, fluid structure interaction, for instance.

The main difficulty arising in adopting this method is the definition of boundary conditions on each subdomain. Actually, the internal boundaries, in contrast to boundaries of the global domain, are fictitious, and the physics of the problem doesn't provide boundary conditions. Two strategies, both iterative, can be adopted. The first one consists in doing a domain decomposition with partial overlapping of the subdomains, and using the previous iteration solution on neighboring subdomains to define the boundary conditions on the current subdomain. The second strategy consists in partitioning the global domain into nonoverlapping subdomains and imposing continuity conditions at the interfaces.

Once the decomposition strategy is selected, the solution method on each subdomain is the same as on the global domain, with now a reasonably small number of unknowns. To fix the ideas, consider the simple example of a scalar equation solved by finite differences on a structured mesh that can be decomposed into P subdomains of the same size N. The numerical treatment will require solving P linear systems costing $O(N^2)$ on each subdomain, that is, a total cost of $O(PN^2)$ per iteration, instead of $O(P^2N^2)$ for the global problem. The added cost of the new method comes from the iterative nature of the algorithm, and therefore the size and the number of the subdomains must be carefully selected in order to ensure the competitiveness of the algorithm.

In any case, even if the computing time increases compared to the initial global scheme, we always gain the crucial advantage of being able to fit the problem in the computer's memory.

Last but not the least, even though this advantage cannot be illustrated within the scope of a MATLAB project, domain decomposition methods have really found their full worthiness with the development of parallel computing (see, for instance, Smith, Bjørstad, and Gropp (1996)). The solution of a problem set on the subdomains can be distributed to different processors, with a serious hope of computing time speedup as well as memory savings.

To illustrate the principles of the method, this project presents an implementation of the Schwarz method with overlapping on model problems of the 1D and 2D Laplacian

$$\begin{cases} -\Delta u(x) + c(x)u(x) = f(x), & \text{for} \quad x \in \Omega \subset \mathbb{R}^n, \\ u = g, & \text{on} \quad \partial\Omega. \end{cases} \tag{10.1}$$

10.2 1D Finite Difference Solution

In one dimension, the above problem becomes

$$\begin{cases} -u''(x) + c(x)u(x) = f(x), & \text{for} \quad x \in (a,b), \\ u(a) = u_a, \\ u(b) = u_b. \end{cases} \tag{10.2}$$

The finite difference method for second-order boundary value problems such as the one above is described in detail in Lucquin and Pironneau (1996), Le Dret and Lucquin (2016), and we just summarize the main features here. We discretize the interval $[a,b]$ on $n+2$ points $x_i = a + ih$ for $i = 0, \ldots, n+1$ with a uniform step $h = \frac{b-a}{n+1}$. We denote by U the vector formed by the approximation of the solution $u(x)$ at points x_i. We set $U_0 = u_a$ and $U_{n+1} = u_b$ to ensure that the numerical solution satisfies the boundary conditions. The finite difference discretization of the second derivative (as described, for instance, in Chap. 1) leads to the linear system

$$(S) \qquad A_h U = B_h,$$

where A_h is the $n \times n$ tridiagonal matrix

$$A_h = \frac{1}{h^2} \begin{pmatrix} 2 + h^2 c_1 & -1 & 0 & \cdots & \cdots & & 0 \\ -1 & 2 + h^2 c_2 & \ddots & \ddots & \ddots & & \vdots \\ 0 & \ddots & 2 + h^2 c_3 & \ddots & & \ddots & \vdots \\ \vdots & \ddots & & \ddots & \ddots & & 0 \\ \vdots & & \ddots & & \ddots & 2 + h^2 c_{n-1} & -1 \\ 0 & \cdots & & \cdots & 0 & -1 & 2 + h^2 c_n \end{pmatrix},$$

where $c_i = c(x_i)$ and B_h is the following vector in $\mathbb{R}^n$:

$$B_h = \begin{pmatrix} f(a+h) + \dfrac{u_a}{h^2} \\ f(a+2h) \\ \cdot \\ \cdot \\ f(b-h) + \dfrac{u_b}{h^2} \end{pmatrix}.$$

10.3 Schwarz Method in One Dimension

For simplicity, we first decompose the computational domain $[a,b]$ into two subdomains with overlapping: we choose an odd value n and two integer values i_l and i_r symmetric with respect to $\frac{n+1}{2}$ such that $i_l < \frac{n+1}{2} < i_r$. We set $x_l = i_l h$ and $x_r = i_r h$, thus defining two intervals $]a, x_r[$ and $]x_l, b[$ with a nonempty overlap $[a, x_r] \cap [x_l, b] = [x_l, x_r] \neq \emptyset$. We now plan to compute the solution u of the problem (10.2) by solving two problems posed on the subintervals $[a, x_r]$ and $[x_l, b]$:

$$(P_1) \begin{cases} -u_1''(x) + c(x)u_1(x) = f(x), & \text{for} \quad x \in \,]a, x_r[, \\ u_1(a) = u_a, \\ u_1(x_r) = \alpha; \end{cases}$$

and $(P_2) \begin{cases} -u_2''(x) + c(x)u_2(x) = f(x), & \text{for} \quad x \in \,]x_l, b[, \\ u_2(x_l) = \beta, \\ u_2(b) = u_b. \end{cases}$

The solution u_1 (respectively u_2) is expected to be the restriction on the interval $[a, x_r]$ (respectively $[x_l, b]$) of the solution u of the problem set on the full interval $[a, b]$. The two solutions u_1 and u_2 must therefore be identical within the overlapping region $[x_l, x_r]$, which allows us to define the boundary conditions in x_l and x_r:

$$u_1(x_r) = \alpha = u_2(x_r) \quad \text{and} \quad u_2(x_l) = \beta = u_1(x_l).$$

Since we do not know a priori the values of α and β, we solve the two problems iteratively: α is fixed arbitrarily, at first, for instance, by linear interpolation of the global boundary conditions

$$\alpha = \frac{1}{b-a}\left(u_a(b - x_r) + u_b(x_r - a)\right).$$

Then we set $u_2^0(x_r) = \alpha$ and we compute for $k = 1, 2, \ldots$ the solutions u_1^k and u_2^k of the following problems:

$$(P_1) \begin{cases} -u_1''(x) + c(x)u_1(x) = f(x), & \text{for} \quad x \in (a, x_r), \\ u_1(a) = u_a, \\ u_1(x_r) = u_2^{k-1}(x_r); \end{cases}$$

then

$$(P_2) \begin{cases} -u_2''(x) + c(x)u_2(x) = f(x), & \text{for} \quad x \in (x_l, b), \\ u_2(x_l) = u_1^k(x_l), \\ u_2(b) = u_b. \end{cases}$$

We claim that when the overlap region is nonempty, this algorithm converges to the solution u of the global problem (10.2) as $k \longrightarrow \infty$. This result was first proved using the fixed point theorem by Schwarz (1870), in the case $c(x) = 0$. It was rediscovered one century later using a variational formulation approach by Lions (1988). More efficient methods from the algorithmic point of view have been developed since then, which do not require an overlap but impose additional transfer conditions between subdomains. The following paragraphs will illustrate numerically the convergence, after discretization of $(P1)$ and $(P2)$ by finite difference.

10.3.1 Discretization

The problems P_1 and P_2 are solved using finite differences, in the same manner as used for the global problem (10.2) in the previous section. The bounds x_l and x_r have been set so that the two subdomains are of the same size. Denoting by V^k (respectively W^k) the vector of the approximate discrete solution on the subdomain $[a, x_r]$ (respectively $[x_l, b]$), the algorithm for the kth iteration is as follows:

$$
\left\|
\begin{aligned}
&\text{initialization}: \quad W_{i_r}^0 = \alpha \\
&\text{for} \quad k = 1, 2, \ldots, \quad \text{do} \\
&\qquad A_{h,l} V^k = B_{h,l} + \frac{1}{h^2}[u_a, 0, \ldots, 0, W_{i_r}^{k-1}]^T, \\
&\qquad A_{h,r} W^k = B_{h,r} + \frac{1}{h^2}[V_{i_l}^k, 0, \ldots, 0, u_b]^T, \\
&\text{end}
\end{aligned}
\right.
\tag{10.3}
$$

where $A_{h,l}$ (respectively $A_{h,r}$) is the discretization matrix for the operator $-\Delta + cI$ on $]a, x_r[$ (respectively $]x_l, b[$) in $\mathbb{R}^{i_r-1} \times \mathbb{R}^{i_r-1}$:

$$
A_{h,l} = \frac{1}{h^2}
\begin{pmatrix}
2+h^2c_1 & -1 & 0 \cdots & & \cdots & & 0 \\
-1 & 2+h^2c_2 & -1 & 0 & \ddots & & \vdots \\
0 & \ddots & \ddots & \ddots & \ddots & & \vdots \\
\vdots & & \ddots & \ddots & \ddots & & \ddots & 0 \\
\vdots & & \ddots & 0 & -1 & 2+h^2c_{i_r-2} & -1 \\
0 & & \cdots & & \cdots & 0 & -1 & 2+h^2c_{i_r-1}
\end{pmatrix},
$$

and

$$
A_{h,r} = \frac{1}{h^2}
\begin{pmatrix}
2+h^2c_{i_l+1} & -1 & 0 \cdots & & \cdots & & 0 \\
-1 & 2+h^2c_{i_l+2} & -1 & 0 & \ddots & & \vdots \\
0 & \ddots & \ddots & \ddots & \ddots & & \vdots \\
\vdots & & \ddots & \ddots & \ddots & & \ddots & 0 \\
\vdots & & \ddots & 0 & -1 & 2+h^2c_{n-1} & -1 \\
0 & & \cdots & & \cdots & 0 & -1 & 2+h^2c_n
\end{pmatrix}.
$$

The vectors B_l and B_r contain the values of the right-hand-side term f evaluated at the discretization points

$$(B_{h,l})_i = f(x_i), \quad \text{for} \quad i = 1, \ldots, i_r - 1,$$
$$(B_{h,r})_i = f(x_{i+i_l}), \quad \text{for} \quad i = 1, \ldots, n - i_l.$$

The stopping criterion in the iteration loop is obtained by measuring the gap between the two solutions within the overlap region $[x_l, x_r]$, where they should finally coincide:

$$\|e^k\| \le \varepsilon, \quad \text{with} \quad e^k_{i-i_l} = V^k_i - W^k_{i-i_l}, \quad \text{for} \quad i = i_l + 1, \ldots, i_r - 1.$$

In order to test the performance of the method, we can also compare the results with the solution U obtained in Sect. 10.2 using the classical method on the whole domain.

We accordingly compute two vectors e^k_l and e^k_r of components

$$(e^k_l)_i = U_i - V^k_i, \quad \text{for} \quad i = 1, \ldots, i_r - 1,$$
$$(e^k_r)_{i-i_l} = U_i - W^k_{i-i_l}, \quad \text{for} \quad i = i_l + 1, \ldots, n,$$

and we observe the decay of their norm with iterations along with the e_k norm decay.

Exercise 10.1 Write a program to implement Algorithm 10.3. Display in the same window, but with different colors, the solutions V^k, W^k, and U, refreshing the graphics at each iteration k. One should obtain a sequence of graphs as in Fig. 10.2. Represent the evolution of the three errors $\|e^k\|$, $\|e^k_l\|$, and $\|e^k_r\|$ as functions of k in another window as in Fig. 10.1.
A solution of this exercise is proposed in Sect. 10.5 at page 238.

Exercise 10.2 Modify the program of Exercise 10.1 and turn it into a function receiving as input argument the number of points $n_o = i_r - i_l - 1$ in the overlap region. This function computes and returns as output arguments the number of iterations necessary to reach the tolerance error and the computing time. Write a program that calls this function for different values of n_o and analyze the influence of the size of the overlap region on the algorithm's convergence. A solution of this exercise is proposed in Sect. 10.5 at page 239.

Figures 10.1 and 10.2 illustrate the results for a beam of length 1 meter, with a constant stiffness coefficient $c = 10$. The left end of the beam is fixed equal to 10 cm higher than the right end. The beam is subject to its own weight of 1 N/m as well as to an overload of 9 N/m on a 40 cm portion, starting 20 cm from its left end.

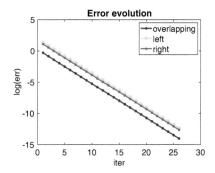

Fig. 10.1 Logarithm of the L^2 norm of the error versus the number of iterations.

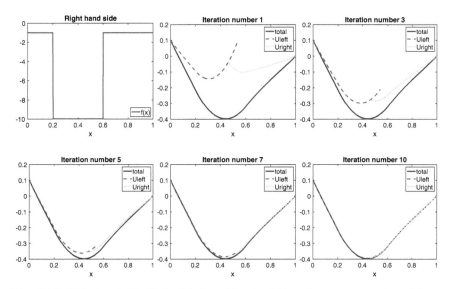

Fig. 10.2 Right-hand side $f(x)$, global, and local solutions for varying number of iterations. $n_0 = 20$, $N = 500$.

10.4 Extension to the 2D Case

We now focus on the 2D problem (10.1) set on a rectangle. We restrict ourselves to the case $c = 0$, thus modeling a steady heat conduction 3D problem in a metallic piece where one dimension is much larger than the two others (see Fig. 10.3). Variations in the temperature will be neglected in this direction. In the first case, we impose on the boundary inhomogeneous Dirichlet conditions:

$$\begin{cases} -\Delta u(x_1, x_2) = F(x_1, x_2), & \text{for} \quad (x_1, x_2) \in \,]a_1, b_1[\times]a_2, b_2[, \\ u(a_1, x_2) = f_2(x_2), & \text{for} \quad x_2 \in \,]a_2, b_2[, \\ u(b_1, x_2) = g_2(x_2), & \text{for} \quad x_2 \in \,]a_2, b_2[, \\ u(x_1, a_2) = f_1(x_1), & \text{for} \quad x_1 \in \,]a_1, b_1[, \\ u(x_1, b_2) = g_1(x_1), & \text{for} \quad x_1 \in \,]a_1, b_1[. \end{cases} \quad (10.4)$$

From a practical point of view this computation models, for instance, a **thermal shock** test on a metallic beam. An experimental setup can consist, for instance, of a null temperature on faces $x_1 = a_1$ and $x_1 = b_1$, that is, $f_2(x_2) = g_2(x_2) = 0$, a temperature of 50°C on the face $x_2 = a_2$, that is, $f_1(x_1) = 50$, and a temperature of 100°C on the face $x_2 = b_2$, that is, $g_1(x_1) = 100$. Furthermore, since no internal heat sources are present, the right-hand side is null: $F(x_1, x_2) = 0$.

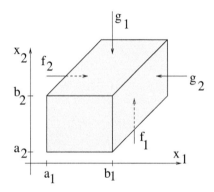

Fig. 10.3 Sketch of the metallic piece subject to a thermal shock.

10.4.1 Finite Difference Solution

We restrict ourselves to the case in which the domain dimensions are such that it is possible to use the same discretization step in both directions x_1 and x_2. We therefore define $h = \frac{b_1 - a_1}{n_1} = \frac{b_2 - a_2}{n_2}$. On this regular grid (see Fig. 10.4), we denote by $u_{i,j} = u(a_1 + ih, a_2 + jh)$ (respectively $f_{i,j}$) the discretized value of the solution $u(x_1, x_2)$ (respectively $F(x_1, x_2)$) at the discretization points. In that case, using Taylor expansions in x_1 and in x_2 in order to approximate partial derivatives in both directions, we obtain the Laplacian discretization with a five-point scheme,

$$\Delta u_{i,j} \approx \frac{4u_{i,j} - u_{i-1,j} - u_{i+1,j} - u_{i,j-1} - u_{i,j+1}}{h^2},$$

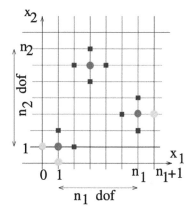

Fig. 10.4 Discretization of the domain interior Ω using $n_1 \times n_2$ points.

which is an $O(h^2)$ approximation if u is smooth enough (i.e., $u \in C^4$). The finite difference solution W is hence a solution of the following linear system:

$$\frac{4u_{i,j} - u_{i-1,j} - u_{i+1,j} - u_{i,j-1} - u_{i,j+1}}{h^2} = f_{i,j}, \qquad (10.5)$$

where the unknowns are the $u_{i,j}$ for $i = 1, \ldots, n_1$ and $j = 1, \ldots, n_2$. In fact, the values of the solution on the boundaries, corresponding to indices $i = 0$ or $i = n_1 + 1$ and $j = 0$ or $j = n_2 + 1$, are set by the boundary conditions

$$u_{0,j} = f_2(a_2 + jh), \quad u_{n+1,j} = g_2(a_2 + jh),$$
$$u_{i,0} = f_1(a_1 + ih), \quad u_{i,n+1} = g_1(a_1 + ih).$$

Each row of the linear system (10.5) has at most five nonzero terms: the diagonal term coefficient is $\frac{4}{h^2}$, and for the off-diagonal terms, corresponding to the neighbors with indices

$$(i+1, j), \quad (i-1, j), \quad (i, j-1), \quad \text{and} \quad (i, j+1),$$

which do not belong to the boundary, the coefficients are equal to $\frac{-1}{h^2}$. These nodes are represented by blue squares in Fig. 10.4.

We can build the matrix using block matrix symbolism: the degrees of freedom (i, j), $(i+1, j)$, and $(i-1, j)$ are neighbors in the grid and also consecutive in the global numbering of the degrees of freedom. The coefficients of the linear system that link nodes belonging to a given row j, for $j = 1, \ldots, n_2$, can therefore be rewritten as a tridiagonal matrix

$$T = \frac{1}{h^2} \begin{pmatrix} 4 & -1 & 0 & \dots & \dots & 0 \\ -1 & 4 & -1 & \ddots & \ddots & \vdots \\ 0 & \ddots & 4 & \ddots & \ddots & \vdots \\ \vdots & \ddots & \ddots & \ddots & \ddots & 0 \\ \vdots & & \ddots & \ddots & 4 & -1 \\ 0 & \dots & \dots & 0 & -1 & 4 \end{pmatrix}. \tag{10.6}$$

The two other neighbors of the node (i, j) in the grid are the nodes $(i, j - 1), (i, j + 1)$, which are n_1 nodes away on each side of the central node in the global numbering, and their connection is ensured through a diagonal matrix $D = \frac{-1}{h^2} I_{n_1 \times n_1}$ on each side of the matrix T. The matrix of the global linear system $AU = B$ is hence a tridiagonal block matrix of size $n_2 \times n_2$, each block being of size $n_1 \times n_1$:

$$A = \begin{pmatrix} T & D & 0 & \dots & 0 \\ D & T & D & \ddots & \vdots \\ 0 & \ddots & \ddots & \ddots & 0 \\ \vdots & \ddots & D & T & D \\ 0 & \dots & 0 & D & T \end{pmatrix}. \tag{10.7}$$

The right-hand side B of the linear system is a vector of size $n_1 \cdot n_2$, which can be built as a matrix to benefit from the numbering of the degrees of freedom associated with the grid. We start by initializing the right-hand-side vector using the right-hand-side function $F(x_1, x_2)$ at the grid nodes:

$$B_{i,j} = f_{i,j}, \quad \text{for} \quad 1 \le i \le n_1, \quad 1 \le j \le n_2. \tag{10.8}$$

If the node (i, j) has a neighbor on the grid boundary (represented by a green circle in Fig. 10.4), the contribution $\frac{u_{i',j'}}{h^2}$ of this neighbor (i', j') in the Laplacian discretization at point (i, j) must be added to the (i, j) right-hand-side coefficient, the value of $u_{i',j'}$ being set by the boundary condition. The boundary condition contributions are therefore added to all terms $B_{1,i}$, $B_{n_1,i}$ for $i = 1, \dots, n_2$ and $B_{i,1}, B_{i,n_2}$ for $i = 1, \dots, n_1$. Beware of the special case occurring at the four corners!

$$\begin{cases} B_{1,i} = B_{1,i} + \dfrac{f_2(a_2 + ih)}{h^2}, \\ B_{n_1,i} = B_{n_1,i} + \dfrac{g_2(a_2 + ih)}{h^2}, \end{cases} \quad \text{for} \quad 1 \le i \le n_1,$$

$$\begin{cases} B_{i,1} = B_{i,1} + \dfrac{f_1(a_1 + ih)}{h^2}, \\ B_{i,n_2} = B_{i,n_2} + \dfrac{g_1(a_1 + ih)}{h^2}, \end{cases} \quad \text{for} \quad 1 \le j \le n_2. \tag{10.9}$$

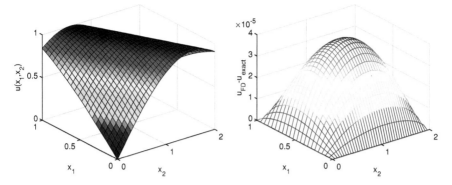

Fig. 10.5 Laplacian problem with Dirichlet boundary conditions. Left: global solution; right: error between exact solution and finite difference approximation.

Exercise 10.3 1. Write a function DDM_F_LaplaceDirichlet to compute the matrix A of size $n_1 n_2 \times n_1 n_2$ of the global linear system. The function computes and returns the matrix A. It receives in its input arguments the lower bounds of the domain, a_1 and a_2; the number of points in each direction, n_1 and n_2; and the discretization step h. The matrix A will be built by blocks using a tridiagonal matrix T and the identity matrix, both of size $n_1 \times n_1$.

2. Write a function DDM_F_RHS2dDirichlet to compute the right-hand-side vector of the linear system. The function returns the vector B. It receives in its input arguments the lower bounds of the domain, a_1 and a_2; the number of points in each direction, n_1 and n_2; and the discretization step h. It calls the function rhs2d(x1,x2), which computes the value of the right-hand-side function $F(x_1, x_2)$.

3. Write a function DDM_F_FD2dDirichlet to compute the finite difference solution. The calling sequence should be

```
function [n2,b2,Solm]=DDM_F_FD2dDirichlet(n1,a1,a2,b1,b2, ...
                       rhs2d, f1,g1,f2,g2)
```

where f1,g1,f2,g2 are the names of the functions defining the boundary conditions.

4. We select a function $u(x_1, x_2) = \sin(x_1 + x_2)$ that is the exact solution of the problem $-\Delta u = f$ with the right-hand-side function $F(x_1, x_2) = 2\sin(x_1 + x_2)$ and with boundary conditions equal to the restrictions of the exact solution on the boundaries:

$$f_1(x_1) = \sin(x_1 + a_2), \quad g_1(x_1) = \sin(x_1 + b_2),$$
$$f_2(x_2) = \sin(a_1 + x_2), \quad g_2(x_2) = \sin(b_2 + x_2).$$

Program the functions DDM_F_rhs2dExact(x1,x2), DDM_F_f1Exact(x1), DDM_F_g1Exact(x1), DDM_F_f2Exact(x2), DDM_F_g2Exact(x2) corresponding to this test case, along with the function DDM_F_U02dExact(x1,x2), which will be used to compute the exact solution at the grid discretization points.

5. Write a program DDM_M_TestFD2d to test the previously defined functions: the size of the domain must be carefully chosen to ensure that the discretization step is the same in both directions. The solution computed with the parameters $a_1 = a_2 = 0$, $b_1 = 1$, $b_2 = 2$, and $n_1 = 20$ is represented in the left panel of Fig. 10.5. Check the computation by displaying the error, that is, the difference between the exact solution $u(x_1, x_2) = \sin(x_1 + x_2)$ and the finite difference solution, as in the right panel of Fig. 10.5.

6. Modify the previous program and adapt it to the thermal shock case (10.4) defined in Sect. 10.4, to obtain the solution displayed in Fig. 10.6. Use first a square domain of size $b_1 - a_1 = b_2 - a_2 = 6$, then a rectangular one of dimensions $b_1 - a_1 = 6$ and $b_2 - a_2 = 20$.

A solution of this exercise is proposed in Sect. 10.5 at page 239.

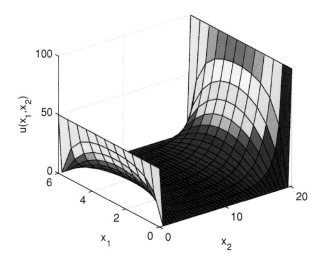

Fig. 10.6 Solution of the thermal shock problem computed with finite differences on the full domain.

10.4.2 Domain Decomposition in the 2D Case

We will now apply the technique described in Sect. 10.3 in the 2D case. The global domain is decomposed into n_s subdomains with an overlap in the direction x_2. Here again, in order to simplify the implementation, we assume that the domain can be discretized with the same step h in both directions. Furthermore, we also impose that all subdomains have the same size $(n_2+1)h$ and that all overlap regions have the same number of grid cells n_o. Figure 10.7 shows an example of a decomposition satisfying these constraints. Note that they are very restrictive and might not be satisfied for an arbitrary domain $[a_1, b_1] \times [a_2, b_2]$. It might be necessary to adjust the total length $b_2 - a_2$. We denote by $u^{s,k}$ the solution on the sth subdomain $[a_1, b_1] \times [a_2^s, b_2^s]$, for

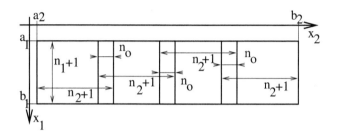

Fig. 10.7 Decomposition into four identical subdomains with constant overlap.

$s = 1, \ldots, n_s$, at iteration number k, for $k = 0, 1, \ldots$. The bounds of the subdomain, a_2^s and b_2^s, are equal to $a_2^1 = a_2$ and $a_2^s = a_2^{s-1} + (n_2 + 1 - n_o)h$ for $k > 1$ and $b_2^s = a_2^s + (n_2 + 1)h$.

With this notation, $u^{s,k}$ is the solution of the problem

$$
\begin{cases}
-\Delta u^{s,k}(x_1, x_2) = F(x_1, x_2), & \text{for } (x_1, x_2) \in (a_1, b_1) \times (a_2^s, b_2^s), \\
u^{k,s}(a_1, x_2) = f_2(x_2), & \text{for } x_2 \in (a_2^s, b_2^s), \\
u^{s,k}(b_1, x_2) = g_2(x_2), & \text{for } x_2 \in (a_2^s, b_2^s), \\
u^{s,k}(x_1, a_2^s) = \begin{cases} f_1(x_1), & \text{for } s = 1, \\ u^{s-1,k}(x_1, a_2^s), & \text{for } s = 2, \ldots, n_s, \end{cases} & \text{for } x_1 \in (a_1, b_1), \\
u^{s,k}(x_1, b_2^s) = \begin{cases} g_1(x_1), & \text{for } s = n_s, \\ u^{s+1,k-1}(x_1, b_2^s), & \text{for } s = 1, \ldots, n_s - 1, \end{cases} & \text{for } x_1 \in (a_1, b_1).
\end{cases}
$$

At the first iteration, the boundary condition on the right end of the subdomain is not defined, except for the last one, where it is the global boundary condition. For all others, it must therefore be arbitrarily fixed, for instance by linear interpolation of the global boundary conditions f_1 and g_1:

$$
u^{s+1,0}(x_1, b_2^s) = ((b_2 - b_2^s)f_1(x_1) + (b_2^s - a_2)g_1(x_1))/(b_2 - a_2).
$$

The iterations stop when the sum (or the maximum) of the error norm on each overlap region is below a given tolerance.

We denote by X_1 the vector of abscissa $a_1 + jh$, for $j = 1, \ldots, n_1$, and by X_2^s the vector of ordinates in the sth subdomain $a_2 + jh + (s-1)(n_2+1-n_o)$, for $j = 1, \ldots, n_2$. The Schwarz algorithm in two dimensions can be written as

Initialization:

$$U_{a_2l} = f_1(X_1), \quad U_{b_2r} = g_1(X_1),$$
$$U^{s,0}_{.,n_o} = ((b_2 - b_2^s)U_{b_2r} + (b_2^s - a_2)U_{a_2l})/(b_2 - a_2), \quad \text{for } s = 2, \ldots, n_s,$$
$$U^s_{a_1} = f_1(X_2^s), \quad U^s_{b_1} = g_1(X_2^s), \quad B^s_{.,.} = F(X_1, X_2^s),$$
$$B^s_{1,.} = B^s_{1,.} + U^s_{a_1}/h^2, \quad B^s_{n_1,.} = B^s_{n_1,.} + U^s_{b_1}/h^2,$$

for $k = 1, 2, \ldots$ do

    for $s = 1, \ldots, n_s$ do

        if $s = 1$, $U^s_{a_2} = U_{a_2l}$ else $U^s_{a_2} = U^{s-1,k}_{.,n_2+1-n_o}$ (10.10)

        if $s = n_s$, $U^s_{b_2} = U_{b_2r}$ else $U^s_{b_2} = U^{s+1,k-1}_{.,n_o}$

        $B^{s,k} = B^s$, $B^{s,k}_{.,1} = B^{s,k}_{.,1} + U^s_{a_2}/h^2$ $B^{s,k}_{.,n_2} = B^{s,k}_{.,n_2} + U^s_{b_2}/h^2$

        solve $AU^{s,k} = B^{s,k}$

        if $s > 1$, $R^s_{.,j} = U^{s,k}_{.,j} - U^{s-1,k}_{.,n_2-n_o+1+j}$,   for $j = 1, \ldots, n_o - 1$

    end

    $E^k = \sup\limits_{s=2,\ldots,n_s} \|R^s\|$

if $E^k < \varepsilon$ end

In this algorithm, A is the matrix resulting from the discretization of the operator $-\Delta$, defined by (10.7). It is here the same matrix for all subdomains.

Exercise 10.4 1. Write a function DDM_F_Schwarz2dDirichlet to implement the above algorithm. The calling syntax should be

```
function [conviter,cpu,mem,n2,b2]=DDM_F_Schwarz2dDirichlet(n1,...
    ns,no,a1,a2,b1,b2,rhs2d,f1,g1,f2,g2,RHS2d,Laplace,n11),
```

where the input parameters are

- n1, the number of cells in direction x_1,
- n11, the number of degrees of freedom in direction x_1,
- ns, the number of subdomains,
- no, the number of cells in the overlap regions in direction x_2,
- Laplace, the name of the function to compute the Laplacian discretization matrix,
- RHS2d, the name of the function to compute the right-hand side,
- rhs2d, the name of the function $F(x_1, x_2)$,
- f1, the name of the function $f1(x_1)$ defining the boundary condition on the edge $x_2 = a_2$,
- g1, the name of the function $g1(x_1)$ defining the boundary condition on the edge $x_2 = b_2$,
- f2, the name of the function $f2(x_2)$ defining the boundary condition on the edge $x_1 = a_1$, and
- g2, the name of the function $g2(x_2)$ defining the boundary condition on the edge $x_1 = b_1$.

The function returns as output arguments

- conviter, the number of iterations necessary to have a maximum error in the overlap regions below the specified tolerance tol,

- cpu, the computing time, and
- mem, the necessary memory.

2. Write a program DDM_M_TestSchwarz2d to test the algorithm with the same function f as in the global case for the following parameter values: $a_1 = a_2 = 0$, $b_1 = 1$, $n_1 = 9$, $b_2 = 30$, $n_o = 10$, and $n_s = 20$.
A solution of this exercise is proposed in Sect. 10.5 at page 241.

Study of the method's performance: we now study the influence of the subdomain size on the convergence speed. We therefore need to estimate, for a given subdomain decomposition, the computing time necessary to achieve the specified accuracy. Computing time is measured within MATLAB using the commands tic and toc at the beginning and end of the script, or part of the script, that is to be monitored. Since it is the elapsed time that is actually measured, this is best done on a single-user computer. Furthermore, in order to be able to compare several configurations, only one parameter should vary. We keep constant the dimensions of the global domain, which imposes constraints on the number of subdomains, their size, and the size of the overlap region.

Exercise 10.5 Fix the parameters $b_1 = 1$, $b_2 = 50$, $n_1 = 9$, and the overlap size $n_o = 4$. We assume that realistic values for the number s of subdomains will run from 5 to 60. For each value of n_s within this range, check whether the decomposition is possible, and, if it is, compute the solution using the function DDM_F_Schwarz2. Display the performance in computing time, memory, and number of iterations, as a function of parameters n_o and n_s. Analyze the influence of the overlap size on the algorithm's convergence.
A solution of this exercise is proposed in Sect. 10.5 at page 241.

10.4.3 Implementation of Realistic Boundary Conditions

More realistic heat conduction test cases require the implementation of additional boundary conditions besides the Dirichlet one that we have used so far. Let us consider, for instance, the temperature field within a bus bar like the one sketched in Fig. 10.8. The electric field produces heat at the uniform rate $F(x_1, x_2) = q = 10^6$ W/m^3. The following temperatures are imposed on the electrodes: 40°C on the left end and 10°C on the right end, thanks to a cooling liquid circulation device. The two lateral faces of the bar as well as the bottom face are insulated, meaning that a Neumann boundary condition has to be imposed. On the upper face, we impose a Fourier, or Robin, boundary condition in order to model the natural convection-driven cooling phenomenon. The thermal transfer coefficient is equal to $c_{th} = 75$ W/m^2 and the outside temperature is 0°C. The thermal diffusivity coefficient of the alloy is equal to $k = 20$ W/m K. Solving for $\tilde{u} = u - u_{ext}$, problem (10.1) becomes in this particular case

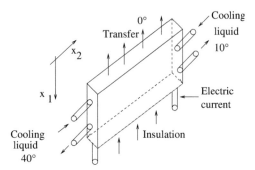

Fig. 10.8 Bus bar sketch.

$$\begin{cases} -k\Delta u(x_1, x_2) = q, & \text{for} \quad x \in \Omega, \\ u = 40, \quad \text{on} \quad x_2 = a_2, \\ u = 10, \quad \text{on} \quad x_2 = b_2, \\ \dfrac{\partial u}{\partial n} = 0, \quad \text{on} \quad x_1 = b_1, \\ \dfrac{\partial u}{\partial n} + c_{\text{th}} u = 0, \quad \text{on} \quad x_1 = a_1. \end{cases} \tag{10.11}$$

Here $\frac{\partial}{\partial n}$ denotes the derivative with respect to the normal vector to the surface. In order to discretize these Neumann and Fourier boundary conditions, we introduce in the system the degrees of freedom corresponding to these nodes on the faces $x_1 = a_1$ and $x_1 = b_1$. There are now $n_1 + 2$ nodes in the x_1 direction instead of the n_1 in the Dirichlet case. For the nodes where the Neumann condition applies, we write

$$u_{n_1+1,j} = u_{n_1+2,j},$$

which eliminates the outside node reference $u_{n_1+2,j}$ in the Laplacian discretization at nodes of indices $(n_1 + 1, j)$, giving eventually

$$\frac{3u_{n_1+1,j} - u_{n_1,j} - u_{n_1+1,j-1} - u_{n_1+1,j+1}}{h^2} = f_{n_1+1,j} = \frac{q}{k},$$

On the other hand, the Fourier condition is discretized as

$$u_{0,j} - u_{-1,j} + hc_{\text{th}} u_{0,j} = 0,$$

which eliminates the reference to nodes $(-1, j)$ in the Laplacian discretization at nodes of indices $(0, j)$, leading eventually to

$$\frac{3u_{0,j} - hc_{\text{th}} u_{0,j} - u_{1,j} - u_{0,j-1} - u_{0,j+1}}{h^2} = f_{0,j}.$$

Exercise 10.6 1. Modify Algorithm (10.10) to take into account the Fourier and Neumann boundary conditions.
2. Implement a function DDM_F_LaplaceFourier that builds the linear system tridiagonal block matrix.
3. Implement functions DDM_F_RHS2dFourier, DDM_F_f1BB, DDM_F_g1BB, and DDM_F_rhs2dBB for the bus bar problem.
4. Modify function DDM_F_FinDif2d and script DDM_M_TestFinDif2d in order to treat this problem with the global finite difference algorithm.
5. Modify function DDM_F_Schwarz2d and script DDM_M_TestSchwarz2d in order to treat this problem with the Schwarz algorithm.

Solutions of this exercise are proposed in Sect. 10.5 at page 240 for the global solution and at page 243 for the domain decomposition solution.

The global treatment provides the solution displayed in Fig. 10.9. The influence of the Fourier condition is clearly seen on the boundary at $x_1 = 0$, where the solution decreases toward the outside temperature value.

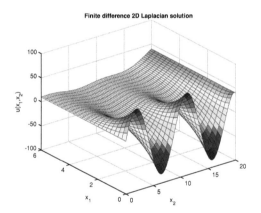

Fig. 10.9 Temperature in the bus bar computed with finite differences on the full domain.

10.4.4 Possible Extensions

The first extension from the point of view of the domain decomposition is of course to adapt the implementation to a decomposition on subdomains of different sizes. The degrees of freedom bookkeeping is then more complicated, and each subdomain requires the computation of a specific matrix. Should these matrices be computed once and for all and stored in memory, or computed again at each iteration? What is the influence of this strategic issue on the computing time?

Another extension of the project can be the domain decomposition in both directions, which will enable the treatment of problems set on domains with complex geometry. The storage of the solution and the connection of the degrees of freedom in the overlapping regions require a rigorous implementation.

10.5 Solutions and Programs

Solution of Exercises 10.1 **and** 10.2

The function DDM_F_FunSchwarz1d implements the Schwarz algorithm in the case of two subdomains of the same size. This constraint greatly simplifies the implementation, since the matrices of the local linear systems have the same dimensions for both subdomains. In the case where the coefficient $c(x)$ is constant, the matrix is exactly the same in both subdomains. The enforcement of the constraint requires a careful translation of the mathematical indices into MATLAB (once again, keep in mind that the indices of an array in MATLAB start from 1). One method consists in setting the number of space steps in the global domain to an even number, and therefore the number of discretization points, including the edges at $x = a$ and b, to an odd number n_x. The space step is denoted by $h = (b - a)/(n_x - 1)$. Then the (even) number of space steps within the overlap is fixed to $2n_o$, with the parameter n_o sent as input to the function. From these data, the position x_l of the left side of the right-hand-side subdomain is computed:

$$x_l = 0.5(a + b) - n_o h,$$

along with the position x_r of the right edge of the left-hand-side subdomain:

$$x_r = 0.5(a + b) + n_o h.$$

Eventually, the number of space steps in each subdomain is equal to

$$i_l = i_r = (n_x + 1)/2 + n_o - 1.$$

Once these parameters are set, the finite difference matrix for the subdomains, of size $i_g - 1 \times i_g - 1$, can be computed. Two right-hand-side vectors are also defined. The influence of the boundary conditions at points x_r and x_l is not included at this stage since they vary at each iteration.

The function DDM_F_FunSchwarz1d uses the function DDM_F_rhs1d to compute the right-hand side.

The output parameters of the function are the number of iterations required to reach the convergence tolerance, and the computing time, estimated using the tic and toc MATLAB commands.

To answer Exercise 10.1, the function is called once with $n_o = 10$ from the script DDM_M_CallSchwarz1d. The parameter detail is set to 1 so that the solution is displayed at each iteration, and the evolution of the error as a function of iterations is also displayed once convergence has been reached.

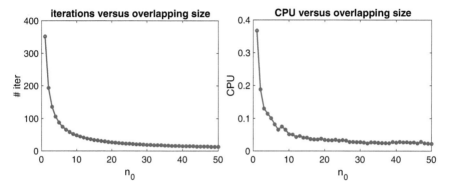

Fig. 10.10 Left: number of iterations. Right: computing time performance of the decomposition versus the size of the overlapping region.

The script DDM_M_PerfSchwarz1d does the performance study required in Exercise 10.2. It calls the function DDM_F_FunSchwarz1d for all values n_o between 1 and $n_x/10$, and stores the corresponding number of iterations and computing time in arrays. A graph of the number of iterations and computing time as a function of the overlap size is displayed in Fig. 10.10. The method converges faster as the overlap increases, and the computing time for each iteration does not vary that much; therefore, the overall computing time decreases as the size of the overlap region increases.

Solution of Exercise 10.3

To implement the 2D problem, it is interesting to preserve the double numbering of the discretization nodes associated with the Cartesian grid for the graphical representation of the solution, the boundary conditions, and the right-hand-side implementation. The global numbering of the degrees of freedom in a column vector can be used only to solve the linear systems. MATLAB can easily convert an $n_1 \times n_2$ array containing the unknowns $u_{i,j}$ into an array u_k with $k = 1, \ldots, n_1 \times n_2$, and conversely, as in the following script:

```
% If size(tab)=[n1,n2]
col=tab(:); % size(col)=[n1*n2,1] and col((j-1)*n1+i)=tab(i,j)
% inversely if  size(col)=[n1*n2,1]
tab=zeros(n1,n2);
tab(:)=col;
```

We first give MATLAB programming solutions for functions: the finite difference matrix for the 2D Laplacian operator in the case of Dirichlet boundary conditions is built by the function DDM_F_LaplaceDirichlet. In the test case proposed in question 5, the exact solution

$$u(x_1, x_2) = \sin(x_1 + x_2)$$

is programmed in file DDM_F_U02dExact.m. The boundary conditions compatible with the exact solution are programmed in files DDM_F_f1Exact.m, DDM_F_g1Exact.m, DDM_F_f2Exact.m, and DDM_F_g2Exact.m. The right-hand-side function

$$f(x_1, x_2) = -\Delta u(x_1, x_2) = 2\sin(x_1 + x_2)$$

is programmed in the function DDM_F_rhs2dExact.m. The right-hand side of the linear system in the case of Dirichlet boundary conditions is assembled by the function DDM_F_RHS2dDirichlet. It uses the right-hand-side function as specified in (10.8) and the boundary conditions as specified in (10.9). The computation of the finite difference solution is performed by the function DDM_F_FD2dDirichlet. It receives in its input arguments the functions DDM_F_f1Exact, DDM_F_g1Exact, DDM_F_f2Exact, DDM_F_g2Exact, and DDM_F_rhs2dExact, whose local names are respectively f1, g1, f2, g2, and rhs2d. It is able to treat other test cases and other boundary conditions. In the present case of inhomogeneous Dirichlet boundary conditions on all the boundaries, the number of degrees of freedom in the x_1 (respectively x_2) direction is equal to n_1 (respectively n_2), that is, the number of inside nodes in this direction.

The test case proposed in question 5 of Exercise 10.3 is treated in the first part of the calling script DDM_M_TestFinDif2d. The finite difference solution is compared with the exact solution by displaying their difference.

To compute the solution of the thermal shock described in Fig. 10.3, for which the exact solution is not known, call the function DDM_F_FD2dDirichlet with the functions DDM_F_rhs2dCT, DDM_F_f1CT, DDM_F_g1CT, and DDM_F_f2CT as input arguments in order to compute the right-hand side. Eventually, the last computation performed in the script DDM_M_TestFinDif2d.m corresponds to the bus bar problem of Exercise 10.6. It is actually done by the function DDM_F_FinDif2dFourier. The Laplacian matrix is computed by DDM_F_LaplaceFourier, where Neumann boundary conditions on the edge $x_1 = a_1$ and Fourier boundary conditions on the edge $x_1 = b_1$ are handled. Different functions DDM_F_RightHandSideFourier and DDM_F_rhs2dBB are used to compute the right-hand side. The inhomogeneous Dirichlet conditions, defined on edges parallel to x_1, are taken into account using functions DDM_F_f1BB and DDM_F_g1BB.

Solution of Exercise 10.4

Algorithm (10.10) corresponding to the Schwarz method in the case of Dirichlet boundary conditions on all four edges is programmed in the function
DDM_F_Schwarz2dDirichlet below and is tested in the script
DDM_M_TestSchwarz2dDirichlet for the two examples treated in the previous exercise. An interesting programming feature is the array RHS indexed by
the subdomain, which contains the right-hand side of the linear system for
the corresponding subdomain. This array is initialized with the contribution
of the right-hand-side function $f(x_1, x_2)$ as well as the Dirichlet boundary
conditions on the global domain edges. The array is used in the subsequent
iterations, in the loop on subdomains, to initialize the right-hand-side vector
Rhsm of the local linear system. The contribution of the Dirichlet boundary
condition on internal edges, which depends on the solution on neighboring
subdomains, is then added before the linear system is solved.
The matrix Lapl of the linear system is the same for all subdomains and does
not depend on the solution; it is therefore computed once and for all outside
the loop on iterations.
    Figure 10.11 displays the solutions after 1 and 4 iterations for the first test
case, where the exact solution is known.

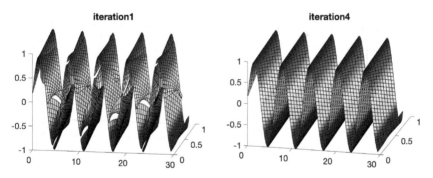

Fig. 10.11 Solutions computed on 20 subdomains, $n_0 = 5$, after 1 iteration (left) and 4
iterations (right).

Solution of Exercise 10.5

To analyze the convergence, we propose the script DDM_M_Perf.m, which also
calls the function DDM_F_Schwarz2dDirichlet. The first test case is considered, with this time a larger domain in the x_2 direction: $b_2 - a_2 = 50$. The
width of the domain remains equal to 1 and is discretized with 11 cells, leading to $n_1 = 10$ degrees of freedom in the x_1 direction. The size of the overlap
region is fixed to 10 cells in the x_2 direction. All "reasonable" values for the
number of subdomains are tried in a loop, from $n_s = 5$ subdomains (which

corresponds to 117 degrees of freedom per subdomain in the x_2 direction) to $n_s = 60$ subdomains (which corresponds to 18 degrees of freedom per subdomain). Since the discretization step must be the same in both directions x_1 and x_2, some configurations are impossible: the step h is fixed by the number of points in the x_1 direction, $n_1 = 10$, that is, $h = (b_1 - a_1)/(n_1 + 1)$. Therefore, the number of internal points in the x_2 direction is also fixed to $n_2^{\text{total}} = (b_2 - a_2)/h$, with the constraint that n_2 must be an integer. Furthermore, the number of points in the x_2 direction in one subdomain, n_2, taking into account the overlap region, must satisfy $n_s(n_2 + 1) = n_2^{\text{total}} - n_o(1 - n_s)$, while being also an integer. Only the decomposition configurations leading to a total length of 50 with a 0.2% tolerance are considered.

The function DDM_F_Schwarz2dDirichlet is called for these allowable configurations, this time with the input argument detailed set equal to 0 to inhibit some of the intermediate graphical outputs. On the other hand, output parameters return the number of iterations, the computing time, the memory size, and the number of points per subdomain n_2. Figure 10.12 shows

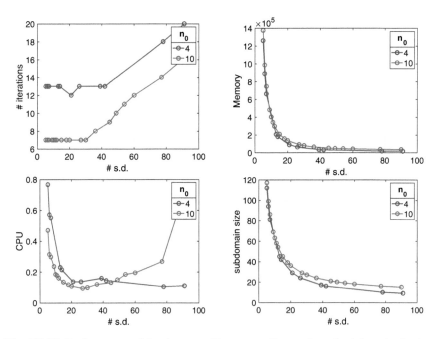

Fig. 10.12 Performances of the decomposition versus the number of subdomains for two different sizes of the overlap.

the results obtained for two different sizes of the overlap regions: $n_o = 4$ and $n_o = 10$ cells in the x_2 direction. A comparison of the computing time curve with the number of iterations curve is particularly interesting. One would expect that the number of iterations should increase with the number of subdomains, but since the computing time necessary for each subdomain decreases along with their size, the evolution of the global computing time is less predictable. The simulations indicate that the optimal number of subdomains might depend on the size of the overlap. For a small overlap $n_0 = 4$, it seems that the CPU and memory decrease monotonously as the number of subdomains increases. For a larger overlap $n_0 = 10$, there is an optimal number of 27 subdomains, where the method is the fastest. It is then also faster than with $n_0 = 4$ but slightly more demanding in memory.

For comparison, the direct computation of the global solution on 9×499 degrees of freedom would require 1.45 seconds of computing time, which is more than the decomposition method requirement in the worst configuration case. The memory necessary to store the matrix of a global linear system on the order of 2.10^7 is also prohibitive.

Solution of Exercise 10.6

We now denote by X_1 the vector containing the $n_1 + 2$ abscissa $a_1 + jh$, $j = 0, \ldots, n_1+1$, including a_1 and b_1. The vectors X_2^i of Algorithm (10.10) are unchanged. The Laplacian matrix is modified in order to take into account the Neumann and Fourier boundary conditions, using the block representation (10.7). The T and D matrices are now of dimension $(n_1 + 2) \times (n_1 + 2)$, and the matrix T is different from the previous one (10.6) in the first and last rows

$$T = \frac{1}{h^2} \begin{pmatrix} 3 - hc_{\text{th}} & -1 & 0 & \cdots & \cdots & 0 \\ -1 & 4 & -1 & \ddots & \ddots & \vdots \\ 0 & \ddots & 4 & \ddots & \ddots & \vdots \\ \vdots & & \ddots & \ddots & \ddots & \vdots \\ \vdots & & & \ddots & 4 & -1 \\ 0 & \cdots & \cdots & 0 & -1 & 3 \end{pmatrix}.$$

Since the outside temperature is equal to $0°C$, there is no contribution of the Fourier boundary condition on the right-hand side. The Schwarz algorithm is now as follows:

initialization :
$$U_{a_2 l} = f_2(X_1), \quad U_{b_2 r} = g_2(X_1)$$
$$U^{i,0}_{.,n_o} = (iU_{b_2 r} + (n_s - s)U_{a_2 l})/n_s, \quad \text{for} \quad s = 2, \ldots, n_s$$
$$B^s_{.,.} = F(X_1, X^s_2), \quad \text{for} \quad s = 1, \ldots, n_s$$
  for $k = 1, 2, \ldots,$ do
      for $s = 1, \ldots, n_s$ do
          if $s = 1,$ $U^s_{a_2} = U_{a_2 l}$ else $U^s_{a_2} = U^{s-1,k}_{.,n_2+1-n_o}$
          if $s = n_s,$ $U^s_{b_2} = U_{b_2 r}$ else $U^s_{b_2} = U^{s+1,k-1}_{.,n_o}$ (10.12)
          $B^{s,k} = B^s, \quad B^{s,k}_{.,1} = B^{s,k}_{.,1} + U^s_{a_2}/h^2 \quad B^{s,k}_{.,n_2} = B^{s,k}_{.,n_2} + U^s_{b_2}/h^2$
          solve $AU^{s,k} = B^{s,k}$
          if $s > 1,$ $R^s_{.,j} = U^{s,k}_{.,j} - U^{s-1,k}_{.,n_2-n_o+1+j}, \quad \text{for } j = 1, \ldots, n_o - 1$
      end of s
      $E^k = \sup_{s=2,\ldots,n_s} \|R^s\|$
  if $E^k < \varepsilon$ end of k

This algorithm is programmed in the function DDM_F_Schwarz2dFourier and tested in the script DDM_M_BusSchwarz2d.m.

Chapter References

This project provided the general idea of domain decomposition, which is to decompose a global problem into suitable subproblems of smaller complexity. In practice, however, domain decomposition methods are mostly implemented for complex shapes discretized with finite element methods. Nonoverlapping algorithms as in Quarteroni and Valli (1999) are then preferable, and a large number of such techniques use the so-called *mortar* formulation (see Wohlmuth, 2001).

P.L. Lions, On the alternating Schwarz method I, in *First International Symposium on Domain Decomposition Methods for Partial Differential Equations*, ed. by R. Gowinski, G.H. Golub, G.A. Meurant, J. Périaux (SIAM, Philadelphia, 1988), pp. 1–42

H. Le Dret, B. Lucquin, *Partial Differential Equations: Modeling, Analysis and Numerical Approximation* (Birkhäuser/Springer, Cham, 2016)

B. Lucquin, O. Pironneau, *Introduction to Scientific Computing* (Willey, Chichester, 1998)

A. Quarteroni, A. Valli, *Domain Decomposition Methods for Partial Differential Equations*, Numerical Mathematics and Scientific Computation (The Clarendon Press, Oxford University Press, New York, 1999)

H.A. Schwarz, *Gesammelte Mathematische Abhandlungen*, vol. 2 (Springer, Berlin, 1890). First published in Vierteljahrsschrift Naturforsch. Ges. (Zurich, 1870)

B.F. Smith, P.E. Bjørstad, W.D. Gropp, *Domain Decomposition, Parallel Multilevel Methods for Elliptic Partial Differential Equations* (Cambridge University Press, Cambridge, 1996)

B.I. Wohlmuth, *Discretization Methods and Iterative Solvers Based on Domain Decomposition*, Lecture Notes in Computational Science and Engineering, vol. 17 (Springer, Berlin, 2001)

Chapter 11
Geometrical Design: Bézier Curves and Surfaces

Project Summary

Level of difficulty:	2
Keywords:	Bézier curves, Bézier surfaces
Application fields:	Computer-aided geometric design, geometric modeling, computational graphics

11.1 Introduction

Many fields in the computational science area need descriptions of complex objects: virtual reality, computational graphics, geometric modeling, computer-aided geometric design (CAGD). These descriptions are commonly obtained using basic elements: points, curves, surfaces, and volumes. Elementary tools used to handle these elements are mathematical functions such as polynomials and rational functions, which allow easy graphical representation in many situations: union of objects, intersection, complement.

The very first studies in geometrical design go back to the sixties and were related to industrial projects. For example, J. Ferguson (Boeing) and S. Coons (Ford) in the United States, P. de Casteljau (Citroën) and P. Bézier (Renault) in France were pioneers in the discipline. This chapter gives an introduction to geometrical design by studying some properties of the so-called Bézier curves and surfaces. More details can be found in Bézier (1986), de Casteljau (1985), Farin and Hansford (2000), Farin (1996) and Fergusson (1997), for example.

© The Author(s), under exclusive license to Springer Nature Switzerland AG 2023   247
I. Danaila et al., *An Introduction to Scientific Computing*,
https://doi.org/10.1007/978-3-031-35032-0_11

11.2 Bézier Curves

Let $n \geq 2$ and $m > 0$ be integers and $t \in [0, 1]$ a parameter; consider $m + 1$ points in $\mathbb{R}^n$: $P_0, P_1, \ldots, P_m$ (distinct or not) and define the point $P(t)$ by

$$P(t) = \sum_{k=0}^{m} \binom{m}{k} t^k (1-t)^{m-k} P_k, \tag{11.1}$$

where $\binom{m}{k} = \frac{m!}{k!(m-k)!}$ is the binomial coefficient. The *Bézier curve* $\mathcal{B}_m$ with control points $P_0, P_1, \ldots, P_m$ is the trajectory described by $P(t)$ as t goes from 0 to 1. The polynomials $B_m^k(t) = \binom{m}{k} t^k (1-t)^{m-k}$ are the *Bernstein polynomials* of degree m, with the following properties:

$$\begin{cases} \forall t \in [0, 1],\, 0 < B_m^k(t) < 1,\, \sum_{k=0}^{m} B_m^k(t) = 1, \\[2mm] B_m^k(0) = 0,\text{ for } 0 < k \leq m,\; B_m^0(0) = 1, \\[2mm] B_m^k(1) = 0,\text{ for } 0 \leq k < m,\; B_m^m(1) = 1. \end{cases} \tag{11.2}$$

It follows from (11.1) and (11.2) that $P(0) = P_0$ and $P(1) = P_m$. Generally, P_0 and P_m are the only control points on the curve $\mathcal{B}_m$. Definition (11.1) allows one to represent, exactly and in a condensed form, a great diversity of curves in $\mathbb{R}^n$. Figure 11.1 (left) displays an example of a Bézier curve defined in $\mathbb{R}^2$ with five control points. Note that the order in which the control points are considered in (11.1) will dictate the shape of the curve: for example, in Figs. 11.1 (left) and (right) the same control points are used, but P_0 and P_4 have been interchanged. More generally, attempting to change any control point will result in the entire curve being modified.

Since Bernstein polynomials are linearly independent functions, two Bézier curves coincide for the same value of m when they share the same control points. Nevertheless, it is important to note that the same Bézier curve admits different representations of type (11.1), corresponding to different values of m. For example, consider two points P_0 and P_1 and define Q_1 to be the midpoint of $P_0 P_1$; the line segment $P_0 P_1$ is defined by either

$$P(t) = \sum_{k=0}^{1} \binom{1}{k} t^k (1-t)^{1-k} P_k \qquad (m = 1),$$

or

$$Q(t) = \sum_{k=0}^{2} \binom{2}{k} t^k (1-t)^{2-k} Q_k \qquad (m = 2),$$

with $t \in [0,1]$, $Q_0 = P_0$, and $Q_2 = P_1$. If we introduce the new control points $R_0 = P_0$, $R_1 = (2P_0 + P_1)/3$, $R_2 = (P_0 + 2P_1)/3$, and $R_3 = P_1$, the Bézier curve corresponding to the definition

$$R(t) = \sum_{k=0}^{3} \binom{3}{k} t^k (1-t)^{3-k} R_k \qquad (m=3),$$

with $t \in [0,1]$ is still the line segment $P_0 P_1$ (check that $P(t) = Q(t) = R(t)$).

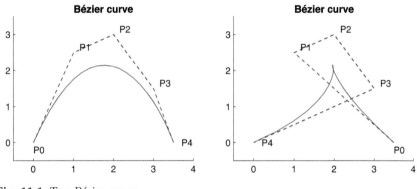

Fig. 11.1 Two Bézier curves.

Remark 11.1 The control point P_k, for any $1 < k < m$, does not generally belong to the Bézier curve; it is, however, possible to introduce another type of Bézier curve, called a *Bézier interpolation curve*, which contains all its control points. Both kinds of curves belong to the family of *spline curves*. Spline curves are generated using the general definition (11.3), in which functions f_k are polynomials of degree m:

$$P(t) = \sum_{k=0}^{m} f_k(t) P_k. \tag{11.3}$$

Many other curves (B-splines, NURBS) are defined by way of such a formula (Coons (1974), Hoschek and Lasser (1997), or Piegl and Tiller (1995)). In (11.3) the blending functions f_k may be polynomials (of degree $p \neq m$), rational functions, etc. All the corresponding curves are entirely defined by setting the control points and choosing the associated functions. Note that a curve may be also defined "piecewise", as the union of distinct curves sharing the same endpoints. In this chapter, we shall limit our study to Bézier curves defined by formula (11.1).

11.3 Basic Properties of Bézier Curves

In this section we study some properties of Bézier curves, which are relevant for practical applications.

11.3.1 Convex Hull of the Control Points

According to (11.1), the point $P(t)$ is defined as the barycenter of the $m + 1$ control points P_k, with corresponding weights $B_m^k(t)$. If follows from the first relationship in (11.2) that $P(t)$ belongs to the convex hull of the control points. We may see in Fig. 11.2 (left) that a Bézier curve lies entirely within the convex hull of the control points. Note that this convex hull contains the polygon $P_0 P_1 \ldots P_m P_0$, which is commonly referred as the *control polygon*. From a more general point of view, it is worth noting that in many situations the control polygon is not convex (as can be seen from Fig. 11.2 (right)).

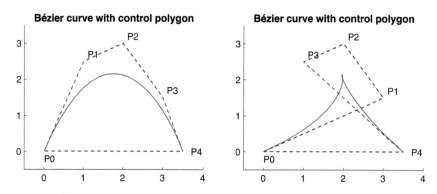

Fig. 11.2 Control polygon. Left: convex; right: nonconvex.

11.3.2 Multiple Control Points

When defining a Bézier curve, it is not necessary to use distinct control points. This allows us to create more or less complicated shapes, closed or open curves, as displayed in Figs. 11.3 and 11.4.

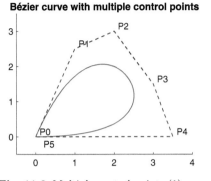

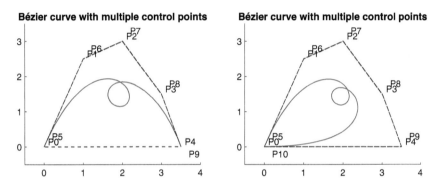

Fig. 11.3 Multiple control points (1).

Fig. 11.4 Multiple control points (2).

11.3.3 Tangent Vector to a Bézier Curve

Let $P(t)$ be the point of the Bézier curve $\mathcal{B}_m$ corresponding to the value t of the parameter. The tangent vector to $\mathcal{B}_m$ at $P(t)$ is defined by

$$\boldsymbol{\tau}(t) = \frac{d}{dt}P(t) = \sum_{k=0}^{m} \frac{d}{dt}B_m^k(t)P_k. \tag{11.4}$$

It follows from definition (11.2) that

$$\frac{d}{dt}B_m^k(t) = \begin{cases} -m(1-t)^{m-1}, & \text{if } k = 0, \\[2mm] m(1-mt)(1-t)^{m-2}, & \text{if } k = 1, \\[2mm] \binom{m}{k}(k-mt)t^{k-1}(1-t)^{m-k-1}, & \text{if } 1 < k < m-1, \\[2mm] mt^{m-2}(m-1-mt), & \text{if } k = m-1, \\[2mm] mt^{m-1}, & \text{if } k = m. \end{cases}$$

Consequently, when $P_0 \neq P_1$, the tangent vector at P_0 is $\boldsymbol{\tau}(0) = m\overrightarrow{P_0 P_1}$. The Bézier curve $\mathcal{B}_m$ is tangent at P_0 to the edge $P_0 P_1$ of the control polygon. Similarly, if $P_{m-1} \neq P_m$, then $\boldsymbol{\tau}(1) = m\overrightarrow{P_{m-1} P_m}$ and the curve is tangent at P_m to the edge $P_{m-1} P_m$. This property is illustrated in the previous figures.

Remark 11.2 More generally, it can be proved that

$$\frac{d}{dt} B_m^k(t) = m(B_{m-1}^{k-1}(t) - B_{m-1}^k(t)).$$

This formula may be useful to compute the tangent vector $\boldsymbol{\tau}(t)$.

11.3.4 Junction of Bézier Curves

We address now the problem of linking two Bézier curves. Consider the Bézier curve defined with $m + 1$ control points P_0, P_1, ..., P_m and another curve defined with $m' + 1$ control points P_0', P_1', ..., $P_{m'}'$. We are interested in studying how these two curves connect, and more particularly how this junction looks on a display. This is an important problem in CAGD, where maximum quality in the rendering of pictures is expected.

In order to get a $\mathcal{C}^0$ connection (that is, a continuous junction of the two curves) we have to impose a basic condition: $P_m = P_0'$. Then, since the first curve $\mathcal{B}_m$ is tangent to the line segment $P_{m-1} P_m$ at P_m and the second curve $\mathcal{B}_{m'}'$ is tangent to $P_0' P_1'$ at P_0', a tangential connection of the curves is obtained if and only if the three points P_{m-1}, P_m (or P_0'), and P_1' lie on a straight line. This condition, which is called the $\mathcal{G}^1$ continuity condition, is generally sufficient to get a satisfactory layout. Nevertheless, for a better rendering, it is natural to ask for more, namely a $\mathcal{C}^1$ continuity condition. This will be satisfied if the tangent vector $\boldsymbol{\tau}$ passes continuously from the first curve to the second one. We know that the tangent vector to $\mathcal{B}_m$ at P_m is $m\overrightarrow{P_{m-1} P_m}$, while for $\mathcal{B}'_m$ the tangent vector at P_0' is $m\overrightarrow{P_0' P_1'}$. The $\mathcal{C}^1$ continuity condition is then satisfied when $\overrightarrow{P_{m-1} P_m} = \overrightarrow{P_0' P_1'}$. This is equivalent to saying that $P_m = P_0'$ is the midpoint of the line segment $P_{m-1} P_1'$.

Figure 11.5 shows an example of a $\mathcal{G}^1$ junction (left), together with an example of a $\mathcal{C}^1$ junction (right). The actual difference is not visible here, but it is clear from Fig. 11.5 (right) that P_4 (point $P_4 \equiv P_0'$) is the midpoint of the line segment $P_3 P_1'$, while this is not true in Fig. 11.5 (left). Distinguishing between these two kinds of junction is important when one has to handle evenly spaced points on the curve $\mathcal{B} = \mathcal{B}_m \cup \mathcal{B}_m'$.

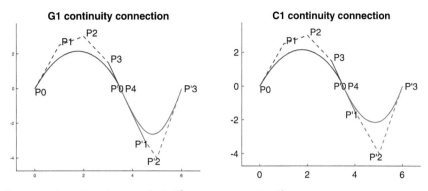

Fig. 11.5 Junction of curves. Left $\mathcal{G}^1$ continuity; right: $\mathcal{C}^1$ continuity.

11.3.5 Generation of the Point $P(t)$

Although the point $P(t)$ is exactly defined from (11.1), the effective construction of $P(t)$ for a given value of the parameter $t \in [0, 1]$ in this way is time-consuming. Furthermore, since the calculation of high degree polynomials values is not an accurate process, the point resulting from (11.1) will be generally different from the actual $P(t)$. Fortunately, there is another way, cheap and accurate, to obtain $P(t)$ using the recurrence between Bernstein polynomials:

$$B_{m+1}^k(t) = tB_m^{k-1}(t) + (1 - t)B_m^k(t). \qquad (11.5)$$

This result is proved by writing

$$t\,B_m^{k-1}(t) + (1 - t)B_m^k(t) = t\binom{m}{k-1}t^{k-1}(1-t)^{m-k+1} + (1-t)\binom{m}{k}t^k(1-t)^{m-k}$$

$$= (\binom{m}{k-1} + \binom{m}{k})t^k(1-t)^{m-k+1}$$

$$= \binom{m+1}{k}t^k(1-t)^{m+1-k} = B_{m+1}^k(t).$$

This property is useful for displaying Bézier curves: for a chosen value of $t \in [0, 1]$, the points P_q^p, for $p = 0, 1, \ldots, m$ and $q = p, p + 1, \ldots, m$, are successively defined by

$$
\begin{aligned}
&\text{initialization: for } p = 0, \text{ do} \\
&\quad \text{for } q = 0, 1, \ldots, m, \text{ do} \\
&\qquad P_q^0 = P_q \\
&\quad \text{end do} \\
&\text{end do} \\
&\text{construction: for } p = 1, 2, \ldots, m, \text{ do} \\
&\quad \text{for } q = p, p+1, \ldots, m, \text{ do} \\
&\qquad P_q^p = t P_{q-1}^{p-1} + (1-t) P_q^{p-1}. \\
&\quad \text{end do} \\
&\text{end do}
\end{aligned} \tag{11.6}
$$

We shall prove now that $P_m^m = P(t)$. We first note that in (11.6), at step p, any of the $m - p$ points P_q^p is defined as the barycenter of P_{q-1}^{p-1} and P_q^{p-1}, which are the two points obtained in the previous step. Using mathematical induction on p, we prove that P_q^p satisfies, for $p = 0, 1, \ldots, m$ and $q = p, p+1, \ldots, m$, the relationship

$$
P_q^p = \sum_{j=0}^{m} B_p^j P_j.
$$

This is trivial when $p = 0$ because of the definition of the points P_q^0 ($q = 0, 1, \ldots, m$). We then assume the property to be satisfied to rank $p - 1$ (included) and prove it for rank p. For $q = p, p+1, \ldots, m$, we write

$$
P_q^p = t P_{q-1}^{p-1} + (1-t) P_q^{p-1} = \sum_{j=0}^{m} (t B_{q-1}^j + (1-t) B_q^j) P_j = \sum_{j=0}^{m} B_q^j P_j.
$$

The relation is also satisfied for rank p, and then for any value of $p \le m$. When $p = m$, this relation leads to

$$
P_m^m = \sum_{j=0}^{m} B_m^j P_j = P(t).
$$

It follows that any point $P(t)$ of the Bézier curve $\mathcal{B}_m$ with control points $P_0, P_1, \ldots, P_m$ can be built by means of algorithm (11.6), which is called *de Casteljau's algorithm*. The computational cost to generate $P(t)$ in this way is equivalent to performing $m + \cdots + 1 = m(m+1)/2$ linear combinations; this is much cheaper (and more accurate) than the use of formula (11.1). The construction process is displayed in Fig. 11.6.

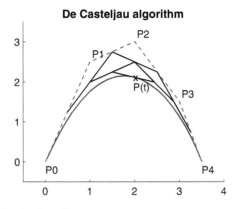

Fig. 11.6 De Casteljau's algorithm.

11.4 Generation of Bézier Curves

It is now time to deal with a few examples. We shall see how easy it is to construct Bézier curves using de Casteljau's algorithm.

Exercise 11.1 1. Choose $m + 1$ points P_0, P_1, ..., P_m in $\mathbb{R}^2$.
2. Write a general procedure generating a point $P(t)$ of the Bézier curve with control points P_0, P_1, ..., P_m, using de Casteljau's algorithm (11.6), for any value $t \in [0, 1]$.
3. Compute and display the corresponding Bézier curve.
4. Repeat the experiment for different values of m and different sets of control points.
5. Check the different continuity conditions.

A solution of this exercise is proposed in Sect. 11.10 at page 265.

Remark 11.3 One may want to construct the curve displayed in Fig. 11.1 (left). In this particular case, $m = 4$ and the control points are $P_0 = (0, 0)$, $P_1 = (1, 2.5)$, $P_2 = (2, 3)$, $P_3 = (3, 1.5)$, $P_4 = (3.5, 0)$.

11.5 Splitting Bézier Curves

Let $\mathcal{B}_m$ be the Bézier curve defined by $m+1$ control points $P_0, P_1, \ldots, P_m$. Let θ be a given value in $[0, 1]$. The point $P(\theta)$ of $\mathcal{B}_m$ is associated with θ by the algorithm (11.6). We successively construct the points P_q^p, for $p = 0, 1, \ldots, m$ and $q = p, p + 1, \ldots, m$, and finally set $P(\theta) = P_m^m$. Consider now the Bézier curve $\tilde{\mathcal{B}}_m$ defined by the $m+1$ control points $P_0^0, P_1^1, \ldots, P_m^m$. The point $\tilde{P}(\tilde{t})$ on this curve is defined by

$$\tilde{P}(\tilde{t}) = \sum_{k=0}^{m} \binom{m}{k} \tilde{t}^k (1-\tilde{t})^{m-k} P_k^k, \tag{11.7}$$

where the parameter $\tilde{t}$ belongs to the interval $[0,1]$. Note that $\tilde{P}(0) = P_0$ and $P(\theta) = P_m^m$. We are going to prove now that the Bézier curve $\tilde{\mathcal{B}}_m$, with ending points P_0 and $P(\theta)$, is the part of the curve $\mathcal{B}_m$ obtained when the parameter t covers the interval $[0, \theta]$. We first check that the points P_q^p generated by the algorithm (11.6) satisfy, for $1 \le p \le m$ and $p \le q \le m$,

$$P_q^p = \sum_{k=0}^{p} \binom{p}{k} \theta^k (1-\theta)^{p-k} P_{q-k}. \tag{11.8}$$

This result is obtained by mathematical induction on p. Using definition (11.6) we obtain, when $p = 1$ and $1 \le q \le m$,

$$P_q^1 = \theta P_{q-1}^0 + (1-\theta) P_q^0 = \sum_{k=0}^{1} \binom{1}{k} \theta^k (1-\theta)^{1-k} P_{q-k}.$$

We assume that the property is true up to the value $p-1$. Then we write, for $p \le q \le m$,

$$P_q^p = \theta P_{q-1}^{p-1} + (1-\theta) P_q^{p-1}$$

$$= \sum_{k=0}^{p-1} \binom{p-1}{k} \theta^{k+1} (1-\theta)^{p-1-k} P_{q-1-k} + \sum_{k=0}^{p-1} \binom{p-1}{k} \theta^k (1-\theta)^{p-k} P_{q-k}$$

$$= \sum_{k=0}^{p-2} \binom{p-1}{p-1-k} \theta^{k+1} (1-\theta)^{p-1-k} P_{q-1-k} + \theta^p P_{q-p}$$

$$+ (1-\theta)^p P_q + \sum_{k=1}^{p-1} \binom{p-1}{k} \theta^k (1-\theta)^{p-k} P_{q-k}.$$

We obtain

$$P_q^p = \sum_{k=1}^{p-1} \binom{p-1}{p-k} \theta^k (1-\theta)^{p-k} P_{q-k} + \theta^p P_{q-p}$$

$$+ (1-\theta)^p P_q + \sum_{k=1}^{p-1} \binom{p-1}{k} \theta^k (1-\theta)^{p-k} P_{q-k}$$

$$= \theta^p P_{q-p} + \sum_{k=1}^{p-1} \left(\binom{p-1}{p-1-(k-1)} + \binom{p-1}{k} \right) \theta^k (1-\theta)^{p-k} P_{q-k} + (1-\theta)^p P_q.$$

Finally,

$$P_q^p = \sum_{k=0}^{p} \binom{p}{k} \theta^k (1-\theta)^{p-k} P_{q-k}.$$

The point $\tilde{P}(\tilde{t})$ of $\tilde{\mathcal{B}}_m$, defined by the $m+1$ control points $P_0^0, P_1^1, \ldots, P_m^m$, satisfies

$$\tilde{P}(\tilde{t}) = \sum_{k=0}^{m} \binom{m}{k} \tilde{t}^k (1-\tilde{t})^{m-k} P_k^k$$

$$= \sum_{k=0}^{m} \binom{m}{k} \tilde{t}^k (1-\tilde{t})^{m-k} \sum_{l=0}^{k} \binom{k}{l} \theta^l (1-\theta)^{k-l} P_{k-l}. \tag{11.9}$$

Let p an integer ($0 \le p \le m$). By gathering in this sum all the terms related to P_p, we obtain

$$\tilde{P}(\tilde{t}) = \sum_{p=0}^{m} \sum_{l=0}^{m-p} \binom{m}{p+l} \binom{p+l}{p} \theta^{p+l} (1-\theta)^{m-p-l} \tilde{t}^p (1-\tilde{t})^l P_p.$$

Then we recall that $\binom{m}{p+l}\binom{p+l}{p} = \binom{m}{p}\binom{m-p}{l}$, and note that

$$\theta^{p+l}(1-\theta)^{m-p-l}\tilde{t}^p(1-\tilde{t})^l = (\theta\tilde{t})^p(\theta-\theta\tilde{t})^l(1-\theta)^{m-p-l}.$$

The formula (11.9) is then written as

$$\tilde{P}(\tilde{t}) = \sum_{p=0}^{m} \binom{m}{p} (\theta\tilde{t})^p \sum_{l=0}^{m-p} \binom{m-p}{l} (\theta-\theta\tilde{t})^l (1-\theta)^{m-p-l} P_p$$

$$= \sum_{p=0}^{m} \binom{m}{p} (\theta\tilde{t})^p (1-\theta\tilde{t})^{m-p} P_p. \tag{11.10}$$

For any given value of θ in $[0,1]$, the product $\theta\tilde{t}$ lies in $[0,\theta]$ when the parameter $\tilde{t}$ covers $[0,1]$. When $\tilde{t} \in [0,1]$ the point $\tilde{P}(\tilde{t})$ defined by (11.9) or (11.10) covers the Bézier curve with ending points P_0 and $P(\theta)$.

How do we get the complementary part of the curve? By a reverse ordering of the control points and by changing θ into $1 - \theta$. Actually, the point $P(\theta)$ of the Bézier curve with control points $P_0, P_1 \ldots, P_m$ is defined according to

$$P(\theta) = \sum_{k=0}^{m} \binom{m}{k} \theta^k (1-\theta)^{m-k} P_k = \sum_{k=0}^{m} \binom{m}{k} (1-\theta)^k \theta^{m-k} P_{m-k}.$$

Then $P(\theta)$ is the same point as the point $Q(1-\theta)$ of the Bézier curve with control points $P_m, P_{m-1}, \ldots, P_0$. In order to obtain the part of the curve with ending points $P(\theta)$ and P_m, we first generate the points $Q_0^0, Q_1^1, \ldots, Q_m^m$ associated with the Bézier curve with control points $P_m, P_{m-1}, \ldots, P_0$. Then we set

$$\tilde{Q}(t) = \sum_{k=0}^{m} \binom{m}{k} t^k (1-t)^{m-k} Q_k^k.$$

Remark 11.4 This result may be generalized to B-spline curves for which more properties can be established in relation to basic operations such as moving, removing and inserting a control point.

11.6 Intersection of Bézier Curves

We address now the problem of finding the intersection of two Bézier curves $\mathcal{B}_m$ and $\mathcal{B}'_{m'}$ defined in $\mathbb{R}^2$ by their control points. We describe the two curves by their generic points

$$P(t) = \sum_{k=0}^{m} \binom{m}{k} t^k (1-t)^{m-k} P_k,$$

$$P'(t') = \sum_{k'=0}^{m'} \binom{m'}{k'} (t')^{k'} (1-t')^{m'-k'} P'_{k'}.$$

(11.11)

Now, is it possible to find two values t and t' such that $P(t) = P'(t')$? How do we compute them when they exist? According to the theory of algebraic geometry, one can deduce implicit representations $f(x, y)$ and $f'(x, y)$ of both curves $\mathcal{B}_m$ and $\mathcal{B}'_{m'}$. But within the corresponding formulation, searching for a possible common point is equivalent to finding the roots of a polynomial of degree $m + m'$. This is a too complicated and time-consuming way to get the solution. We propose here a method based on particular properties of Bézier curves. We proceed as follows. Since any Bézier curve $\mathcal{B}_m$ is entirely contained in the convex hull $\mathcal{E}_m$ of its control points, we know that the intersection set $\mathcal{B}_m \cap \mathcal{B}'_{m'}$ is empty when the convex hulls $\mathcal{E}_m$ and $\mathcal{E}'_{m'}$ do not intersect. Conversely, both curves $\mathcal{B}_m$ and $\mathcal{B}'_{m'}$ may intersect when the convex hulls intersect. In order to obtain a more accurate view of the problem in this case, we can split both curves into two parts and check the intersection of the corresponding convex hulls. Splitting a Bézier curve $\mathcal{B}_m$ into two subcurves $\mathcal{B}_m^1$ and $\mathcal{B}_m^2$ with their associated control points sends us back to the previous section. We denote by $\mathcal{B}_m^1$ the curve corresponding to the part of the curve $\mathcal{B}_m$ obtained for $t \in [0, 0.5]$, while $\mathcal{B}_m^2$ corresponds to the part of the curve $\mathcal{B}_m$ obtained for $t \in [0.5, 1]$. The control points of both curves $\mathcal{B}_m^1$ and $\mathcal{B}_m^2$ are defined by (11.8). We proceed then by successive iterations as long as there exist two intersecting convex hulls. The corresponding algorithm is the following:

$$
\begin{Vmatrix}
\text{initialization :} \\
\quad \mathcal{E}_1 = \mathcal{E}_m, \; \mathcal{E}'_1 = \mathcal{E}'_{m'} \\
\text{iterations: while } \mathcal{E}_k \cap \mathcal{E}'_k \neq \emptyset \text{ , do} \\
\quad \text{split: } \mathcal{B}_k = \mathcal{B}_{k_1} \cup \mathcal{B}_{k_2} \\
\quad \text{associate: } \mathcal{E}_{k_1}, \; \mathcal{E}_{k_2} \\
\quad \text{split: } \mathcal{B}'_{k'} = \mathcal{B}'_{k'_1} \cup \mathcal{B}'_{k'_2} \\
\quad \text{associate: } \mathcal{E}'_{k'_1}, \; \mathcal{E}'_{k'_2} \\
\quad \text{check: } \mathcal{E}_{k_i} \cap \mathcal{E}'_{k'_j} \qquad (*) \\
\text{end do}
\end{Vmatrix}
\qquad (11.12)
$$

The intersection of two bounded convex sets is a bounded convex set, so $\mathcal{E}_{k_i} \cap \mathcal{E}'_{k'_j}$ is convex and contained in both $\mathcal{E}_{k_i}$ and $\mathcal{E}'_{k'_j}$; thus its "size" is decreasing as the algorithm (11.12) evolves. This algorithm converges in the following way: Either all intersections are empty and then curves $\mathcal{B}_m$ and $\mathcal{B}'_{m'}$ do not intersect, or there exists at least one intersection whose size is vanishing as (11.12) proceeds; then the convex intersection is shrinking to a point common to both curves. Note that algorithm (11.12) is able to cope with multiple intersections (Fig. 11.7).

In order to check automatically whether sets $\mathcal{E}_k$ and $\mathcal{E}'_k$ intersect, we need to compute the convex hull of $m+1$ given points in $\mathbb{R}^2$. Since $\mathcal{E}_k$ is bounded by a convex polygon $\mathcal{P}_k$, its computer representation is a *mesh* of $\mathcal{P}_k$. This mesh $\mathcal{M}_k$ may be any set of triangles whose union is equal to $\mathcal{P}_k$. Their vertices are the control points and they satisfy the classical rule that the intersection of two distinct triangles is either empty or reduced to a common vertex or a common edge. We can then check whether $\mathcal{E}_k$ and $\mathcal{E}'_k$ intersect by testing the intersection of all pairs of triangles $(T_{k,i}, T'_{k,j}) \in \mathcal{M}_k \times \mathcal{M}'_k$.

Although this method is correct, it is too tedious for our purpose. For the sake of simplicity, we proceed as follows: Each convex hull $\mathcal{E}_k$ is embedded in a rectangle $R_k = [x^k_m, x^k_M] \times [y^k_m, y^k_M]$ defined by the extreme values of the coordinates of the control points. Then we replace in (11.12) the line quoted by $(*)$ by *check* $R_{k_i} \cap R_{k'_j}$. Since $\mathcal{E}_k \subset R_k$, this may slow down the convergence of (11.12), but the modified algorithm is very simple to implement (the intersection of two rectangles, when it exists, is a rectangle that is easy to compute).

Stopping criterion: it is necessary to stop the iterations in algorithm (11.12). An efficient way to check the accuracy of the result is to set an acceptable smallest size σ of the rectangle $R = R_k \cap R_{k'}$. When the length (or the width) of R is smaller than σ, we define the common point $S = \mathcal{B}_m \cap \mathcal{B}'_{m'}$ as the intersection of both diagonals $M_1 M_2\ M'_1 M'_2$ of R. Finally, once S has been spotted, it remains to set it on the Bézier curve $\mathcal{B}_m$; in other words, we have to compute the corresponding value of the parameter t such that

$$S = S(t) = \sum_{k=0}^{m} \binom{m}{k} t^k (1-t)^{m-k} P_k.$$

This value is obtained by a linear approximation:

$$t \approx t(M_2) \frac{x(S) - x(M_1)}{x(M_2) - x(M_1)} + t(M_1) \frac{x(M_2) - x(S)}{x(M_2) - x(M_1)}. \qquad (11.13)$$

We proceed in the same way to compute the value of t':

$$S = S'(t') = \sum_{k'=0}^{m'} \binom{m'}{k'} (t')^{k'} (1-t')^{m'-k'} P'_{k'},$$

$$\qquad\qquad (11.14)$$

$$t' \approx t'(M'_2) \frac{x(S) - x(M'_1)}{x(M'_2) - x(M'_1)} + t'(M'_1) \frac{x(M'_2) - x(S)}{x(M'_2) - x(M'_1)}.$$

Remark 11.5 We use ordinates $y(M_1)$ and $y(M_2)$ in (11.13) when $x(M_1) = x(M_2)$.

Remark 11.6 The stopping criterion may be modified by computing the distance between the curve and the straight line $M_1 M_2$, instead of the size of the rectangle R. This is obtained via an approximation of the curvature. For example, if (x_k, y_k) are the coordinates of sampling points of $\mathcal{B}_m$, we may use a value of h, defined by

$$h = \max_{k}(|x_{k-1} - 2x_k + x_{k+1}|, |y_{k-1} - 2y_k + y_{k+1}|).$$

11.6.1 Implementation

Exercise 11.2 1. Compute and display two Bézier curves $\mathcal{B}_m$ and $\mathcal{B}'_{m'}$ defined in $\mathbb{R}^2$.
2. Implement the algorithm (11.12) in its simplified formulation. Check the intersection of the curves.
3. Compute the values of t and t' according to (11.13) and (11.14). Display the corresponding points $S(t)$ and $S'(t')$. Compare to the values obtained by (11.12).

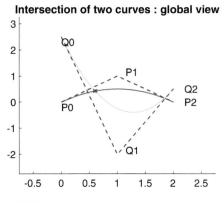

Fig. 11.7 Intersection of Bézier curves.

A solution of this exercise is proposed in Sect. 11.10 at page 266. See also Fig. 11.7.

11.7 Bézier Surfaces

Similar to Bézier curves, a pleasant and easy-to-handle representation of surfaces is obtained using two parameters t_1 and t_2. Let m_1 and m_2 be two positive integers, and consider $(m_1 + 1) \times (m_2 + 1)$ control points $P_{k_1,k_2} \in \mathbb{R}^3$. For all $(t_1, t_2) \in [0,1] \times [0,1]$, we define the point $P(t_1, t_2)$ by the relation

$$P(t_1, t_2) = \sum_{k_1=0}^{m_1} \sum_{k_2=0}^{m_2} \binom{m_1}{k_1}\binom{m_2}{k_2} t_1^{k_1}(1-t_1)^{m_1-k_1} t_2^{k_2}(1-t_2)^{m_2-k_2} P_{k_1,k_2}.$$
(11.15)

As (t_1, t_2) ranges in $[0,1] \times [0,1]$, the corresponding point $P(t_1, t_2)$ covers a surface, referred to as a *Bézier patch* by CAGD specialists (see Fig. 11.8). A Bézier surface is then the union of the Bézier patches (see Fig. 11.9).

Remark 11.7 There is no assumption made on the layout of the points P_{k_1,k_2} in using definition (11.15). Nevertheless, this layout has a significant influence on the final rendering of the generated surface. In order to modify the surface by moving some control points, it is easier to use points P_{k_1,k_2} situated on a rectangular grid. The resulting surface is called a *rectangular* patch, opposed to a *triangular* patch, whose control points are situated on a triangular grid and the generating point is defined by

$$P(t_1, t_2, t_3) = \sum_{k_1+k_2+k_3=m} \frac{m!}{k_1! k_2! k_3!} t_1^{k_1} t_2^{k_2} t_3^{k_3} P_{k_1,k_2,k_3}.$$
(11.16)

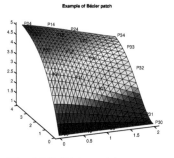

Fig. 11.8 Bézier patch.

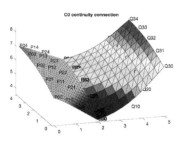

Fig. 11.9 $\mathcal{C}^0$ continuity.

11.8 Basic properties of Bézier Surfaces

11.8.1 Convex Hull

The point $P(t)$ is the barycenter of the $(m_1 + 1) \times (m_2 + 1)$ control points P_{k_1,k_2} with the corresponding weights $B_{m_1}^{k_1}(t_1)B_{m_2}^{k_2}(t_2)$. It follows from (11.2) that $P(t)$ lies in the convex hull of the points P_{k_1,k_2}, which is a volume with polygonal faces.

11.8.2 Tangent Vector

The tangent plane to the Bézier surface at the point $P(t_1, t_2)$ is defined by the two tangent vectors $\boldsymbol{\tau}_1(t_1, t_2) = \frac{d}{dt_1}P(t_1, t_2)$ and $\boldsymbol{\tau}_2(t_1, t_2) = \frac{d}{dt_2}P(t_1, t_2)$. Consequently, when $P_{0,0} \neq P_{1,0}$ and $P_{0,0} \neq P_{0,1}$, the Bézier surface is tangent to the triangle $P_{1,0}P_{0,0}P_{0,1}$ at point $P_{0,0}$. The same property holds for vertices $P_{m_1,0}$, P_{0,m_2} and P_{m_1,m_2}.

11.8.3 Junction of Bézier Patches

Let $\mathcal{S}_{m_1,m_2}$ be the Bézier patch defined by the $(m_1 + 1) \times (m_2 + 1)$ control points P_{k_1,k_2} $(1 \leq k_1 \leq m_1, 1 \leq k_2 \leq m_2)$, and $\mathcal{S}'_{m_1,m_2}$ the Bézier patch defined by the $(m'_1 + 1) \times (m'_2 + 1)$ control points $P'_{k'_1,k'_2}$ $(1 \leq k'_1 \leq m'_1,$ $1 \leq k'_2 \leq m'_2)$. We address now the problem of connecting these two patches.

The simplest junction corresponds to so-called $\mathcal{C}^0$ continuity, supposing that there exists a common curve located on the rim of the surfaces. Such a curve corresponds to an extreme value of t_1 or t_2 (namely 0 or 1). According to definition (11.15) and properties (11.2), such a curve is a Bézier curve. The

$\mathcal{C}^0$ continuity is then satisfied when the corresponding Bézier curves fit; this is true, for example, when the control points are identical on both curves.

Now we look further for a better rendering of the junction, and suppose that $P_{m_1,k_2} = P_{0,k_2}$ for $k_2 = 0, 1, \ldots, m_2$. A $\mathcal{G}^1$ connection is obtained when vectors $\boldsymbol{\tau}_1(1, t_2) = \frac{d}{dt_1} P(1, t_2)$ and $\boldsymbol{\tau}'_1(0, t'_2) = \frac{d}{dt'_1} P'(0, t'_2)$ are collinear at any common point. This condition is equivalent to saying that $P_{m_1,k_2} = P'_{0,k_2}$ is the midpoint of $P_{m_1-1,k_2} P'_{1,k_2}$ for $k_2 = 0, 1, \ldots, m_2$. Furthermore, when the tangent vectors $\boldsymbol{\tau}_2(1, t_2) = \frac{d}{dt_2} P(1, t_2)$ and $\boldsymbol{\tau}'_2(0, t'_2) = \frac{d}{dt'_2} P'(0, t'_2)$ are identical at any common point, a $\mathcal{C}^1$ connection is realized. Examples illustrating the different connections are displayed in Figs. 11.9 and 11.10.

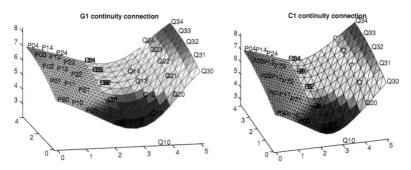

Fig. 11.10 Junction of patches. Left: $\mathcal{G}^1$ continuity; right: $\mathcal{C}^1$ continuity.

11.8.4 Construction of the Point $P(t)$

According to definition (11.15), any point $P(t_1, t_2)$ may be computed by an algorithm similar to (11.6). We write

$$P(t_1, t_2) = \sum_{k_1=0}^{m_1} \binom{m-1}{k-1} t_1^{k_1} (1-t_1)^{m_1-k_1} P_{k_1}, \qquad (11.17)$$

with

$$P_{k_1} = \sum_{k_2=0}^{m_2} \binom{m_2}{k_2} t_2^{k_2} (1-t_2)^{m_2-k_2} P_{k_1,k_2}. \qquad (11.18)$$

Let k_1 be a given integer. According to (11.17)-(11.18), the point P_{k_1} appears to be the point $P_{k_1}(t_2)$ of the Bézier curve defined by the $m_2 + 1$ control points $P_{k_1,0}, P_{k_1,1}, \ldots, P_{k_1,m_2}$. It is then possible to calculate its coordinates by means of de Casteljau's algorithm (11.6). Once the $m_1 + 1$

points P_{k_1} are computed, a new application of (11.6) generates the point $P(t_1, t_2)$ on the Bézier patch. The whole patch is then defined as the union of all rectangular faces generated by vertices $P(t_1, t_2)$, $P(t_1 + \Delta t_1, t_2)$, $P(t_1 + \Delta t_1, t_2 + \Delta t_2)$, and $P(t_1, t_2 + \Delta t_2)$ (here Δt_1 and Δt_2 are the sampling step sizes for t_1 and t_2). The corresponding construction algorithm (11.19) is called the *de Boor–Coox algorithm*:

$$
\begin{aligned}
&\text{for } p_1 = 0, \text{ initialization:} \\
&\quad \text{for } q_1 = 0, 1, \ldots, m_1, \text{ do} \\
&\qquad \text{construction of point } P_{q_1} \text{ by (11.6)} \\
&\qquad \text{for } p_2 = 0, \text{ initialization:} \\
&\qquad\quad \text{for } q_2 = 0, 1, \ldots, m_2, \text{ do} \\
&\qquad\qquad P^0_{q_1, q_2} = P_{q_1, q_2} \\
&\qquad\quad \text{end do} \\
&\qquad \text{end do} \\
&\qquad \text{for } p_2 = 1, 2, \ldots, m_2, \text{ do} \\
&\qquad\quad \text{for } q_2 = p_2, p_2 + 1, \ldots, m_2, \text{ do} \\
&\qquad\qquad P^{p_2}_{q_1, q_2} = t_2 P^{p_2-1}_{q_1, q_2-1} + (1 - t_2) P^{p_2-1}_{q_1, q_2}. \\
&\qquad\quad \text{end do} \\
&\qquad \text{end do: set } P_{q_1} = P_{q_1, m_2} \\
&\qquad \text{end of computation of } P_{q_1}, \text{ set: } P^0_{q_1} = P_{q_1} \\
&\quad \text{end do} \\
&\text{for } p_1 = 1, 2, \ldots, m_1, \text{ do} \\
&\quad \text{for } q_1 = p_1, p_1 + 1, \ldots, m_1, \text{ do} \\
&\qquad P^{p_1}_{q_1} = t_1 P^{p_1-1}_{q_1-1} + (1 - t_1) P^{p_1-1}_{q_1}. \\
&\quad \text{end do} \\
&\text{end do: set } P(t_1, t_2) = P^{m_1}_{m_1}
\end{aligned}
\tag{11.19}
$$

11.9 Construction of Bézier Surfaces

Exercise 11.3 1. Choose $(m_1 + 1) \times (m_2 + 1)$ points $P_{0,0}, P_{1,0}, \ldots, P_{m_1,m_2}$ in $\mathbb{R}^3$.
2. Write a general procedure generating the point $P(t_1, t_2)$ of the Bézier patch with control points $P_{0,0}, P_{1,0}, \ldots, P_{m_1,m_2}$, using the de Boor–Coox algorithm for any value $(t_1, t_2) \in [0,1] \times [0,1]$.
3. Compute and display the corresponding Bézier patch.
4. Repeat the computation for different values of (m_1, m_2) and different sets of control points.
5. Check the different continuity conditions.

A solution of this exercise is proposed in Sect. 11.10 at page 266.

Exercise 11.4 (for the brave) Extend the method for computing the intersection of two curves to compute the intersection of Bézier surfaces. Write and check the corresponding procedure.

11.10 Solutions and Programs

11.10.1 Solution of Exercise 11.1

The script **CAGD_M_ex1.m** defines a set of control points, and builds and displays the resulting Bézier curve, as shown in Fig. 11.1 (left). This procedure calls the function **CAGD_F_cbezier**, which computes a set of sampling points, and the function **CAGD_F_casteljau**, which builds a point according to de Casteljau's algorithm (11.6).

```
function [x,y]=CAGD_F_casteljau(t,XP,YP)
%%
%%    Construction of a point of a Bezier curve
%%    according to de Casteljau's algorithm
%%
m=size(XP,2)-1;
xx=XP;yy=YP;
for kk=1:m
xxx=xx; yyy=yy;
for k=kk:m
xx(k+1)=(1-t)*xxx(k)+t*xxx(k+1);
yy(k+1)=(1-t)*yyy(k)+t*yyy(k+1);
end
end
x=xx(m+1);y=yy(m+1);
```

The procedure CAGD_M_ex1 then calls the function CAGD_F_tbezier, which displays the Bézier curve and its control points. The control polygon, as shown in Fig. 11.2, is available for calling procedures CAGD_M_ex1b and CAGD_F_pbezier. Exchanging the points P_1 and P_3 will have as result the generation of a different curve, with a nonconvex control polygon (see procedure CAGD_M_ex1c).

Procedures CAGD_M_connectCC0, CAGD_M_connectCG1, CAGD_M_connectCC1 plot examples of C^0, G^1, and C^1 continuity.

11.10.2 Solution of Exercise 11.2

The procedure CAGD_M_ex2 defines two sets of control points, and builds and displays both corresponding Bézier curves. Each curve is then located within a rectangle by the function CAGD_F_drectan. A possible intersection of these rectangles is then tested by the function CAGD_F_rbezier. When this is successful, the function CAGD_F_dbezier is used; this procedure is an implementation of algorithm (11.12), seeking iteratively the intersection of both curves. When the rectangular intersection is small enough, the coordinates of the intersection point are computed; then an approximation of the corresponding value of the parameter t is obtained from (11.13) and (11.14).

Finally, the procedure calls the function CAGD_F_cbezier, which computes a sampling of points of a Bézier curve, using the functions CAGD_F_casteljau and CAGD_F_tbezier. The resulting curve and control points are then displayed.

11.10.3 Solution of Exercise 11.3

The procedure CAGD_M_ex3 defines a set of control points, and builds and displays the corresponding Bézier patch, as represented in Fig. 11.8 (you can use the "Rotate 3D" button to obtain a global view of the surface). This procedure calls the functions CAGD_F_sbezier, which gives a sampling of points of a Bézier surface; CAGD_F_coox; and CAGD_F_ubezier, which finally displays both the surface and its control points. The function CAGD_F_coox builds one point of a Bézier surface according to the de Boor–Coox algorithm (11.19).

```
function [x,y,z]=CAGD_F_coox(t1,t2,XP,YP,ZP)
%%
%%   Construction of a point on a Bzier surface
%%   according to the de Boor--Coox algorithm
```

```
%%
np1=size(XP,1);np2=size(XP,2);
xx1=zeros(np1,1);yy1=zeros(np1,1);zz1=zeros(np1,1);
for k1=1:np1
xx2=zeros(np2,1);yy2=zeros(np2,1);zz2=zeros(np2,1);
for k2=1:np2
xx2(k2)=XP(k1,k2);yy2(k2)=YP(k1,k2);zz2(k2)=ZP(k1,k2);
end
[x,y,z]=CAGD_F_cast3d(t2,xx2,yy2,zz2);
xx1(k1)=x;yy1(k1)=y;zz1(k1)=z;
end
[x,y,z]=CAGD_F_cast3d(t1,xx1,yy1,zz1);
```

The procedures CAGD_M_connectSC0, CAGD_M_connectSG1, and CAGD_M_connectSC1 display examples of $\mathcal{C}^0$ (respectively $\mathcal{G}^1$ and $\mathcal{C}^1$) and to 11.10. The obtained results correspond to Figs. 11.9 and 11.10.

Chapter References

P. Bézier, *Courbes et Surfaces, Mathématiques et CAO*, vol. 4 (Hermes, Paris, 1986)

P. de Casteljau, *Formes à Pôles, Mathématiques et CAO*, vol. 2 (Hermes, Paris, 1985)

S.A. Coons, *Surface Patches and B-splines Curves, CAGD* (1974)

G. Farin, D. Hansford, *The Essentials of CAGD* (AK Peters, 2000)

G. Farin, *Curves and Surfaces for CAGD: A Practical Guide*, 4th edn. (Academic Press, New York, 1996)

J. Ferguson, Multivariable curve interpolation. J. Assoc. Comput. Mach. (1964)

J. Hoschek, D. Lasser, *Fundamentals of Computer Aided Geometric Design* (Peters, Massachusetts, 1997)

L. Piegl, W. Tiller, *The NURBS Book* (Springer, Berlin, 1995)

Chapter 12
Gas Dynamics: The Riemann Problem and Discontinuous Solutions: Application to the Shock Tube Problem

Project Summary

Level of difficulty: 3

Keywords:	Nonlinear hyperbolic systems, Euler equations for gas dynamics, centered schemes: Lax–Wendroff, MacCormack; upwind schemes: Godunov, Roe
Application fields:	Shock tube, supersonic flows

The interest in studying the shock tube problem is threefold. From a fundamental point of view, it offers an interesting framework to introduce some basic notions about nonlinear hyperbolic systems of partial differential equations (PDEs). From a numerical point of view, this problem constitutes, since the exact solution is known, an inevitable and difficult test case for any numerical method dealing with discontinuous solutions. Finally, there is a practical interest, since this model is used to describe real shock tube experimental devices.[1]

12.1 Physical Description of the Shock Tube Problem

The fundamental idea of the shock tube is the following: consider a long one-dimensional (1D) tube, closed at its ends and divided into two equal regions by a thin diaphragm (see Fig. 12.1). Each region is filled with the same gas, but with different thermodynamic parameters (pressure, density, and temperature). The region with the highest pressure is called the *driven*

---

[1] The first shock tube facility was built in 1899 by Paul Vieille to study the deflagration of explosive charges. Nowadays, shock tubes are currently used as low-cost high-speed wind tunnels, in which a wide variety of aerodynamic or aeroballistic topics are studied: supersonic aircraft flight, gun performance, asteroid impacts, shuttle atmospheric entry.

© The Author(s), under exclusive license to Springer Nature Switzerland AG 2023   269
I. Danaila et al., *An Introduction to Scientific Computing*,
https://doi.org/10.1007/978-3-031-35032-0_12

section of the tube, while the low-pressure part is the *working section*. The gas being initially at rest, the sudden breakdown of the diaphragm generates a high-speed flow, which propagates in the working section (this is the place where the model of a free-flying object, such as a supersonic aircraft, will be placed).

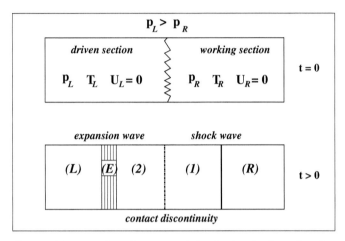

Fig. 12.1 Sketch of the initial configuration of the shock tube ($t = 0$) and waves propagating in the tube after the diaphragm breakdown ($t > 0$).

Let us get into a more detailed analysis of the problem. Consider (Fig. 12.1) that the left part of the tube is the driven section, defined by the pressure p_L, the density ρ_L, the temperature T_L, and the initial velocity $U_L = 0$. Similarly, the parameters of the (right part) working section are $p_R < p_L$, ρ_R, T_R, and $U_R = 0$.

At time $t = 0$ the diaphragm breaks, generating a process that naturally tends to equalize the pressure in the tube. The gas at high pressure expands through an *expansion (or rarefaction) wave* and flows into the working section, pushing the gas of this part. The rarefaction is a continuous process and takes place inside a well-defined region (the expansion fan) that propagates to the left (region (E) in Fig. 12.1); the width of the expansion fan grows in time.

The compression of the low-pressure gas generates a *shock wave* propagating to the right. The expanded gas is separated from the compressed gas by a *contact discontinuity*, which can be regarded as a fictitious membrane traveling to the right at constant speed. At this point of our simplified description, we just note that some of the physical functions defining the flow in the tube ($p(x), \rho(x), T(x)$, and $U(x)$) are discontinuous across the shock wave and the contact discontinuity. These discontinuities, which cause the difficulty of the problem, will be described in greater detail in the following sections.

12.2 Euler Equations of Gas Dynamics

To simplify the mathematical description of the shock tube problem we consider an infinitely long tube (to avoid reflections at the tube ends) and neglect viscous effects in the flow. We also suppose that the diaphragm is completely removed from the flow at $t = 0$. Under these simplifying hypotheses, the compressible flow in the shock tube is described by the one-dimensional (1D) Euler system of PDEs (see, for instance, Hirsch 1988; LeVeque 1992)

$$\frac{\partial}{\partial t} \underbrace{\begin{pmatrix} \rho \\ \rho U \\ E \end{pmatrix}}_{W(x,t)} + \frac{\partial}{\partial x} \underbrace{\begin{pmatrix} \rho U \\ \rho U^2 + p \\ (E+p)U \end{pmatrix}}_{F(W)} = 0, \tag{12.1}$$

where ρ is the density of the gas and E the total energy:

$$E = \rho e + \frac{\rho}{2} U^2, \tag{12.2}$$

with e the internal energy per unit mass. To close this system of equations, we need to write the constitutive law of the gas (or equation of state). Considering the ideal gas model, the equation of state is

$$p = \rho \mathcal{R} T, \tag{12.3}$$

and the internal energy is proportional to the temperature:

$$e = C_v T, \tag{12.4}$$

with $C_v = \mathcal{R}/(\gamma - 1)$ the heat coefficient at constant volume. The constants $\mathcal{R}$ and γ characterize the thermodynamic properties of the gas ($\mathcal{R}$ is the universal gas constant divided by the molecular mass and γ is the ratio of specific heat coefficients). Combining (12.2), (12.3) and (12.4) leads to the following new expression for the total energy:

$$E = \frac{p}{\gamma - 1} + \frac{\rho}{2} U^2. \tag{12.5}$$

It is also useful to define the local speed of sound a, the Mach number $\mathcal{M}$, and the total enthalpy H:

$$a = \sqrt{\gamma \mathcal{R} T} = \sqrt{\gamma \frac{p}{\rho}}, \quad \mathcal{M} = \frac{U}{a}, \quad H = \frac{E+p}{\rho} = \frac{a^2}{\gamma - 1} + \frac{1}{2} U^2. \tag{12.6}$$

Considering the column vector of unknowns $W = (\rho, \rho U, E)^t$, the Euler system of equations (12.1) can be written in the following *conservative* form:

$$\frac{\partial W}{\partial t} + \frac{\partial}{\partial x} F(W) = 0, \tag{12.7}$$

with the initial condition (we denote by x_0 the abscissa of the diaphragm):

$$W(x,0) = \begin{cases} (\rho_L, \rho_L U_L, E_L), & x \leq x_0, \\ (\rho_R, \rho_R U_R, E_R), & x > x_0. \end{cases} \tag{12.8}$$

The vector W contains the conserved variables and $F(W)$ the flux-functions. Note that with this choice of the vector of unknowns W, the pressure is not an unknown, since it can be derived from (12.5) using the components of W.

Definition 12.1 The nonlinear hyperbolic system of PDEs (12.7) and piecewise constant initial condition (12.8) define the Riemann problem.

The mathematical analysis of the Euler system of PDEs usually considers its *quasilinear* form:[2]

$$\frac{\partial W}{\partial t} + A \frac{\partial W}{\partial x} = 0, \tag{12.9}$$

with the Jacobian matrix

$$A = \frac{\partial F}{\partial W} = \begin{pmatrix} 0 & 1 & 0 \\ \frac{1}{2}(\gamma - 3)U^2 & (3 - \gamma)U & \gamma - 1 \\ \frac{1}{2}(\gamma - 1)U^3 - UH & H - (\gamma - 1)U^2 & \gamma U \end{pmatrix}. \tag{12.10}$$

It is interesting to note that the matrix A satisfies the following remarkable relation:

$$AW = F(W). \tag{12.11}$$

Furthermore, we can easily calculate its eigenvalues

$$\lambda^0 = U, \quad \lambda^+ = U + a, \quad \lambda^- = U - a, \tag{12.12}$$

and the corresponding eigenvectors

$$v^0 = \begin{pmatrix} 1 \\ U \\ \frac{1}{2}U^2 \end{pmatrix}, \quad v^+ = \begin{pmatrix} 1 \\ U + a \\ H + aU \end{pmatrix}, \quad v^- = \begin{pmatrix} 1 \\ U - a \\ H - aU \end{pmatrix}. \tag{12.13}$$

We conclude that the Jacobian matrix A is diagonalizable, i.e. it can be decomposed as $A = P \Lambda P^{-1}$, where

$$\Lambda = \begin{pmatrix} U - a & 0 & 0 \\ 0 & U & 0 \\ 0 & 0 & U + a \end{pmatrix}, \quad P = \begin{pmatrix} 1 & 1 & 1 \\ U - a & U & U + a \\ H - aU & \frac{1}{2}U^2 & H + aU \end{pmatrix}. \tag{12.14}$$

---

[2] The reader who has already explored Chap. 1 of this book may notice that this form is similar to that of the convection equation. The underlying idea is here to generalize the analysis of characteristics in the case of a system of PDEs.

We can easily verify that

$$P^{-1} = \begin{pmatrix} \frac{1}{2}\left(\alpha_1 + \frac{U}{a}\right) & -\frac{1}{2}\left(\alpha_2 U + \frac{1}{a}\right) & \frac{\alpha_2}{2} \\ 1 - \alpha_1 & \alpha_2 U & -\alpha_2 \\ \frac{1}{2}\left(\alpha_1 - \frac{U}{a}\right) & -\frac{1}{2}\left(\alpha_2 U - \frac{1}{a}\right) & \frac{\alpha_2}{2} \end{pmatrix}, \tag{12.15}$$

where $\alpha_1 = (\gamma - 1)U^2/(2a^2)$ and $\alpha_2 = (\gamma - 1)/a^2$.

Definition 12.2 The system (12.9) is said hyperbolic if the matrix A is diagonalizable with real eigenvalues.

The hyperbolic character of the system (12.9) has important consequences on the propagation of the information in the flow field. Certain quantities, called *invariants*,[3] are transported along particular curves in the plane (x, t), called *characteristics*. From a numerical point of view, this suggests a simple way to calculate the solution at any point $P(x, t)$ by gathering all the information transported through the characteristics starting from P and going back to regions where the solution is already known (imposed by the initial condition, for example).

The general form of the equation defining a characteristic is $dx/dt = \lambda$, where λ is an eigenvalue of the Jacobian matrix A. Since the corresponding invariant r is constant along the characteristic, it satisfies

$$\frac{dr}{dt} = \frac{\partial r}{\partial t} + \frac{\partial r}{\partial x}\frac{dx}{dt} = 0, \quad \text{or} \quad \frac{\partial r}{\partial t} + \lambda\frac{\partial r}{\partial x} = 0. \tag{12.16}$$

In the simplest case of the convection equation $\partial u/\partial t + c\,\partial u/\partial x = 0$, with constant transport velocity c, there exists a single characteristic curve, which is the line $x = ct$, and the corresponding invariant is the solution itself, $r = u$ (see also Chap. 1). From (12.12) we infer that the system (12.9) has three distinct characteristics:

$$C^0 : \frac{dx}{dt} = U, \quad C^+ : \frac{dx}{dt} = U + a, \quad C^- : \frac{dx}{dt} = U - a. \tag{12.17}$$

The invariants can be generally presented as differential relations:

$$dr^0 = dp - a^2 d\rho = 0,\ dr^+ = dp + \rho a\,dU = 0,\ dr^- = dp - \rho a\,dU = 0, \tag{12.18}$$

which have to be integrated along the corresponding characteristic curves.

It is possible to integrate exactly relations (12.18) for flows with constant entropy (also called isentropic flows). Entropy is a fundamental thermodynamic concept that, roughly speaking, measures irreversibility: a reversible process has constant entropy, while irreversible processes cause the entropy to increase. Entropy s per unit mass can be related to internal energy e by

---

[3] For a rigorous analysis of hyperbolic systems of PDEs and related definitions (in particular the definition of Riemann invariants), the reader can refer to Godlewski and Raviart (1996), Hirsch (1988), LeVeque (1992).

the thermodynamic law: $de = Tds - pd(1/\rho)$. Recalling (12.4) and (12.3), we obtain that for an ideal gas the entropy variation during its evolution, starting from a reference state A, is $s - s_A = C_v \ln\left(\frac{p/p_A}{(\rho/\rho_A)^\gamma}\right)$. For an isentropic flow, the entropy remains unchanged ($s = s_A$), and we deduce that the ratio p/ρ^γ is constant. In this case, relations (12.18) become

$$r^0 = p/\rho^\gamma, \quad r^+ = U + \frac{2a}{\gamma - 1}, \quad r^- = U - \frac{2a}{\gamma - 1}. \tag{12.19}$$

The above relations will be useful to link states of regions (L) and (2) through the expansion wave (see Fig. 12.1), when deriving the exact solution of the shock tube problem. Another important relation will be needed to link states of regions (1) and (R), located before and after the shock wave discontinuity. These are the so-called jump relations, or Rankine–Hugoniot conditions. In the general case of a discontinuity surface Σ propagating with velocity $\mathbf{U}_\Sigma$ and separating solutions W_+ and W_- on each side of the surface, these relations link the value of the jump $(W_+ - W_-)$ of the solution to the flux jump $(F(W_+) - F(W_-))$. In the particular case of the 1D Euler equations (12.7), the jump equations simply become $U_\Sigma (W_+ - W_-) = F(W_+) - F(W_-)$ (see Godlewski and Raviart (1996) for the mathematical proof of Rankine–Hugoniot conditions in the general case of 3D Euler equations and a general parametric description of the discontinuity surface Σ). We apply below the jump conditions to the shock wave between regions (1) and (R), considering that the shock propagates with constant and positive velocity U_s. Recalling that $U_R = 0$, we obtain:

$$\begin{cases} U_s(\rho_R - \rho_1) & = -\rho_1 U_1, \\ U_s(-\rho_1 U_1) & = p_R - (p_1 + \rho_1 U_1^2), \\ U_s\left(\frac{p_R - p_1}{\gamma - 1} - \frac{1}{2}\rho_1 U_1^2\right) & = -\left(\frac{\gamma}{\gamma - 1}p_1 + \frac{1}{2}\rho_1 U_1^2\right)U_1, \end{cases} \tag{12.20}$$

or, in the equivalent form:

$$\frac{\rho_R}{\rho_1} = 1 - \frac{U_1}{U_s}, \tag{12.21}$$

$$p_R - p_1 = -\rho_1 U_1 U_s \left(1 - \frac{U_1}{U_s}\right) = -\rho_R U_1 U_s, \tag{12.22}$$

$$p_R - p_1 + \gamma\frac{U_1}{U_s}p_1 = \frac{\gamma - 1}{2}\rho_1 U_1^2 \left(1 - \frac{U_1}{U_s}\right) = \frac{\gamma - 1}{2}\rho_R U_1^2. \tag{12.23}$$

If the Mach number of the shock wave is defined as $\mathcal{M}_s = U_s/a_R$, after some algebraic manipulations[4] of the system (12.21)-(12.23), we obtain the following Rankine–Hugoniot equations:

---

[4] It is possible to extract U_1 from (12.22) and replace its expression in (12.23). After diving by $(p_1 - p_R)$ and using that $\rho_R U_s^2 = \gamma\mathcal{M}_s^2 p_R$ we obtain Eq. (12.24).

$$\frac{p_1}{p_R} = \frac{2\gamma}{\gamma+1}\mathcal{M}_s^2 - \frac{\gamma-1}{\gamma+1}, \tag{12.24}$$

$$\frac{U_1}{a_R} = \frac{2}{\gamma+1}\left(\mathcal{M}_s - \frac{1}{\mathcal{M}_s}\right), \tag{12.25}$$

$$\frac{\rho_R}{\rho_1} = \frac{2}{\gamma+1}\frac{1}{\mathcal{M}_s^2} + \frac{\gamma-1}{\gamma+1}. \tag{12.26}$$

12.2.1 Dimensionless Equations

When building numerical applications, we usually prefer to remove physical units from equations and work with dimensionless variables. This simplifies the problem formulation and may reduce computational round-off errors. Physical variables in previous equations are non-dimensionalized (or scaled) using a reference state defined by the parameters of the working section:

$$\rho^* = \rho/\rho_R, \ U^* = U/a_R, \ a^* = a/a_R, \ T^* = T/(\gamma T_R),$$
$$p^* = p/(\rho_R a_R^2) = p/(\gamma p_R), \ E^* = E/(\rho_R a_R^2), \ H^* = H/a_R^2. \tag{12.27}$$

We also nondimensionalize space and time variables as $x^* = x/L$, $t^* = t/(L/a_R)$, where L is the length of the tube.

The Euler equations for dimensionless variables (denoted by the star superscript) keep the same differential form as previously:

$$\frac{\partial}{\partial t^*}\underbrace{\begin{pmatrix}\rho^* \\ \rho^* U^* \\ E^*\end{pmatrix}}_{W^*(x^*,t^*)} + \frac{\partial}{\partial x^*}\underbrace{\begin{pmatrix}\rho^* U^* \\ \rho^* U^{*2}+p^* \\ (E^*+p^*)U^*\end{pmatrix}}_{F^*(W^*)} = 0. \tag{12.28}$$

Dimensionless total energy E^* and total enthalpy H^* become

$$E^* = \frac{p^*}{\gamma-1} + \frac{\rho^*}{2}U^{*2}, \quad H^* = \frac{(a^*)^2}{\gamma-1} + \frac{1}{2}U^{*2}. \tag{12.29}$$

Differences with respect to previous physical equations appear in the equation of state

$$p^* = \rho^* T^*, \tag{12.30}$$

and in the definition of the speed of sound

$$a^* = \sqrt{\gamma\frac{p^*}{\rho^*}} = \sqrt{\gamma T^*}. \tag{12.31}$$

To simplify notation we drop the star superscript in subsequent equations; only dimensionless variables will be considered in the following sections.

12.2.2 Exact Solution

The exact solution of the shock tube problem follows the physical and mathematical descriptions given in previous sections. The tube is separated (see Fig. 12.1) into four uniform regions, i.e. with constant parameters (pressure, density, temperature, and velocity): the left (L) and right (R) regions (which keep the parameters imposed by the initial condition) and two intermediate regions, denoted by subscripts 1 and 2.

It is important to identify these regions in the (x, t) plane (see Fig. 12.2). All the waves are centered at the initial position of the diaphragm ($t = 0, x = x_0$). Since the shock and the contact discontinuity propagate in uniform zones, they have constant velocities and hence are displayed as lines in the (x, t) diagram. The expansion wave extends through the new zone (E), the expansion fan, in which the flow parameters vary continuously (see below). We recall that the shock wave and the contact discontinuity propagate to the right, while the expansion fan moves to the left.

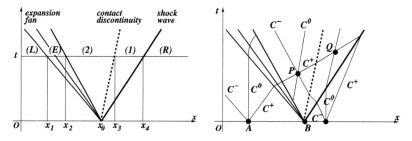

Fig. 12.2 Diagram in the (x, t) plane of the exact solution of the shock tube problem (left). Characteristics used to calculate the exact solution (right).

We start the calculation of the exact solution by writing the dimensionless parameters of the (L) and (R) regions (which are in fact the input parameters for a computer program):

$$\text{Region (R):} \quad \rho_R = 1, \ p_R = 1/\gamma, \ T_R = 1/\gamma, \ a_R = 1, \ U_R = 0, \quad (12.32)$$

$$\text{Region (L):} \quad \rho_L, \ p_L, \ T_L, \ a_L, \ U_L = 0 \text{ (given quantities).} \quad (12.33)$$

We then use the jump relations across the discontinuities and take into account the propagation of the information along the characteristics, as follows:

1. The shock wave implies the discontinuity of all the parameters of the flow. The jump between regions (1) and (R) is described by the Rankine–Hugoniot relations (12.24)-(12.26). Note that these equations are already written in the non-dimensional form corresponding to our scaling (with U_1/a_R becoming non-dimensional U_1). We recall that the velocity U_s of the shock is constant and $\mathcal{M}_s = U_s/a_R$.

2. The contact discontinuity is in fact a discontinuity of the density function, the pressure, and the velocity being continuous. Hence

$$U_2 = U_1, \quad p_2 = p_1. \tag{12.34}$$

3. We now link the parameters of region (2) to those of region (L). For this purpose, we consider a point P inside the region (2) and draw the characteristics passing through this point (see Fig. 12.2). We notice that only C^0 and C^+ characteristics will cross the expansion fan to search the information in region (L). Using expressions (12.19) for the invariants r^0 and r^+ and taking into account that $U_L = 0$, we obtain

$$\frac{\rho_2}{\rho_L} = \left(\frac{p_2}{p_L}\right)^{1/\gamma}, \quad U_2 = \frac{2}{\gamma - 1}(a_L - a_2). \tag{12.35}$$

4. Finally, we combine the previous relations to obtain an implicit equation for the unknown $\mathcal{M}_s$. The detailed calculation follows:

$$\mathcal{M}_s - \frac{1}{\mathcal{M}_s} \overset{(12.25)}{=} \frac{\gamma+1}{2}U_1 \overset{(12.34)}{=} \frac{\gamma+1}{2}U_2 \overset{(12.35)}{=} a_L\frac{\gamma+1}{\gamma-1}\left(1 - \frac{a_2}{a_L}\right).$$

Since

$$\frac{a_2}{a_L} \overset{(12.31)}{=} \left(\frac{p_2\,\rho_L}{p_L\,\rho_2}\right)^{1/2} \overset{(12.35)}{=} \left(\frac{p_2}{p_L}\right)^{\frac{\gamma-1}{2\gamma}} \overset{(12.34)}{=} \left(\frac{p_1}{p_L}\right)^{\frac{\gamma-1}{2\gamma}},$$

we replace p_1/p_L from (12.24) and finally get the following *compatibility equation*:

$$\mathcal{M}_s - \frac{1}{\mathcal{M}_s} = a_L\frac{\gamma+1}{\gamma-1}\left\{1 - \left[\frac{p_R}{p_L}\left(\frac{2\gamma}{\gamma+1}\mathcal{M}_s^2 - \frac{\gamma-1}{\gamma+1}\right)\right]^{\frac{\gamma-1}{2\gamma}}\right\}. \tag{12.36}$$

Once this implicit nonlinear equation is solved (using an iterative Newton method, for example), the obtained value of $\mathcal{M}_s$ will be used in previous relations to determine all the parameters of uniform regions (1) and (2).

To complete the exact solution, we need to determine the extent of each region (i.e. calculate the values of the abscissas x_1, x_2, x_3, x_4 in Fig. 12.2) for a given time value t. We proceed as follows:

- The expansion fan (E) is left-bounded by the C^- characteristic starting from the point B, considered to belong to region (L), i.e. the line of slope $dx/dt = -a_L$. The right bound of the expansion fan is the C^- characteristic starting from the same point B, but considered this time to belong to region (2), i.e. the line of slope $dx/dt = U_2 - a_2$. The values of x_1 and x_2 are consequently

$$x_1 = x_0 - a_L t, \quad x_2 = x_0 + (U_2 - a_2)t. \qquad (12.37)$$

Consider now a point (x, t) inside the region (E), i.e. $x_1 \leq x \leq x_2$. Since this point belongs to a C^- characteristic starting from B, necessarily $(x - x_0)/t = U - a$. Using the C^+ characteristic coming from region (L), we also get that $a + (\gamma - 1)U/2 = a_L$. Combining these two relations and remembering that the flow is isentropic, we can conclude that the exact solution inside the expansion fan is

$$U = \frac{2}{\gamma + 1}\left(a_L + \frac{x - x_0}{t}\right), \quad a = a_L - (\gamma - 1)\frac{U}{2}, \quad p = p_L\left(\frac{a}{a_L}\right)^{\frac{2\gamma}{\gamma - 1}}.$$
$$(12.38)$$

- The contact discontinuity is transported at constant velocity $U_2 = U_1$, so

$$x_3 = x_0 + U_2 t. \qquad (12.39)$$

- Since the shock wave also propagates at constant dimensionless velocity $U_s = \mathcal{M}_s$, we finally obtain

$$x_4 = x_0 + \mathcal{M}_s t. \qquad (12.40)$$

Remark 12.1 The exact solution $W(x, t)$ of the shock tube problem depends only on the ratio x/t, as one would have expected from the characteristics analysis of the Euler system of PDEs.

Exercise 12.1 Write a MATLAB function to compute the exact solution of the shock tube problem. The definition header of the function will be as follows:

```
function [uex,xREG]=HYP_F_shock_tube_exact(x,x0,t)
% Input arguments:
%  x        vector of dimension n containing space discretization
%  x0       initial position x_0 of the diaphragm
%  t        time at which the solution is computed
% Output arguments:
%       uex  array of dimension (3,n) containing exact solution
%               uex(1,1:n)  the density rho
%               uex(2,1:n)  the velocity U
%               uex(3,1:n)  the pressure p
%       xREG abscissas separating regions L,E,2,1,R
```

Write a main program HYP_M_shock_tube_analytical that defines the parameters of the problem and calls the function HYP_F_shock_tube_exact. Consider $x \in [0, 1]$, $x_0 = 0.5$, and a regular (equidistant) grid with $M = 81$ computational points. Define physical parameters corresponding to those used by Sod (see also Hirsch 1988): $\gamma = 1.4$, $\rho_L = 8$, $p_L = 10/\gamma$. Plot the dimensionless exact solution ($\rho(x)$, $U(x)$, and $p(x)$) at time $t = 0.2$.

Hint: define all physical parameters as global variables; use the MAT-LAB built-in function `fzero` to solve the compatibility equation (12.36).

The expected result is displayed in Fig. 12.3. This solution was obtained using the MATLAB program presented in Sect. 12.4 at page 290.

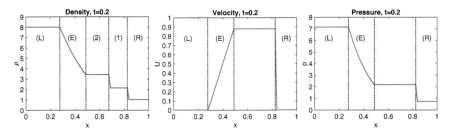

Fig. 12.3 Exact solution of the shock tube problem (Sod's data) at time $t = 0.2$.

12.3 Numerical Solution

The first idea one would have in mind when attempting to numerically solve the Euler system of PDEs (12.28) is to use *elementary* discretization methods discussed in Chap. 1 for scalar PDEs, for example, an Euler or a Runge–Kutta method for the time integration and centered finite differences for the space discretization. We shall see, however, that such methods are not appropriate to compute discontinuous solutions, since they generate nonphysical oscillations. This drawback of the space-centered schemes for computing the shock tube problem will be illustrated using the more sophisticated Lax–Wendroff and MacCormack schemes. We shall also give a quick description of upwind schemes that take into account the hyperbolic character of the system and allow a better numerical solution. Results using Roe's upwind scheme will be finally discussed.

12.3.1 Lax–Wendroff and MacCormack Centered Schemes

The space-centered schemes were historically the first to be used to solve hyperbolic systems. The two most popular schemes, the Lax and Wendroff scheme and the MacCormack scheme, are still used in some industrial numerical codes. We shall apply these schemes to solve the Euler system (12.28) written in the conservative form

$$\frac{\partial W}{\partial t} + \frac{\partial}{\partial x} F(W) = 0. \tag{12.41}$$

We use a regular (or equidistant) discretization of the domain of definition of the problem $(x, t) \in [0, 1] \times [0, T]$:

- in space

$$x_j = (j - 1)\delta x, \quad \delta x = \frac{1}{M - 1}, \quad j = 1, 2, \ldots, M, \tag{12.42}$$

- and in time

$$t^n = (n - 1)\delta t, \quad \delta t = \frac{T}{N - 1}, \quad n = 1, 2, \ldots, N. \tag{12.43}$$

For both schemes, the numerical solution W_j^{n+1} (at time t_{n+1} and space position x_j) is computed in two steps (a predictor and a corrector step) following the formulas displayed in Fig. 12.4.

We discuss in the following some remarkable features of these schemes.

1. (*Boundary values.*) From the schematic representation of the predictor and corrector steps in Fig. 12.4, we notice that only the components $j = 2, \ldots, (M - 1)$ of the solution are calculated. The remaining components for $j = 1$ and $j = M$ need to be prescribed by appropriate boundary conditions. Since the tube is assumed infinite, we impose $W_1^n = W_L$ and $W_M^n = W_R$ at any time level t_n. Practically, this is equivalent to leaving unchanged the first and last components of the solution vector. Meanwhile, it is obvious that the computation must stop before one of the waves (expansion or shock) hits the boundary.

2. (*Propagation of information.*) The predictor step of the Lax–Wendroff scheme computes an intermediate solution at mid-points $(j + \frac{1}{2})$, for $j = 1, \ldots, (M - 1)$, using forward finite differences. These intermediate values are then used in the centered finite difference scheme of the corrector step.

The MacCormack scheme combines backward differences for the predictor step with forward differences for the corrector step. We can show in fact that the idea behind this scheme is the following Taylor expansion:

$$W_j^{n+1} = W_j^n + \left(\overline{\frac{\partial W}{\partial t}}\right)_j \delta t, \tag{12.44}$$

where

$$\left(\overline{\frac{\partial W}{\partial t}}\right)_j = \frac{1}{2}\left[\left(\frac{\partial W}{\partial t}\right)_j^n + \left(\frac{\partial \tilde{W}}{\partial t}\right)_j\right] = \frac{1}{2}\left[\frac{\tilde{W}_j - W_j^n}{\delta t} - \frac{F(\tilde{W}_j) - F(\tilde{W}_{j-1})}{\delta x}\right]$$

is an approximation of the first derivative in time.

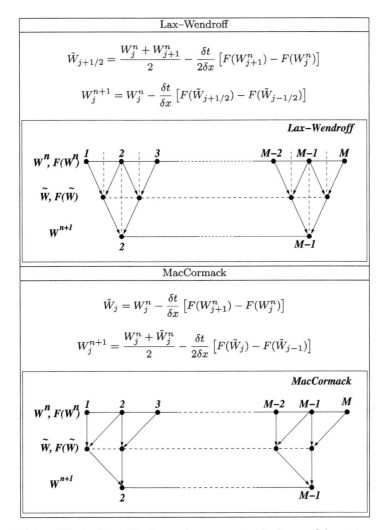

Fig. 12.4 Lax–Wendroff and MacCormack space-centered schemes. Schematic representation of their predictor and corrector steps.

In conclusion, the information is searched on both sides of the computed point j. The information propagation along characteristics is not taken into account, since no distinction is made between upstream and downstream influences. We shall see that this lack of physics in these numerical schemes will generate unwanted (nonphysical) oscillations of the solution.

3. (*Accuracy*) Both schemes use a three-point stencil $(j-1, j, j+1)$ to reach second-order accuracy in time and space.

4. (*Stability*) Both schemes are explicit and consequently subject to stability conditions. Similar to the (scalar) convection equation (see Chap. 1),

we can write the stability (or CFL[5]) condition in the general form

$$\max_{1\leq i\leq 3} (|\lambda_i|) \frac{\delta t}{\delta x} \leq 1,$$

where λ_i, $i = 1, 2, 3$, are the eigenvalues of the Jacobian matrix $\partial F/\partial W$, regarded here as propagation speeds of the corresponding characteristic waves $(dx/dt = \lambda)$. Using (12.17), we obtain the stability condition

$$(|U| + a) \frac{\delta t}{\delta x} \leq 1. \tag{12.45}$$

For numerical applications, this condition is used to compute the time step

$$\delta t = \text{cfl} \cdot \frac{\delta x}{|U| + a}, \quad \text{with} \quad \text{cfl} < 1. \tag{12.46}$$

Exercise 12.2 For the same physical and numerical parameters as in the previous exercise, compute the numerical solution of the shock tube problem at $t = 0.2$ using Lax–Wendroff and MacCormack centered schemes. Compare the exact solution and comment on the results. Hints:
• set an array w(1:3,1:M) to store the discrete values of the vector $W = (\rho, \rho U, E)^t$ of conservative variables;
• using (12.46) with cfl = 0.95, compute the time step in a separate function function dt = HYP_F_calc_dt(w,dx,cfl);
• write a function to compute $F(W)$;
• use vector programming to translate the formulas in Fig. 12.4 into MAT-LAB program lines (avoid loops!); for example, the predictor step of the Lax–Wendroff scheme will be coded in a single line:

```
wtilde=0.5*(w(:,1:M-1)+w(:,2:M))-0.5*dt/dx*(F(:,2:M)-F(:,1:M-1));
```

• for each scheme, superimpose numerical and exact solutions for (ρ, U, p) as in Fig. 12.5.
A solution of this exercise is proposed in Sect. 12.4 at page 290.

Numerical results displayed in Fig. 12.5 for both schemes, show good accuracy in smooth regions, whereas unwanted oscillations appear at the interfaces between different regions of the solution. The contact discontinuity is also poorly captured. The MacCormack scheme seems to better capture the shock discontinuity, but it introduces higher-amplitude oscillations at the end of the expansion wave where the flow is strongly accelerated.

---

[5] Courant–Friedrichs–Lewy.

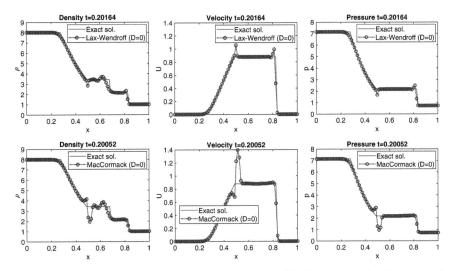

Fig. 12.5 Numerical results for the shock tube problem (Sod's parameters) using centered schemes. Lax–Wendroff scheme (up) and MacCormack scheme (down).

12.3.1.1 Artificial Dissipation

The oscillations generated by the centered schemes around discontinuities can be damped by adding a supplementary term to the initial equation (12.41):

$$\frac{\partial W}{\partial t} + \frac{\partial}{\partial x} F(W) - \delta x^2 \frac{\partial}{\partial x}\left(D(x)\frac{\partial W}{\partial x}\right) = 0. \qquad (12.47)$$

The mathematical form of this term is inspired by the heat equation (discussed in Chap. 1). The idea is to simulate the effect of a physical dissipation (or diffusion) process which is well known to have a smoothing effect[6] on the solution. Since the dissipation term is proportional to the gradient $\partial W/\partial x$ of the solution, the smoothing will be important in regions with sharp gradients (as the shock discontinuity) where numerical oscillations are expected to disappear.

The coefficient $D(x)$, also called *artificial viscosity* by analogy with Navier–Stokes equations (see Chap. 15), has to be positive to ensure a stabilizing effect[7] on the numerical solution. Moreover, its value has to be chosen such that the influence of the artificial term is negligible (i.e. of an order greater than or equal to the truncation error) in the smooth regions of the solution.

Several methods have been suggested to prescribe the artificial viscosity $D(x)$ and to modify classical centered schemes accordingly (see, for instance, Hirsch (1988), Fletcher (1991)). We illustrate the simplest technique, which

---

[6] This smoothing effect is nicely illustrated for the heat equation in Chap. 1, Exercise 1.10.

[7] The heat equation with negative diffusivity has physically unstable solutions!

considers a constant coefficient $D(x) = D$ and presents Eq. (12.47) in the conservative form (12.41) with a modified flux $F^*(W)$:

$$\frac{\partial W}{\partial t} + \frac{\partial}{\partial x}F^*(W) = 0, \quad \text{where} \quad F^*(W) = F(W) - D\delta x^2 \frac{\partial W}{\partial x}. \quad (12.48)$$

In order to use the same three-points stencil to define the schemes, the new vector $F^*(W)$ will be discretized

- using backward differences in the predictor step

$$F^*(W_j) = F(W_j) - (D\delta x)(W_j - W_{j-1}), \quad (12.49)$$

- and forward differences in the corrector step

$$F^*(\tilde{W}_j) = F(\tilde{W}_j) - (D\delta x)(\tilde{W}_{j+1} - \tilde{W}_j). \quad (12.50)$$

Exercise 12.3 Modify the previous program by adding an artificial dissipation term to both Lax–Wendroff and MacCormack schemes. Use (12.49)–(12.50) to modify the flux $F(W)$. Discuss the effect of the value of the artificial viscosity D (take $0 \leq D \leq 10$). What is the influence of D on the value of the time step?

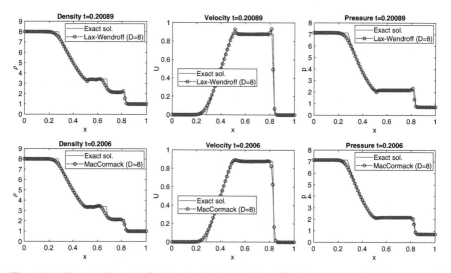

Fig. 12.6 Numerical results for the shock tube problem (Sod's parameters) using centered schemes with artificial dissipation. Lax–Wendroff scheme (up) and MacCormack scheme (down).

The results obtained with an artificial dissipation term are displayed in Fig. 12.6. Numerical oscillations are reduced near the shock and expansion waves, but large dissipation is also introduced in other regions of the solution. In particular, the contact discontinuity (see the graph for $\rho(x)$) is considerably smeared. Increasing the value of D allows one to completely remove the oscillations, but the overall accuracy is not satisfactory. More sophisticated methods have been proposed (see the references at the end of the chapter) to render the dissipation more selective with respect to the nature of discontinuities, but the general tradeoff between damping the oscillations and overall accuracy suggests that the artificial dissipation does not bring a real solution to the problem. A different approach, including more physics in the numerical approximation, is presented in the next section.

12.3.2 Upwind Schemes (Roe's Approximate Solver)

The origin of the numerical oscillations generated by the centered schemes discussed in the previous section comes from complete ignorance of the hyperbolic character of the Euler system of PDEs, and, in particular, the propagation of the information along characteristics. These important (physical) features will be considered in building upwind schemes.

Physical information can be introduced at different levels of the numerical approximation. We distinguish between:

1. *flux splitting upwind schemes*, which use different directional discretization of the flux $F(W)$, depending on the sign of the eigenvalues λ of the Jacobian matrix (12.10); since λ corresponds to the propagation speed of the associated characteristic, these schemes include only the information on the direction of propagation of waves (up- or downstream);
2. *Godunov-type schemes*, which introduce a higher level of physical approximation by considering a discretization based on the exact solution of the Riemann problem at each interface between computational points; when the local Riemann problem is solved approximately, we talk about Riemann solvers.

The following sections present the basic principle of Godunov schemes and the Riemann approximate solver of Roe.

12.3.2.1 Godunov-Type Schemes

The basic principle of a Godunov-type scheme is the following: the solution W^n is considered to be piecewise constant over each grid cell defined as the interval $]x_{j-1/2}, x_{j+1/2}[$; this allows us to define locally a Riemann problem at each interface between the cells; each local Riemann problem is solved **exactly** to calculate the solution W^{n+1} at the next time level.

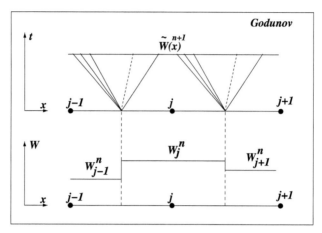

Fig. 12.7 Principle of a Godunov-type scheme.

More precisely, the numerical solution is advanced from time level $t_n = n\delta t$ to $t_{n+1} = t_n + \delta t$ in three steps (see Fig. 12.7):

Step 1. Using the known values W_j^n, define the piecewise constant function

$$W^n(x) = W_j^n, \quad x \in \,](j - 1/2)\delta x, (j + 1/2)\delta x[. \tag{12.51}$$

Step 2. Calculate the solution function $\tilde{W}^{n+1}(x)$, $x \in \,](j - 1/2)\delta x, (j + 1/2)\delta x\,[$ by gathering the exact solutions of the two Riemann problems defined at interfaces $\left(j - \frac{1}{2}\right)$ and $\left(j + \frac{1}{2}\right)$. This step requires that the waves issued from the two neighboring Riemann problems not intersect. This implies that the time step should be limited such that

$$\max_j \left(|U| + a\right)_{j+1/2}^n \frac{\delta t}{\delta x} \leq \frac{1}{2}. \tag{12.52}$$

Step 3. Obtain the solution $W^{n+1}(x)$, which is also a piecewise constant function, by averaging $\tilde{W}^{n+1}(x)$ over each cell:

$$W_j^{n+1} = \frac{1}{\delta x} \int_{(j-1/2)\delta x}^{(j+1/2)\delta x} \tilde{W}^{n+1}(x)dx. \tag{12.53}$$

We can show that the Godunov scheme can be written in the following *conservative* form:

$$\frac{W_j^{n+1} - W_j^n}{\delta t} + \frac{\Phi(W_j^n, W_{j+1}^n) - \Phi(W_j^n, W_{j-1}^n)}{\delta x} = 0, \tag{12.54}$$

where the flux vector is generally defined as

$$\Phi(W_j^n, W_{j+1}^n) = F(\tilde{W}_{j+1/2}^{n+1}). \tag{12.55}$$

The advantage of the conservative form is that it is valid over the entire domain of definition of the problem, even though the solution is discontinuous. This form is also used to derive approximate Riemann solvers. The exact form of the flux vector will be presented in the next section for the Roe solver.

12.3.2.2 Roe's Approximate Solver

The approximate solver of Roe is based on a simple and ingenious idea: the Riemann problem (12.9) at interface $\left(j + \frac{1}{2}\right)$ is replaced by the *linear* Riemann problem

$$\frac{\partial \tilde{W}}{\partial t} + A_{j+1/2} \frac{\partial \tilde{W}}{\partial x} = 0, \quad \tilde{W}(x, n\delta t) = \begin{cases} W_j^n, & x \le \left(j + \frac{1}{2}\right)\delta x, \\ W_{j+1}^n, & x > \left(j + \frac{1}{2}\right)\delta x. \end{cases} \tag{12.56}$$

The first question raised by this approach is how to properly define the matrix $A_{j+1/2}$, which depends on W_j^n and W_{j+1}^n. This matrix is a priori chosen such that the following criteria are satisfied.

1. The hyperbolic character of the initial equation is conserved by the linear problem; hence $A_{j+1/2}$ admits a decomposition similar to (12.14):

$$A_{j+1/2} = P_{j+1/2}\, \Lambda_{j+1/2}\, P_{j+1/2}^{-1}. \tag{12.57}$$

   In order to take into account the sign of the propagation speed of characteristic waves, it is useful to define the matrices following:

   - $\text{sign}(A_{j+1/2}) = P_{j+1/2}\,(\text{sign}(\Lambda))\, P_{j+1/2}^{-1}$, where $\text{sign}(\Lambda)$ is the diagonal matrix defined by the signs of the eigenvalues λ_l: $\text{sign}(\Lambda) = \text{diag}(\text{sign}\lambda_l)$.
   - $|A_{j+1/2}| = P_{j+1/2}|\Lambda|P_{j+1/2}^{-1}$, where $|\Lambda| = \text{diag}(|\lambda_l|)$.

2. The linear Riemann problem is consistent with the initial problem, i.e. for all variables u,

$$A_{j+1/2}(u, u) = A(u, u). \tag{12.58}$$

3. The numerical scheme is conservative, i.e. for all variables u and v,

$$F(u) - F(v) = A_{j+1/2}(u, v)(u - v). \tag{12.59}$$

For the practical calculation of the matrix $A_{j+1/2}$, the original idea of Roe was to present the conservative variables W and conservative fluxes $F(W)$ in (12.28) as quadratic forms of the components of the column vector $Z = \sqrt{\rho}(1, U, H)^t = (z_1, z_2, z_3)^t$:

$$W = \begin{pmatrix} z_1^2 \\ z_1 z_2 \\ \frac{1}{\gamma} z_1 z_3 + \frac{\gamma-1}{2\gamma} z_2^2 \end{pmatrix}, \quad F(W) = \begin{pmatrix} z_1 z_2 \\ \frac{\gamma-1}{\gamma} z_1 z_3 + \frac{\gamma-1}{2\gamma} z_2^2 \\ z_2 z_3 \end{pmatrix}. \quad (12.60)$$

Using the following identity, valid for arbitrary quadratic functions f, g,

$$(fg)_{j+1} - (fg)_j = \bar{f}(g_{j+1} - g_j) + \bar{g}(f_{j+1} - f_j), \quad \text{where} \quad \bar{f} = \frac{f_{j+1} + f_j}{2},$$

we can find two matrices $\bar{B}$ and $\bar{C}$ such that

$$\begin{cases} W_{j+1} - W_j = \bar{B}(Z_{j+1} - Z_j), \\ F(W_{j+1}) - F(W_j) = \bar{C}(Z_{j+1} - Z_j). \end{cases} \quad (12.61)$$

This implies that

$$F(W_{j+1}) - F(W_j) = (\bar{C}\,\bar{B}^{-1})(W_{j+1} - W_j), \quad (12.62)$$

which corresponds exactly to (12.59). Consequently, a natural choice for the matrix $A_{j+1/2}$ will be

$$A_{j+1/2} = \bar{C}\,\bar{B}^{-1}. \quad (12.63)$$

A remarkable property of this matrix (the reader is invited to derive it as an exercise!) is that it can be calculated from (12.10) by replacing the variables (ρ, U, H) with the corresponding *Roe's averages*

$$\bar{\rho}_{j+1/2} = R_{j+1/2}\rho_j, \ \bar{U}_{j+1/2} = \frac{R_{j+1/2}U_{j+1} + U_j}{1 + R_{j+1/2}}, \ \bar{H}_{j+1/2} = \frac{R_{j+1/2}H_{j+1} + H_j}{1 + R_{j+1/2}},$$

$$\bar{a}_{j+1/2}^2 = (\gamma - 1)\left(\bar{H}_{j+1/2} - \frac{\bar{U}_{j+1/2}^2}{2}\right), \quad \text{where } R_{j+1/2} = \sqrt{\frac{\rho_{j+1}}{\rho_j}}. \quad (12.64)$$

It is also remarkable that eigenvalue and eigenvector formulas (12.12) and (12.13) still apply to $A_{j+1/2}$ if one uses the corresponding Roe's averaged variables. This considerably simplifies the calculation of matrices $\text{sign}(A_{j+1/2})$ and $|A_{j+1/2}|$, which accounts for the popularity of Roe's approximate solver.

Once the matrix $A_{j+1/2}$ is defined, the upwinding in Roe's scheme follows the general principle of first-order upwind schemes applied to *linear* systems (see, for instance, Hirsch (1988) for more details). The flux in the general conservative form (12.54) becomes for Roe's solver

$$\Phi(W_j^n, W_{j+1}^n) = \frac{1}{2}\left\{F(W_j^n) + F(W_{j+1}^n) - \text{sign}(A_{j+1/2})[F(W_{j+1}^n) - F(W_j^n)]\right\}, \quad (12.65)$$

or, if we use (12.59),

$$\Phi(W_j^n, W_{j+1}^n) = \frac{1}{2}\left\{F(W_j^n) + F(W_{j+1}^n) - |A|_{j+1/2}[W_{j+1}^n - W_j^n]\right\}. \quad (12.66)$$

To summarize, Roe's scheme will be used in the form

$$W_j^{n+1} = W_j^n - \frac{\delta t}{\delta x} \left[\Phi(W_j^n, W^{rem} n_{j+1}) - \Phi(W_j^n, W_{j-1}^n) \right],$$ (12.67)

with the flux Φ given by (12.66); the matrix $|A_{j+1/2}| = P_{j+1/2}|A|P_{j+1/2}^{-1}$ will be calculated using Roe's averages (12.64) in Eqs. (12.14) and (12.15).

Remark 12.2 Roe's scheme is first-order accurate in time and space.

Exercise 12.4 Use Roe's scheme (12.67) to solve numerically the shock tube problem (Sod's parameters). Compare to the numerical results previously obtained using centered schemes.

The results obtained using Roe's scheme are displayed in Fig. 12.8. Compared to centered schemes, the numerical solution is smooth, without oscillations. The shock wave is accurately and sharply captured, but the scheme proves too dissipative around the contact discontinuity, which is strongly smeared.

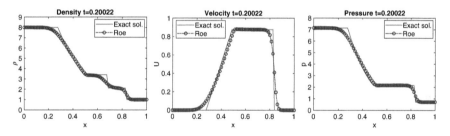

Fig. 12.8 Numerical computation of the shock tube problem (Sod's parameters) using Roe's approximate solver.

More accurate Riemann solvers can be derived in the framework of Godunov-type schemes by increasing the space accuracy. For example, we can use piecewise linear functions in steps 1 and 3 of the Godunov scheme to obtain solvers of second order in space. Several other approaches have been suggested in the literature to include more physics in the numerical discretization, leading to other classes of numerical methods, including TVD (total variation diminishing) and ENO (essentially nonoscillatory) schemes, which are now currently used to solve hyperbolic systems of PDEs. The reader who wishes to pursue the study of upwind schemes beyond this introductory presentation is referred to more specialized texts such as Fletcher (1991), Hirsch (1988), LeVeque (1992), Saad (1998).

12.4 Solutions and Programs

The exact solution of the shock tube problem for a given time value t is computed in the script HYP_M_shock_tube_analytical.m. It calls the function HYP_F_shock_tube_exact.m. The compatibility relation (12.36) is implemented as an implicit function (i.e. $f(x) = 0$) in HYP_F_mach_compat.m; this function is used as the first argument of the MATLAB built-in function fzero to compute the root corresponding to the value of $\mathcal{M}_s$. The final solution, containing the discrete values for (ρ, U, p), is computed according to relations in Sect. 12.2.2. Note the use of the MATLAB built-in function find to compute the abscissas x separating the different regions of the solution.

The main program resulting from successively solving all the exercises of this project is HYP_M_shock_tube.m. After defining the input data (which are the parameters of regions (L) and (R)) as global variables, the space discretization is built and the solution is initialized using Sod's parameters. Three main arrays are used for the computation:

usol(1:3,1:M) to store the nonconservative variables $(\rho, U, p)^t$,
w(1:3,1:M) for the conservative vector $W = (\rho, \rho U, E)^t$,
and F(1:3,1:M) for the conservative fluxes $F(W)$.

The program allows one to choose among three numerical schemes: Lax–Wendroff, MacCormack, and Roe. When a centered scheme is selected, the value of the artificial dissipation is requested. The numerical solution is superimposed on the exact solution using the function implemented in the script HYP_F_plot_graph.m. The most important functions called from the main program are

- HYP_F_trans_usol_w: computes $W = (\rho, \rho U, E)^t$ from $usol = (\rho, U, p)^t$;
- HYP_F_trans_w_usol: computes $usol = (\rho, U, p)^t$ from $W = (\rho, \rho U, E)^t$;
- HYP_F_trans_w_f: computes $F = (\rho U, \rho U^2 + p, (E + p)U)^t$ from $W = (\rho, \rho U, E)^t$;
- HYP_F_calc_dt: computes $\delta t = \text{cfl·}\delta x/(|U| + a)$ from $W = (\rho, \rho U, E)^t$.

All these functions are written with a concern for transparency with respect to mathematical formulas. For this purpose, the vector programming capabilities of MATLAB were used. Let us explain in detail this technique for the predictor step of the Lax–Wendroff scheme (see Fig. 12.4):
- the flux $F(W)$ is computed from W values for all $j = 1, \ldots, M$ components

```
F = HYP_F_trans_w_f(w);
```

- the artificial dissipation vector is added following (12.49); we use the MATLAB built-in function diff to compute differences $W_j - W_{j-1}$; these differences are computed along the rows of the array w and only for $j \geq 2$; according to the left-boundary conditions, the artificial dissipation vector will be completed by zeros for $j = 1$:

```
F = F-Ddx*[zeros(3,1) diff(w,1,2)];
```

- the intermediate solution $\tilde{W}$ is computed only for the components $j = 1, \ldots, M-1$:

```
wtilde=0.5*(w(:,1:M-1)+w(:,2:M))-0.5*dt/dx*(F(:,2:M)-F(:,1:M-1));
```

A similar MATLAB code will be written for the corrector step, having in mind that for this step, right-boundary conditions apply, and consequently, only the components $j = 2, \ldots M-1$ of W^{n+1} are computed:

```
Ftilde = HYP_F_trans_w_f(wtilde);
Ftilde=Ftilde-Ddx*[diff(wtilde,1,2) zeros(3,1)];
w(:,2:M-1)=w(:,2:M-1)-dt/dx*(Ftilde(:,2:M-1)-Ftilde(:,1:M-2));
```

Particular attention was devoted to the implementation of Roe's scheme, which requires a separate function HYP_F_flux_roe to compute the conservative flux Φ. To reduce memory storage, the flux at the interface $\left(j + \frac{1}{2}\right)$ is computed using this once (and once is not habit!) a for loop and several local variables that can be easily identified from mathematical relations. Note also that the analytical form (12.15) for $P_{j+1/2}^{-1}$ was used instead of the (time-consuming) MATLAB built-in function inv, which calculates the inverse of a matrix.

Chapter References

C.A.J. Fletcher, *Computational Techniques for Fluid Dynamics*. (Springer-Verlag, 1991)

E. Godlewski, P.-A. Raviart, *Numerical Approximation of Hyperbolic Systems of Conservation Laws*. (Springer-Verlag, 1996)

C. Hirsch, *Numerical Computation of Internal and External Flows*. (John Wiley & Sons, 1988)

R. LeVeque, *Numerical Methods for Conservation Laws*. (Birkhäuser, 1992)

M. Saad, *Compressible Fluid Flow*. (Pearson Education, 1998)

Chapter 13
Optimization Applied to Model Fitting

Project Summary

Level of difficulty:	2
Keywords:	Optimization, minimization, descent, gradient, Hessian, parameter identification
Application fields:	Biology, epidemiology

13.1 Principle and examples of continuous optimization

What is optimization? From the mathematical point of view, it consists in finding the extrema of a numerical function defined on a set E. Starting from there, an optimization problem can be very abstract and purely mathematical or model a real problem of everyday life, going through all the degrees of complexity in the most varied scientific fields.

The set E can be continuous or discrete. For example, $J_1(x) = ||x||^2$ with $x \in \mathbb{R}^2$ and $J_2(n) = n_1^2 + n_2^2$ with $n \in \mathbb{Z}^2$ both admit a minimum in $(0,0)$, but the proof of the result calls on different properties. More generally, combinatorics problems, consisting in minimizing functions defined on subsets of $\mathbb{Z}^d$, belong to the field of "discrete optimization", which is not discussed in this chapter, where we tackle exclusively continuous optimization problems defining on $\mathbb{R}^n$ or subsets of $\mathbb{R}^n$. An optimization problem can have one or several objectives. For instance, a class of classical problems in economics consists of maximizing profits while minimizing production costs. The set E can be of finite or infinite dimension. For example, a shape optimization problem consisting in finding the shape of an object which maximizes its resistance to a constraint is posed in the set of bounded parts of $\mathbb{R}^3$. This set is of infinite dimension, in the general case it is not even a vector space.

The practical resolution of such a problem consists in seeking an approximate solution in a vector space of finite dimension judiciously designed. The search for the solution of a partial differential equation can also be formulated as a minimization problem in the functional space where one seeks the solution of the equation. The numerical resolution may consist in approximating solution by finite or spectral elements, the corresponding minimization problem becoming at the same time of finite dimension.

Finally, let us mention two main families of numerical methods available to solve a continuous optimization problem in finite dimension: deterministic methods and stochastic methods. These are all iterative methods, with properties of convergence, precision, and speed, that will influence their selection for a given problem. Deterministic methods are in general based on the regularity of the function to be minimized in order to move in the space of unknowns in directions that will cause it to decrease globally. These are fast methods, but can be trapped around local minima, depending on the starting point. Moreover, they require either knowledge of the gradient of the function, or the implementation of an approximation method to calculate it. Stochastic minimization methods are more robust, do not make regularity assumptions on the function to be minimized, and are better able to find a global minimum in the case of local minima. On the other hand, they are very slow and computationally intensive compared to descent methods.

In this chapter, we recall some theoretical results concerning unconstrained optimization and we illustrate them on the problem of identifying parameters by adjustment to the data, for a simple epidemic model.

13.2 Mathematical overview in the case $J : \mathbb{R}^n \to \mathbb{R}$

In this section, we summarize the theoretical results underlying the numerical methods that we plan to use. We first introduce the canonical continuous optimization problem in finite dimensions. We seek the solution of

$$(P) \qquad \inf_{x \in \mathbb{R}^n} J(x)$$

where $J : \mathbb{R}^n \longrightarrow \mathbb{R}$ is a smooth function. If (P) has a solution $x^\star \in \mathbb{R}^n$ such that $J(x^\star) = \min_{\mathbb{R}^n} J(x)$, then $x^\star$ is a global minimizer. A vector $x^\star$ is a local minimizer if there exists an open subset $V \subset \mathbb{R}^n$ such that $x^\star \in V$ and $J(x^\star) = \min_V J(x)$.

Let us suppose that J is Fréchet differentiable on V: for all $x, h \in V$, such that $x + h \in V$

$$J(x + h) = J(x) + \langle \nabla J(x), h \rangle + o(\|h\|), \tag{13.1}$$

where $\langle .,. \rangle$ denotes the canonical scalar product on $\mathbb{R}^n$, $\|.\|$ the associated norm and $\nabla J(x)$ the gradient of J at x

$$\nabla J(x) = \left(\frac{\partial J(x)}{\partial x_i} \right)_{i=1,\ldots,n}.$$

From definition (13.1) we can derive the first-order necessary optimality condition: if $J(x^\star) = \min_V J(x)$ then $\nabla J(x^\star) = 0_{\mathbb{R}^n}$ (see, for instance, Nocedal-Wright (2006) for a proof).

If J is twice differentiable on V, for all $x \in V$, h such that $x + h \in V$

$$J(x+h) = J(x) + \langle \nabla J(x), h \rangle + \frac{1}{2} \langle HJ(x)h, h \rangle + o(\|h\|^2), \qquad (13.2)$$

where $HJ(x)$ is the Hessian matrix of J at x

$$HJ(x) = \left(\frac{\partial^2 J(x)}{\partial x_i \partial x_j} \right)_{i,j=1,\ldots,n}.$$

Hence we can infer that if $J(x^\star) = \min_V J(x)$ then the Hessian matrix $HJ(x^\star)$ is positive semidefinite and J is locally convex on a neighborhood of $x^\star$. On the other hand, if $\nabla J(x^\star) = 0$ and if

- either $HJ(x^\star)$ is symmetric positive definite,
- or $HJ(x)$ is positive semidefinite on a neighborhood of $x^\star$,

then $x^\star \in \mathbb{R}^n$ is a local minimizer of J.

For a more detailed presentation of necessary or sufficient optimality conditions see, for instance, Nocedal-Wright (2006).

13.3 Numerical methods

Many iterative numerical methods have been developed to compute an approximation of $x^\star$. The basic principle consists in defining a sequence x_k, $k = 0, \ldots$, starting from a guess value x_0. Relying on the first order necessary condition $\nabla J(x^\star) = 0_{\mathbb{R}^n}$, the approximation of $x^\star$ is the first value x_k such that $\|\nabla J(x_k)\| \leq \varepsilon$. We present two classes of iterative methods: descent and Newton-like methods.

13.3.1 Descent algorithms

The so-called "descent" iterative methods consist in exploring the search space following descent directions $(d_k)_k$.

Algorithm 13.1 *Descent algorithm*
Input *function* $J : \mathbb{R}^n \longrightarrow \mathbb{R}$ *and its gradient* $\nabla J : \mathbb{R}^n \longrightarrow \mathbb{R}^n$,
required precision: $\varepsilon > 0$,
maximum number of iterations: $k_{\max}$,
initial guess for the solution: $x_0 \in \mathbb{R}^n$.
Result *Approximation of* $x^\star$ *such that* $J(x^\star) = \min_{x \in \mathbb{R}^n} J(x)$.
Initialization $k = 0$,
While $\|\nabla J(x_k)\| > \varepsilon$ *and* $k < k_{\max}$,
    Choose descent direction d_k *(such that* $\langle \nabla J(x_k), d_k \rangle < 0$*)*,
    Choose step α_k *in direction* d_k, *such that* $J(x_k + \alpha_k d_k) \leq J(x_k)$,
    $x_{k+1} = x_k + \alpha_k d_k$,
    $k \leftarrow k + 1$,
end While
Return x_k.

A descent direction for J at x is a vector d for which it exists $\alpha_d > 0$ such that,

$$J(x + \alpha d) \leq J(x), \quad \text{for all } 0 \leq \alpha \leq \alpha_d.$$

From (13.1), we obtain that a descent direction d is such that $\langle d, \nabla J(x) \rangle < 0$. The most evident choice for d_k is therefore the opposite of the gradient. It will be our choice in numerical examples. However it is not the optimal direction in terms of algorithmic efficiency, as proven by the famous conjugated gradient method (Hestenes 1952).

The step α_k in direction d_k is chosen in order to make the function decrease, $J(x_k + \alpha_k d_k) \leq J(x_k)$. In the general case, choosing the step α_k also requires an iterative algorithm. We will use here the simplest one, the "backtracking line search", based on Armijo's rule.

Algorithm 13.2 *Backtracking linesearch (BLS) algorithm.*
Input *Function J, current position x, descent direction d, coefficients $\tau \in$*]0, 1[*and* $\omega \in$]0, 1[.
Result α *such that* $J(x + \alpha d) < J(x)$.
Initialization $k = 0$, *initial guess* $\alpha_0 = 1$.
While $J(x + \alpha_k d) > J(x) + \omega \alpha_k \langle d, \nabla J(x) \rangle$,
    Choose $\alpha_{k+1} = \tau \alpha_k$,
    $k \leftarrow k + 1$,
end While
Return $\alpha = \alpha_k$.

13.3.2 Newton-like algorithms

The other family of methods is based on the famous Newton algorithm, generally used to find the zeros of a nonlinear system of equations, which is indeed what is required to solve $\nabla J(x^\star) = 0$. As illustrated in Fig. 13.1, in the scalar case, at each iteration, x_{k+1} is defined as the intersection of the abscissa axis with the tangent to the graph of J' at x_k, which is a linear operation. A generalization of this method in the vector case, the Newton Raphson algorithm, consists in solving at each iteration the linear system $HJ(x_k)(x_{k+1} - x_k) = -\nabla J(x_k)$.

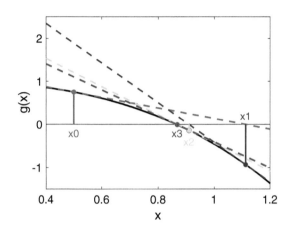

Fig. 13.1 First four iterations of the Newton algorithm to find the zero of $J'(x) = g(x) = \cos(x) - x^3$, starting from $x_0 = 0.5$.

Algorithm 13.3 *Newton Raphson algorithm.*
Input *Function $J(x)$, gradient $\nabla J(x)$, Hessian matrix $HJ(x)$, tolerance ε, max number of iterations $k_{\max}$.*
Result *Approximation of $x^\star$ such that $\nabla J(x^\star) = 0$.*
Initialization $k = 0$, x_0.
While $\|\nabla J(x_k)\| > \varepsilon$ *and* $k \le k_{\max}$,
    Solve $HJ(x_k)d_k = -\nabla J(x_k)$,
    $x_{k+1} = x_k + d_k$,
    $k \leftarrow k + 1$,
end While
Return x_k.

This method is very efficient because its convergence speed is quadratic. However, in the general case, it converges only if the initial guess is close enough to the (unknown) solution. Furthermore, computing the Hessian can be costly, cumbersome, and sometimes impossible. Quasi-Newton alternatives are implemented in modern toolboxes. They consist in approximating the Hessian matrix, or better, its inverse, like in the following Broyden-Fletcher-Goldfarb-Shanno (BFGS) variant (Fletcher 2000).

Algorithm 13.4 *Quasi-Newton algorithm.*
Input *Function $J(x)$, gradient $\nabla J(x)$, tolerance ε, max number of iterations*
$k_{\max}$.
Result *Approximation of $x^\star$ such that $\nabla J(x^\star) = 0$.*
Initialization $k = 0$, x_0, $g_0 = J(x_0)$, $B_0 = I_{n \times n}$.
While $\|g_k\| > \varepsilon$ *and* $k \leq k_{\max}$,
    Set descent direction $d_k = -B_k g_k$.
    Choose step α_k in direction d_k with BLS algorithm (13.2).
    Update $x_{k+1} = x_k + \alpha_k d_k$ and $g_{k+1} = \nabla J(x_{k+1})$.
    Compute $s_k = g_{k+1} - g_k$.

$$\text{Update } B_{k+1} = B_k + \frac{s_k s_k^T}{\langle d_k, s_k \rangle} - \frac{B_k d_k d_k^T B_k}{\langle d_k, B_k d_k \rangle}.$$

    $k \leftarrow k + 1$.
end While
Return x_k.

Exercise 13.1 1. Write a MATLAB function
    `function [alpha,iter]=OPT_F_BLS(J,x,y,d,g)`, where $y = J(x)$ and
    $g = \nabla J(x)$.
    `OPT_F_BLS` computes the step α at point x in the descent direction d (such
    that $\langle d, g \rangle < 0$, using the BLS Algorithm 13.2 with $\tau = 0.5$ and $\omega = 10^{-4}$.
    The return argument `iter` is the number of iterations.
2. Write a MATLAB function
    `[xstar,iter,fcall]=OPT_F_Descent(J, x0,eps, maxiter)`.
    The definition of function J is expected to be `function [y,g]=J(x)`,
    where y is the value of the function J and g its gradient ∇J at point
    x.
    The descent direction is set as the opposite of the gradient. The step in
    the descent direction is computed using `OPT_F_BLS` function. The return
    arguments `iter` and `fcall` are the number of iterations and the number
    of calls to the function J.
3. Write a MATLAB function
    `function [xstar,iter]=OPT_F_Newton(J, x0,eps, maxiter)`.
    The definition of function J is now expected to be `function [y,g,H]=J(x)`,
    where $y = J(x)$, $g = \nabla J(x)$ and $H = HJ(x)$.

13.4 Application to fitting of model parameters

13.4.1 General Least-Square problem

In this paragraph, we show how the adjustment of the parameters of a model
leads to solving an optimization problem. We define a model as a family of
functions $\mathcal{M}$ whose elements f_x depend on a parameter $x \in \mathbb{R}^n$. For instance,

$$\mathcal{M} = \left\{ f_x : \mathbb{R} \to \mathbb{R}, t \longmapsto x\,t \right\},$$

is the set of scalar linear functions parametrized by $x \in \mathbb{R}$. Suppose that we have at our disposal a dataset $\{(t_i, y_i)\}_{1 \leq i \leq d}$ representative of the behavior described by our model $\mathcal{M}$. This means that there exists a function $f_\alpha \in \mathcal{M}$ such that

$$\forall\, i = 1 \ldots d, \qquad y_i = f_x(t_i),$$

where $t_i \in \mathbb{R}^m$ and $y_i \in \mathbb{R}^p$, $i = 1 \ldots d$ for a given parameter $x \in \mathbb{R}^n$. Indeed in practice the data will rather satisfy $y_i = f_x(t_i) + \varepsilon_i$ with an error $(\varepsilon_i)_i$ arising from acquisition procedures or from the imperfect adequacy of the model. The so-called "model fitting" problem consists in finding the function in $\mathcal{M}$ which "best" reproduces the dataset $\{(t_i, y_i)\}_{1 \leq i \leq d}$. It can be formalized by defining the least square distance $J(x)$ between the dataset and the set $\{(t_i, f_x(t_i))\}_{1 \leq i \leq d}$ for a given parameter $x \in \mathbb{R}^n$

$$J(x) = \sum_{i=1}^{d} \| f_x(t_i) - y_i \|_2^2.$$

The Euclidean norm $\|.\|_2$ measures the distance between one data point and its approximation by the model. Fitting the model will consist in minimizing $J(x)$ with respect to $x \in \mathbb{R}^n$. We therefore seek $x^\star$ such that

$$J(x^\star) = \min_{x \in \mathbb{R}^n} J(x).$$

Exercise 13.2 In order to solve the problem numerically using the algorithms programmed in the previous section, the gradient and the Hessian of the function $J(x) = \sum_{i=1}^{d} \| f_x(t_i) - y_i \|_2^2$ are required.

1. Assume that the model function f_x is sufficiently smooth in x. Compute $\nabla J(x)$ and $HJ(x)$ in terms of $\left(\dfrac{\partial f_x(t)}{\partial x_k} \right)_{k=1,\ldots,n}$ and $\left(\dfrac{\partial^2 f_x(t)}{\partial x_k \partial x_j} \right)_{k,j=1,\ldots,n}$.
   In the scalar model case $m = p = 1$, write simpler expressions in terms of $\nabla_x f_x(t)$ and $H_x f_x(t)$.
2. In the scalar model case $m = p = 1$, we assume that
   The outputs of `function [yx,gx,Hx]=fx(t,x)` are $f_x(t)$, $\nabla_x f_x(t)$ and $H_x f_x(t)$. Write a MATLAB function
   `function [y,g,H]=OPT_F_J(x,fx,tdata,ydata)`
   which computes $J(x)$, $\nabla J(x)$ and $HJ(x)$.

Exercise 13.3 Assume that the model is scalar and polynomial. For example, $f_x(t) = x_0 + x_1 t + x_2 t^2$ in the case $n = 3$.

1. Write the least square minimization problem introducing the matrix $T \in \mathcal{M}_{d,3}(\mathbb{R})$ such that $T_{i,p} = t_i^{p-1}$ for $p = 1 \ldots 3$ and $i = 1, \ldots, d$.
2. Write the linear system giving the solution $x^\star \in \mathbb{R}^3$.

3. Use this example to check the descent algorithm 13.3.

    a. Choose a vector $x \in \mathbb{R}^3$, choose $d = 10$ points in the interval $[-1, 1]$, compute $f_x(t_i)$ for $i = 1, \ldots, d$ and add a randomly generated noise of variance $\sigma = 0.1$.

    b. Compute $x^\star$ provided by question 2.

    c. Compute $x^\star$ using the descent algorithm. You will need to define the MATLAB function `[yx,gx,Hx]=OPT_F_poly_mod(t,x)`.

4. The data in file `OPT_unknown_poly.xlsx` (Excel format) has been generated with an unknown polynomial of degree $p \leq 7$. Read the data using the following syntax and identify the degree and the coefficients of the polynomial.

```
M = readmatrix("OPT_unknown_poly.xlsx");
T_data=M(1,:);
y_data=M(2,:);
```

13.4.2 Epidemic modeling

In this section, we adjust the parameters of two models using datasets recorded during the Plague epidemic in Mumbai between December 17, 1905 and July 21, 1906. This disease was almost always rapidly fatal, and the recorded numbers of deaths $(D_i)_{i=1,\ldots,30}$ are available on a time span of $d = 30$ weeks. The $(t_i, D_i)_{i=1,\ldots,30}$ data points are available in the file `data.xls`. We reproduce in the sequel two numerical experiments detailed by Bacaer (2012).

Exercise 13.4 Read the file `data.xlsx` and display the data points (t_i, D_i) for $i = 1, \ldots, d$, marked with symbols "o". The time axis unit should be weeks.

13.4.2.1 SIR model fitting

Mathematical modeling of epidemics has become famous due to the Covid pandemic. Here we consider a compartmental method, which goes back to the seminal work by Kermack (1927). The population is divided into three compartments whose size varies with time: the susceptible $S(t)$, the infected $I(t)$, and the recovered (or dead) $R(t)$. In this simple model, one cannot catch the disease twice, and the total population—including the deaths—remains constant and equal to $N = S(t) + I(t) + R(t)$. The sizes of the three compartments are governed by an ODE model

$$\begin{cases} \dfrac{dS(t)}{dt} = -\alpha S(t)I(t), \\[2mm] \dfrac{dI(t)}{dt} = \alpha S(t)I(t) - \beta I(t), \\[2mm] \dfrac{dR(t)}{dt} = \beta I(t). \end{cases} \qquad (13.3)$$

The coefficients α and β quantify the contagiousness and the recovery (or death) rate of the disease, respectively. The Cauchy problem for an initial condition (S_0, I_0, R_0) has a unique solution for $t \geq 0$. In fact, it is sufficient to consider the two equations for S and I. Given numerical values for α and β, and the initial condition (S_0, I_0), the system of ODE can be solved and the values $I(t_i)$ multiplied by the β coefficient can be compared to experimental data points D_i, for $i = 1, \dots, d$.

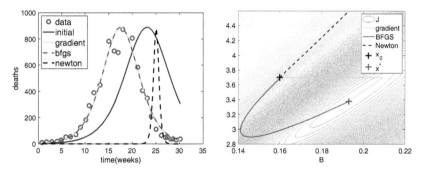

Fig. 13.2 Left: Data points (blue circles o), initial guess (blue) and model solutions obtained using the gradient descent (green), quasi Newton-BFGS (magenta) and Newton (black). Right: Isovalues of $J(x)$ and minimization paths for gradient, BFGS (both converging toward the red cross) and Newton (diverging).

Exercise 13.5 A model, called exponential model in the sequel, has been suggested by Kermack (1927) to reproduce the behavior of $\dfrac{dR}{dt}$:

$$\frac{dR}{dt} \approx E(t; A, B; \varphi) = \frac{A}{\cosh^2(Bt - \varphi)},$$

where A, B, and φ depend in a rather complicated way on the SIR model parameters. The value $A = 890$ is provided. Optimal values for $x = (B, \varphi)$ will be obtained by minimizing

$$J(x) = \sum_{i=1}^{d} \|E(t_i; A; B; \varphi) - D_i\|_2^2.$$

1. Write a MATLAB function `function [yx,gx,Hx]=OPT_F_expmod(t,x)` to compute $E(t; A, B, \varphi)$. The argument x is the column vector containing B and φ. The value of the parameter $A = 890$ is fixed. Arrays `gx` and `Hx` contain $\nabla_x E(t; x)$ and $H_x E(t; x)$, respectively.
2. Compute the optimal solution using the gradient method (13.1) and the Newton method (13.3) (see the left panel of Fig. 13.2):

   a. Use the functions `OPT_F_J.m` from Exercice 13.2 and `OPT_F_expmod.m` to design the objective function.
   b. Use $x_0 = (0.15, 3.6)^T$ as initial guess, do you get a reasonable solution? Start from the solution $x_0 = (0.2, 3.4)^T$ proposed in the reference Bacaer (2012). Why is your solution different?
   c. Minimize instead of $J(x)$ the following functional

$$J_w(x) = \sum_{i=1}^{d} \frac{\|f_x(t_i) - y_i\|_2^2}{\|y_i\|^2}.$$

3. Draw the contour of the objective function on a box around $[0.14, 0.5] \times [2.5, 7]$ (see the right panel of Fig. 13.2). Does it help to understand why the model fitting is so difficult?
4. Implement the quasi-Newton algorithm (13.4).

Exercise 13.6 We now use the ODE system (13.3) to predict the number of deaths per time unit: $D(t; x) = \dfrac{dR(t)}{dt} = \beta I(t)$.

1. Write a MATLAB `function [ya]=OPT_F_SIR(t,x)` computing the solution of the ODE system $(S(t), I(t))$ at time instants stored in vector t. The model parameters are defined as $x = (\alpha, \beta)^T$. Use the built-in MATLAB ODE solver `ode45` (see examples in Chap. 2). Draw $D(t; x)$ as a function of t for different values of x: $(5, 5)^T$, $(4.5, 5)^T$, and $(5.5, 5)^T$.
2. Define the system of ODE satisfied by $(S(t), I(t), \dfrac{\partial S}{\partial \alpha}(t), \dfrac{\partial I}{\partial \alpha}(t), \dfrac{\partial S}{\partial \beta}(t), \dfrac{\partial I}{\partial \beta}(t))$. Modify the function `OPT_F_SIR`.
3. Write a MATLAB `function [J,g]=OPT_F_D_SIR(t,x)` computing $D(t) = \beta I(t)$ and its gradient with respect to $x = (\alpha, \beta)^T$. This function will be used as second input parameter of the fit function `OPT_F_J.m`.
4. Compute the optimal solution using the gradient method and the BFGS method. Use $(S_0, I_0) = (57000, 1)$ as initial condition of the ODE system. Try $x_0 = (10^{-4}; 5)^T$ as initial guess. Do you get a reasonable solution? Is the MATLAB toolbox `fminunc` any better?
5. Normalize the first unknown parameter by S_0, i.e., $x = (\alpha S_0, \beta)$ instead of $x = (\alpha, \beta)$. Making `S0=57000` a global variable, there should be only two lines to modify in function `OPT_F_SIR`. Try with $x_0 = (5; 5)^T$ as initial guess and explain why the three methods give now reasonable answers.

13.5 Solutions and Programs

Solution of Exercise 13.1

See files OPT_F_BLS.m, OPT_F_Descent.m, and OPT_F_Newton.m.

Solution of Exercise 13.2

1. In the general case $x \in \mathbb{R}^n$ and $f_x(t) \in \mathbb{R}^p$. Therefore

$$\nabla J(x) = \left(\frac{\partial J(x)}{\partial x_k}\right)_{k=1,\ldots,n}, \quad \text{where}$$

$$\frac{\partial J(x)}{\partial x_k} = 2\sum_{i=1}^{d}\langle\frac{\partial f_x(t_i)}{\partial x_k}, f_x(t_i) - y_i\rangle_{\mathbb{R}^p}.$$

$$HJ(x) = \left(\frac{\partial^2 J(x)}{\partial x_k \partial x_j}\right)_{k,j=1,\ldots,n}, \quad \text{where}$$

$$\frac{\partial^2 J(x)}{\partial x_k \partial x_j} = 2\sum_{i=1}^{d}\left(\langle\frac{\partial^2 f_x(t_i)}{\partial x_k \partial x_j}, f_x(t_i) - y_i\rangle_{\mathbb{R}^p} + \langle\frac{\partial f_x(t_i)}{\partial x_k}, \frac{\partial f_x(t_i)}{\partial x_j}\rangle_{\mathbb{R}^p}\right).$$

2. In the case $p = m = 1$ the least square function boils down to

$$J(x) = \sum_{i=1}^{d}(f_x(t_i) - y_i)^2.$$

Denoting by $\nabla_x f_x(t) = \left(\frac{\partial f_x(t)}{\partial x_k}\right)_{k=1}^{n}$ and $H_x f_x(t) = \left(\frac{\partial^2 f_x(t)}{\partial x_k \partial x_j}\right)_{k,j=1}^{n}$ the gradient and the Hessian of the model f seen at a fixed time t as a function of x, the gradient and Hessian of J become

$$\nabla J(x) = 2\sum_{i=1}^{d}(f_x(t_i) - y_i)\nabla_x f_x(t_i),$$

$$HJ(x) = 2\sum_{i=1}^{d}\left((f_x(t_i) - y_i)H_x f_x(t_i) + \nabla_x f_x(t_i)\nabla_x f_x(t_i)^T\right).$$

See file OPT_F_J.m for the implementation. Note that if a function J(x) of a single argument x is expected, it can be defined using MATLAB anonymous function method

J(x)=@(x) OPT_F_J(x,fx,t_data,x_data)

Calling J(x) will amount to call OPT_F_J(x,fx,t_data,x_data) with fx,t_data,x_data set to their values at the time of J(x) definition.

Solution of Exercise 13.3

1. The function to be minimized is $J(x) = \|Tx - Y\|^2$ where $T \in \mathcal{M}_{d,3}(\mathbb{R})$ such that $T_{i,p} = t_i^{p-1}$ for $p = 1 \ldots 3$ and $i = 1, \ldots, d$ and $Y \in \mathcal{M}_{d,1}(\mathbb{R})$. Its gradient is $\nabla J(x) = 2T^T(Tx - Y)$ which is zero if $T^T Tx = T^T Y$. The matrix $T^T T$ is positive. It is definite if at least n abscissa t_i are different, assuming of course that $d \geq n$. In that case $x^\star = (T^T T)^{-1} T^T Y$.

2. See files `OPT_M_poly.m` and `OPT_F_poly_mod.m`. Note that MATLAB function `OPT_F_J.m` expects 4 input arguments `x,fx,tdata,ydata` while the algorithm `OPT_F_Descent.m` expects a function of one argument only. This is obtained by defining a function of a single argument `fit_fun(x)` by "fixing" the three remaining arguments of `OPT_F_J` using MATLAB "anonymous" function syntax:

   `fit_fun=@(x) OPT_F_J(x,@OPT_poly_mod,tdata,y)`

3. See file `OPT_M_unknown_poly.m`. The unknown polynomial is of degree 4:

$$p(t) = \frac{t}{2} - \frac{t^2}{2} - t^3 + t^4.$$

Fitting a polynomial of degree higher than 4 will give the same solution.

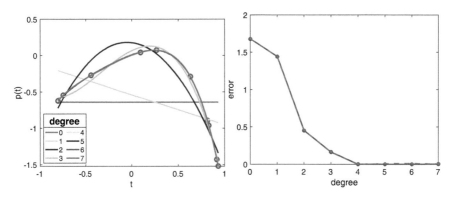

Fig. 13.3 Left: Data points (blue circles o) and best fitted polynomials of degree 0 to 7. Right: Least square error as a function of the polynomial degree.

The left panel of Fig. 13.3 displays the dataset of values of the unknown polynomial at 10 points, along with the best fit polynomials of degree 0 to 7. The last four curves are superimposed and pass through the datapoints. The right panel displays the residual least-square error at the best fit as a function of the polynomial degree. It is zero for degrees higher or equal to 4.

Solution of Exercise 13.4

After reading the MATLAB documentation

```
M = readmatrix("OPT_data.xlsx");
Tdata=M(1,2:31);
Ydata=M(2,2:31);
```

Solution of Exercise 13.5

1. The gradient and Hessian of function

$$E(t;x) = \frac{A}{\cosh^2(Bt - \varphi)},$$

with respect to parameters $x = (B, \varphi)$ are

$$\nabla_x E(t;x) = \frac{-2A \sinh^3(Bt - \varphi)}{\cosh^3(Bt - \varphi)} \begin{pmatrix} t \\ -1 \end{pmatrix},$$

$$H_x E(t;x) = 2A \left(\frac{2}{\cosh^2(Bt - \varphi)} - \frac{3}{\cosh^4(Bt - \varphi)} \right) \begin{pmatrix} t^2 & -t \\ -t & 1 \end{pmatrix}.$$

2. See files OPT_M_exp.m, OPT_F_exp_mod.m, and OPT_F_J_w.m.
3. As seen in Fig. 13.2, the isovalues of the objective function present a narrow valley in the vicinity of the minimum (red cross). The path followed by the Newton method does not find it but goes to another local minimum.

Solution of Exercise 13.6

1. See file OPT_F_SIR.m for the definition of the right-hand side of the ODE model (13.3). Since the MATLAB solver odeint expects a function f(t,y) to solve the ODE system $y'(t) = f(t, y(t))$ it must be defined as an anonymous function

```
SIRty=@(t,y) OPT_F_SIR(t,y,x,1);
```

The fourth argument in OPT_F_SIR calling sequence should be set to 1 if the ODE system including the partial derivatives with respect to x is solved, 0 otherwise and by default.
2. Differentiating the equations of SIR model with respect to α and β leads to

$$
\begin{cases}
\dfrac{dS(t)}{dt} = -\alpha S(t)I(t), \\[2mm]
\dfrac{dI(t)}{dt} = \alpha S(t)I(t) - \beta I(t), \\[2mm]
\dfrac{d}{dt}\dfrac{\partial S(t)}{\partial \alpha} = -S(t)I(t) - \alpha \dfrac{\partial S(t)}{\partial \alpha}I(t) - \alpha S(t)\dfrac{\partial I(t)}{\partial \alpha}, \\[2mm]
\dfrac{d}{dt}\dfrac{\partial I(t)}{\partial \alpha} = S(t)I(t) + \alpha \dfrac{\partial S(t)}{\partial \alpha}I(t) + \alpha S(t)\dfrac{\partial I(t)}{\partial \alpha} + \beta \dfrac{\partial I(t)}{\partial \alpha}, \\[2mm]
\dfrac{d}{dt}\dfrac{\partial S(t)}{\partial \beta} = -\alpha \dfrac{\partial S(t)}{\partial \beta}I(t) - \alpha S(t)\dfrac{\partial I(t)}{\partial \beta}, \\[2mm]
\dfrac{d}{dt}\dfrac{\partial I(t)}{\partial \beta} = \alpha \dfrac{\partial S(t)}{\partial \beta}I(t) + \alpha S(t)\dfrac{\partial I(t)}{\partial \beta} + \beta \dfrac{\partial I(t)}{\partial \beta} + I(t),
\end{cases}
$$

which can be solved starting from the initial condition $(S_0, I_0, 0, 0, 0, 0)^T$.

3. The fit function is

$$
J(x) = \sum_{i=1}^{30} \|\beta I(t_i; \alpha, \beta) - D_i\|_2^2.
$$

In order to use the function OPT_F_J.m and OPT_F_J_w.m we must write a function OPT_F_D_SIR(t;x) computing $\beta I(t; \alpha, \beta)$ and its gradient

$$
\nabla_x D_{SIR}(t; x) = \begin{pmatrix} \beta \dfrac{\partial I(t)}{\partial \alpha} \\[3mm] \beta \dfrac{\partial I(t)}{\partial \beta} + I(t) \end{pmatrix}.
$$

4. The actual fit of the SIR model is computed in the program OPT_M_ode.m which produces Fig. 13.4. Starting with $x_0 = (10^{-4}; 5)^T$, the built-in MAT-LAB solution converges toward an unphysical solution $x = (-0.000054; 5.)^T$. The BFGS quasi-Newton algorithm stops close to the initial guess, and the gradient algorithm marks time close to $x = (0.000095; 5)^T$ which is an acceptable solution.

5. Normalizing α is easily done by introducing the normalization factor as a global variable both in the calling script OPT_M_ode.m and the function OPT_F_SIR. In the calling script just 3 lines should be modified

```
global S0norm
alpha=1.e-4;
beta=5;
S0norm=57000;   % or 1 if no normalization
x0=[alpha*S0norm;beta];
```

In the RHS of the ODE system, the normalization intervenes in the two equations for the corresponding derivatives

$$\frac{d}{dt}\frac{\partial S(t)}{\partial \alpha S_0} = -\frac{S(t)I(t)}{S_0} - \alpha \frac{\partial S(t)}{\partial \alpha S_0}I(t) - \alpha S(t)\frac{\partial I(t)}{\partial \alpha S_0},$$

$$\frac{d}{dt}\frac{\partial I(t)}{\partial \alpha S_0} = \frac{S(t)I(t)}{S_0} + \alpha \frac{\partial S(t)}{\partial \alpha S_0}I(t) + \alpha S(t)\frac{\partial I(t)}{\partial \alpha S_0} + \beta \frac{\partial I(t)}{\partial \alpha S_0}.$$

Therefore the corresponding lines must be modified in OPT_F_SIR

```
global S0norm
alpha=x(1)/S0norm;
...
    DY(3,1)=-S*I/S0norm-alpha*dSda*I-alpha*dIda*S;
```

With this normalization and starting from the same initial guess $x_0 = (10^{-4}S_0; 5)^T$ `fminunc` converges to J(5.66,5.27)=325, the gradient algorithm converges to J(5.44,5.05)=255 and the BFGS quasi-Newton algorithm to J(5.14,4.74)= 217. This last algorithm is therefore now the one achieving the best fit. Its bad behavior in the previous situation is due to initial guess for the Hessian ($= I_{n\times n}$), very far from the true Hessian when the two parameters vary on very different scales. More generally, the three methods now give reasonable answers thanks to the normalization which balances the influence. This example shows that it is useful to compare several methods, since their performance can vary depending on the case.

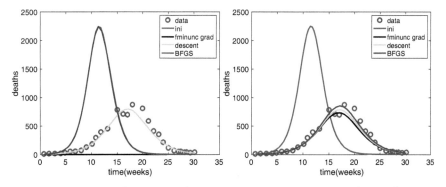

Fig. 13.4 Data points (blue circles o), model solution with initial guess (red line), best fit with `fminunc` (black crosses), gradient (green line), BFGS quasi-Newton (magenta line). Left panel: $x = (\alpha, \beta)$, right panel: $x = (\alpha S_0, \beta)$.

Chapter References

N. Bacaër, The model of Kermack and McKendrick for the plague epidemic in Bombay and the type reproduction number with seasonality. J. Math. Biol. **64**, 403–422 (2012).

S. Boyd, L. Vandenberghe, *Convex Optimization*. (Cambridge University Press, 2004)

J. Nocedal, S.J. Wright, *Numerical Optimization*. (Springer, 2006).

R. Fletcher, *Practical Methods of Optimization*, 2nd edn. (John Wiley & Sons, 2000)

M.R. Hestenes, E. Stiefel, Methods of conjugate gradients for solving linear systems. J. Res. Nat. Bur. Stan. B-49 (1952)

W.O. Kermack, A.G. McKendrick, Contributions to the mathematical theory of epidemics. Proc. R Soc. A **115**, 700–721 (1927)

Chapter 14
Thermal Engineering: Optimization of an Industrial Furnace

Project Summary

Level of difficulty:	2
Keywords:	Finite element method, Laplace differential operator, direct problem, inverse problem
Application fields:	Thermal engineering, optimization

14.1 Introduction

In this chapter we deal with a simple but realistic optimization problem. We have to find the optimal temperature of an industrial furnace in which are made resin pieces, such as car bumpers. The heating system is based on electric resistances, and the first part of this study is to compute the temperature field inside the oven when the values of the resistances are known. This part is called the *direct* problem: the resistances' values are known and the temperature field is unknown. It is important here to emphasize that the mechanical properties of the bumper depend on the temperature during the cooking; so the second part of the study is devoted to computing the resistances' values in order to maintain the bumper temperature at the "*good*" value. This optimization problem is called an *inverse* problem: the temperature is the input and the values of resistances are the outputs.

The computation of the temperature field is performed with the finite element method. Only the main features of this method are recalled here; for further details we refer to Ciarlet (1978), Norrie and de Vries (1973) and Zienkiewicz (1971).

© The Author(s), under exclusive license to Springer Nature Switzerland AG 2023   309
I. Danaila et al., *An Introduction to Scientific Computing*,
https://doi.org/10.1007/978-3-031-35032-0_14

14.2 Formulation of the Problem

For the sake of simplicity we limit the geometry of the problem to elementary shapes: the bumper is a rectangle placed in a rectangular domain Ω representing the oven; the edges are referred to as the boundary $\partial\Omega$ (see Fig. 14.1). This boundary is the union of three nonempty parts: $\partial\Omega_D$, $\partial\Omega_N$, and $\partial\Omega_F$, satisfying the following conditions:

$$\partial\Omega = \partial\Omega_D \cup \partial\Omega_N \cup \partial\Omega_F \text{ and } \partial\Omega_D \cap \partial\Omega_N = \partial\Omega_D \cap \partial\Omega_F = \partial\Omega_F \cap \partial\Omega_N = \emptyset.$$

The partial differential equation arising from the heat diffusion phenomenon in the oven can be written as

$$\begin{cases} \text{Find } T \in V \text{ such that} \\ \text{div} \left[-\mathbb{K} \overrightarrow{\text{grad}} \, T \right] = F \text{ in } \Omega. \end{cases} \tag{14.1}$$

For physical and mathematical reasons, the temperature field has to satisfy some conditions on the wall of the oven. First we impose $T = T_D$ on $\partial\Omega_D$; this is commonly referred to as a *Dirichlet boundary condition*. Another condition rules the thermal flux across $\partial\Omega_N$. This is referred to as a *Neumann boundary condition*. A last condition is devoted to the temperature balance between the inside and the outside of the oven. This *Fourier boundary condition* states that the heat transfer through $\partial\Omega_F$ is proportional to $T - T_F$, the difference between internal and external temperatures.

All these arguments are translated into mathematical terms, and we add them to the formulation of problem (14.1). They are summarized in

$$\begin{cases} T = T_D \text{ on } \partial\Omega_D, \\ \sum_{i,j} \mathbb{K}_{i,j} \dfrac{\partial T}{\partial x_j} \nu_i = f \text{ on } \partial\Omega_N, \\ \sum_{i,j} \mathbb{K}_{i,j} \dfrac{\partial T}{\partial x_j} \nu_i = g(T - T_F) \text{ on } \partial\Omega_F. \end{cases} \tag{14.2}$$

We used here the following notations:

1. T is the temperature in the domain Ω.
2. V is the set of all feasible temperatures.
3. $\mathbb{K} \in \mathbb{R}^{2 \times 2}$ is the thermal conductivity tensor. In a homogeneous isotropic medium, we have $\mathbb{K} = c I_2$, where c is the heat conductivity coefficient, and I_2 is the identity matrix.
4. The volume and surface heat sources are denoted by F and f.
5. The ambient temperature (inside the oven) is set to the value T_D on $\partial\Omega_D$.
6. The outside temperature is set to the value T_F on $\partial\Omega_F$.
7. g is the heat transfer coefficient on $\partial\Omega_F$.
8. $\nu = (\nu_1, \nu_2)^t$ is the outward normal vector on $\partial\Omega$.

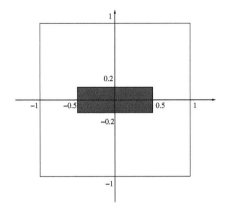

Fig. 14.1 Object and oven.

For the sake of simplicity we assume here perfect thermal insulation, that is, $f = 0$ and $g = 0$. All data necessary to deal with the problem are now determined, and we use a *Green formula*,

$$\forall T, T' \in V, \quad \int_\Omega \text{div} \left[-\mathbb{K} \overrightarrow{\text{grad}} \, T \right] T' dx$$

$$= \sum_{i,j} \int_\Omega \mathbb{K}_{i,j} \frac{\partial T}{\partial x_j} \frac{\partial T'}{\partial x_i} \, dx - \sum_{i,j} \int_{\partial\Omega} \mathbb{K}_{i,j} \frac{\partial T}{\partial x_j} T' \, \nu_i \, ds.$$

Let us introduce now the subspace $V^0 \subset V$ by $V^0 = \{T' \in V, T' \, |_{\partial\Omega_D} = 0\}$, and write the *variational formulation* of problem (14.1):

$$\begin{cases} \text{Find } T \in V^0 + T_D \text{ such that} \\ \forall T' \in V^0 \quad \sum_{K \in \mathcal{T}_h} \int_K (\overrightarrow{\text{grad}} \, T')^t \, \mathbb{K} \overrightarrow{\text{grad}} \, T \, dx = \sum_{K \in \mathcal{T}_h} \int_K T' \, F \, dx. \quad (14.3) \end{cases}$$

It has been proved that problem (14.3) is equivalent to problem (14.1)-(14.2) and has a unique solution T (see Ciarlet 1978).

14.3 Finite Element Discretization

In a real case, the physical data of problem (14.3) are provided by experimental measures; they are not trivial and neither is the geometry of the domain Ω. Consequently, it is not possible to write an explicit solution $T(x, y)$ of problem (14.1)-(14.2). This solution is estimated by the way of an approximation method such as the finite element method. This method uses piecewise polynomial functions defined on triangles or rectangles in 2D formulation (tetrahedra, hexahedra in 3D). In this study we first split the domain Ω into triangular elements, gathered in a triangulation. In the finite element theory a *triangulation* $\mathcal{T}_h$ (or *mesh*) of the domain Ω is a set of triangles satisfying the following properties:

$$\overline{\Omega} = \bigcup_{K \in \mathcal{T}_h} K,$$

$$\forall K, K' \in \mathcal{T}_h \quad K \cap K' = \begin{cases} \emptyset, \\ \text{or a vertex common to } K \text{ and } K', \\ \text{or an edge common to } K \text{ and } K'. \end{cases}$$

Then a finite-dimensional vector subspace $V_h \subset V$ is introduced. A simple example of such a subspace is provided by the so-called "Lagrange finite element" of degree 1. For an arbitrary triangle K in $\mathcal{T}_h$, with vertices A_1, A_2, and A_3, an element T'_h of V_h is defined by

$$T'_h(M) = T'_h(A_1)\lambda_1 + T'_h(A_2)\,\lambda_2 + T'_h(A_3)\lambda_3, \tag{14.4}$$

where $T'_h(A_i)$ is the temperature value at A_i, one of the three vertices of triangle K, while λ_1, λ_2, and λ_3 are the barycentric coordinates of the point M in triangle K.

Remark 14.1 Let K be an arbitrary triangle with vertices A_1, A_2, and A_3. The barycentric coordinates of point M are three real numbers λ_1, λ_2, and λ_3 such that $\lambda_1 + \lambda_2 + \lambda_3 = 1$ and

$$\overrightarrow{OM} = \lambda_1 \overrightarrow{OA_1} + \lambda_2 \overrightarrow{OA_2} + \lambda_3 \overrightarrow{OA_3}.$$

If x, y are the Cartesian coordinates of M, then the barycentric coordinates λ_1, λ_2 are a solution of the linear system

$$\begin{cases} x = x(A_3) + [x(A_1) - x(A_3)]\lambda_1 + [x(A_2) - x(A_3)]\lambda_2, \\ y = y(A_3) + [y(A_1) - y(A_3)]\lambda_1 + [y(A_2) - y(A_3)]\lambda_2, \end{cases} \tag{14.5}$$

where $(x(A_i), y(A_i))$ are the Cartesian coordinates of vertex A_i. The uniqueness of these values is guaranteed if and only if A_1, A_2, and A_3 are not on a straight line.

The definition (14.4) of the approximate temperature T'_h leads to the new relation

$$T'_h(M) = T'_h(A_3) + [T'_h(A_1) - T'_h(A_3)]\lambda_1 + [T'_h(A_2) - T'_h(A_3)]\lambda_2.$$

We introduce then the subspace $V_h^0 \subset V_h$ by

$$V_h^0 = \{T' \in V_h, T'\,|_{\partial \Omega_D} = 0\}.$$

Let T_D be the element of V_h whose components are all zero, except for $T_D(A_i)$, with point A_i located on the boundary $\partial\Omega_D$, whose values come from (14.2). The discrete variational formulation of problem (14.3) is then

$$
\left\{
\begin{array}{l}
\text{Find } T_h \in T_D + V_h^0 \text{ such that} \\[2mm]
\forall T_h' \in V_h^0 \quad \sum_{K \in \mathcal{T}_h} \int_K (\overrightarrow{\mathrm{grad}}\, T_h')^t \, \mathbb{K} \, \overrightarrow{\mathrm{grad}}\, T_h \; dx = \sum_{K \in \mathcal{T}_h} \int_K T_h' \, F \, dx.
\end{array}
\right.
$$

$$(14.6)$$

14.4 Implementation

Formulation (14.6) uses integral calculation on triangles of $\mathcal{T}_h$. Before going further into the details, we examine these terms when K is an arbitrary triangle. One has to compute the value of

$$
\int_K (\overrightarrow{\mathrm{grad}}\, T_h')^t \, \mathbb{K} \, \overrightarrow{\mathrm{grad}}\, T_h \; dx \quad \text{and} \quad \int_K T_h' \, F \, dx.
$$

14.4.1 Matrix Computation

The vector $\overrightarrow{\mathrm{grad}}\, T_h'$ has to be calculated for each T_h' in V_h and each K in $\mathcal{T}_h$. We first write

$$
\frac{\partial T_h'}{\partial \lambda_i} = \frac{\partial T_h'}{\partial x} \times \frac{\partial x}{\partial \lambda_i} + \frac{\partial T_h'}{\partial y} \times \frac{\partial y}{\partial \lambda_i} \quad \text{for } i = 1, 2. \tag{14.7}
$$

Then, using (14.4) and (14.5), we get

$$
\begin{vmatrix}
\dfrac{\partial T_h'}{\partial \lambda_1} = T_h'(A_1) - T_h'(A_3), & \dfrac{\partial x}{\partial \lambda_1} = x(A_1) - x(A_3), & \dfrac{\partial y}{\partial \lambda_1} = y(A_1) - y(A_3), \\[4mm]
\dfrac{\partial T_h'}{\partial \lambda_2} = T_h'(A_2) - T_h'(A_3), & \dfrac{\partial x}{\partial \lambda_2} = x(A_2) - x(A_3), & \dfrac{\partial y}{\partial \lambda_2} = y(A_2) - y(A_3).
\end{vmatrix}
$$

$$(14.8)$$

So a new formulation of (14.7) is

$$
\begin{bmatrix}
T_h'(A_1) - T_h'(A_3) \\
T_h'(A_2) - T_h'(A_3)
\end{bmatrix}
=
\begin{bmatrix}
x(A_1) - x(A_3) & y(A_1) - y(A_3) \\
x(A_2) - x(A_3) & y(A_2) - y(A_3)
\end{bmatrix}
\begin{bmatrix}
\dfrac{\partial T_h'}{\partial x} \\[3mm]
\dfrac{\partial T_h'}{\partial y}
\end{bmatrix}. \tag{14.9}
$$

The matrix determinant in (14.9) is

$$\Delta_K = (x(A_1) - x(A_3)) (y(A_2) - y(A_3)) - (x(A_2) - x(A_3)) (y(A_1) - y(A_3)).$$

Since $|\Delta_K|$ is twice the area of triangle K, matrix (14.9) is invertible when K is not a flat triangle (i.e., the three vertices are not on a straight line). We introduce then an array $[dl \, T_h']_K$ and a matrix B_K by

$$[dl \, T_h']_K = \begin{bmatrix} T_h'(A_1) \\ T_h'(A_2) \\ T_h'(A_3) \end{bmatrix}$$

and

$$B_K = \frac{1}{\Delta_K} \begin{bmatrix} y(A_2) - y(A_3) & y(A_3) - y(A_1) & y(A_1) - y(A_2) \\ x(A_3) - x(A_2) & x(A_1) - x(A_3) & x(A_2) - x(A_1) \end{bmatrix}$$

and write

$$\int_K (\overrightarrow{\text{grad}} \, T_h')^t \, \mathbb{K} \, \overrightarrow{\text{grad}} \, T_h \; dx = [dl \, T_h']_K^t [A_K][dl \, T_h]_K.$$

Matrix $[A_K]$ is the *element matrix*, and is computed by

$$[A_K] = \frac{1}{2} c_K \Delta_K B_K^t B_K.$$

Remark 14.2 The value of c_K, the thermal conductivity coefficient, is different in the air and in the resin, and so depends on K.

14.4.2 Right-Hand-Side Computation

We assume in the following that the heat source function F_r associated with an electrical resistance located at point $P_r(x_r, y_r)$ has the form

$$F_r(x, y) = \frac{F_0}{2} \exp \left[-d^2(x, y) \right], \quad \text{with } d^2(x, y) = \frac{1}{2R_r^2} \left((x - x_r)^2 + (y - y_r)^2 \right),$$

so using the previous notation we may write

$$\pi R_r^2 \int_\Omega F_r \; dx = F_0 \quad \text{and} \quad \int_K T_h' F_r \; dx = [dl \, T_h']_K^t [b_K],$$

where the array $[b_K]$ is the *element right-hand side*, computed by means of the numerical integration formula

$$[b_K] = \frac{\Delta_K}{24} \begin{bmatrix} 2 & 1 & 1 \\ 1 & 2 & 1 \\ 1 & 1 & 2 \end{bmatrix} \begin{bmatrix} F_r(A_1) \\ F_r(A_2) \\ F_r(A_3) \end{bmatrix}.$$

14.4.3 The Linear System

Gathering all these results, we rewrite problem (14.6) in the new form

$$\begin{cases} \text{Find } T_h \in T_D + V_h^0 \text{ such that} \\ \forall T_h' \in V_h^0 \quad \sum_{K \in \mathcal{T}_h} [dl\ T_h']_K^t [A_K][dl\ T_h]_K = \sum_{K \in \mathcal{T}_h} [dl\ T_h']_K^t [b_K]. \end{cases} \quad (14.10)$$

In this formulation, the array $[dl\ T_h']_K$ represents the temperature values of an arbitrary function T_h' in V_h. It is an array whose three components are the temperature values at the vertices of K, an arbitrary triangle of $\mathcal{T}_h$. When we compute the summation in (14.10), element K replaces all triangles of $\mathcal{T}_h$, so all functions T_h' in V_h are taken into account. We then rewrite (14.10) as

$$\begin{cases} \text{Find } T_h \in T_D + V_h^0 \text{ such that} \\ \forall T_h' \in V_h^0 \quad [dl\ T_h']^t\ [A]\ [dl\ T_h] = [dl\ T_h']^t\ [b]. \end{cases} \quad (14.11)$$

Let nv be the number of vertices in triangulation $\mathcal{T}_h$; then $[A]$ is a square matrix of $\mathbb{R}^{nv \times nv}$ and $[b]$ is an array of R^{nv}. Note that

$$[dl\ T_h']^t = [T_h'(A_1), T_h'(A_2), \dots, T_h'(A_{nv})]^t$$

and

$$[dl\ T_h]^t = [T_h(A_1), T_h(A_2), \dots, T_h(A_{nv})]^t$$

are arrays whose nv components are the temperature values at the vertices of $\mathcal{T}_h$. To end, we remark that (14.6) is a linear system with nv equations and nv unknowns:

$$[A]\ [dl\ T_h] = [b]. \quad (14.12)$$

14.5 Boundary Conditions

It is time now to take the boundary conditions into account. The condition $T_h' = 0$ on Ω_D, specified in the definition of the space V_h^0, involves important modifications of the linear system (14.12). For the sake of simplicity, we shall assume in the following lines that the vertices located on Ω_D have the largest indices when the points of triangulation $\mathcal{T}_h$ are ordered. More precisely, the indices of these nv_D vertices are supposed to be $nv - nv_D + 1, nv - nv_D + 2, \dots, nv$. Any element of the finite-dimensional space V_h is then an array of nv real components, and any element of the subspace V_h^0 is an array whose nv_D last components are null. So the linear system (14.12) arising from (14.11) seems to have only $(nv - nv_D)$ rows but nv unknowns! Fortunately, since the solution T_h of problem (14.11) belongs to the space

$T_D + V_h^0$, its nv_D last components are well known and determined by the data T_D associated with the Dirichlet boundary condition $T_h |_{\partial \Omega_D} = T_D$. Finally, the linear system (14.12) has $(nv - nv_D)$ unknowns for the same number of equations! Nevertheless, this treatment has heavy consequences for the computer formulation of (14.12). We write first

$$\begin{bmatrix} A_1 & A_2 \\ A_2^t & A_3 \end{bmatrix} \times \begin{bmatrix} T_h \\ T_{hD} \end{bmatrix} = \begin{bmatrix} b \\ c \end{bmatrix}.$$

The square matrix A_1 is of order $(nv - nv_D)$, A_2 has $(nv - nv_D)$ rows and nv_D columns, and A_3 is a square matrix of order nv_D. When we take the condition $T_h' |_{\partial \Omega_D} = 0$ into account, we see that the nv_D last rows of the linear system vanish. These rows are replaced by the nv_D relations $T_h |_{\partial \Omega_D} = T_{hD} = T_D$, so the linear system is now

$$\begin{bmatrix} A_1 & A_2 \\ 0 & I \end{bmatrix} \times \begin{bmatrix} T_h \\ T_{hD} \end{bmatrix} = \begin{bmatrix} b \\ T_D \end{bmatrix},$$

where I is the identity matrix of order nv_D. For matrix storage reasons it is important to preserve the symmetry of the initial problem. A final modification is then necessary: the matrix A_2 is eliminated in order to obtain

$$\begin{bmatrix} A_1 & 0 \\ 0 & I \end{bmatrix} \times \begin{bmatrix} T_h \\ T_{hD} \end{bmatrix} = \begin{bmatrix} b - A_2 T_D \\ T_D \end{bmatrix},$$

which is the symmetric linear system solved by the computer.

14.5.1 Modular Implementation

When implementing the finite element method, different logical steps have to be taken into account:

1. definition of the triangulation $\mathcal{T}_h$,
2. construction of the linear system (14.12),
3. introduction of the boundary conditions,
4. solution of the modified linear system,
5. visualization of the results.

Any scientific package has to deal with all elements of this list: there exists a distinct procedure corresponding to each step encountered during the implementation. These procedures are called *modules*. Results (output) of the kth-step module are data (input) for $(k + 1)$th-step module. Several modules may exist for the same logical step, in which case they have to share similar formatted input and provide similar formatted output.

14.5.2 Numerical Solution of the Problem

In solving problem (14.11), the very first step is the mesh construction. Numerous packages are devoted to this work, and 2D meshes are easily created. See, for example, the mesh displayed in Fig. 14.2 (left) obtained with the free software *FreeFEM*.[1] There are altogether 304 triangles and 173 vertices. Another mesh, displayed in Fig. 14.2 (right), was computed by the MATLAB "toolbox" PDE-tool, with 1392 triangles and 732 vertices.

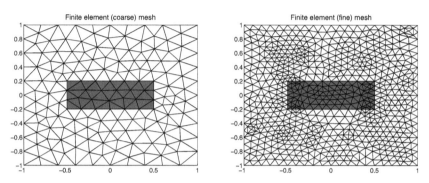

Fig. 14.2 Mesh of the domain. Left: coarse; right: fine.

The mesh description is summarized in the following list

1. Nbpt, Nbtri (two integers): number of vertices (points), number of triangles
2. List of all vertices: for Ns=1,Nbpt

   - Ns, Coorpt[Ns,1], Coorpt[Ns,2], Refpt[Ns] (one integer, two reals, one integer): vertex number, coordinates, and boundary reference;

3. List of all triangles: for Nt=1,Nbtri,

   - Nt, Numtri[Nt,1:3], Reftri[N] (five integers): triangle number, three vertex numbers, and medium reference (air or resin).

Exercise 14.1 1. Create a mesh (or read one of the data files provided with the procedures). Check contents of arrays `Coorpt` and `Numtri`.
2. Compute element matrix and right-hand side for each triangle.
3. Write a procedure that assembles the linear system from element data.
4. Modify the linear system in order to take the boundary conditions into account.[2]
5. Solve the resulting linear system.
6. Visualize the results, plot the isotherm curves.

---

[1] https://freefem.org/
[2] For this experiment $\partial\Omega_D$ is the union of the lines $y = -1$ and $y = 1$, $\partial\Omega_N$ is is the union of the lines $x = -1$ and $x = 1$, and $\partial\Omega_F = \emptyset$.

Hint: Use the following algorithm to assemble A and b:

```
for K=1:Nbtri
   (a) read data for triangle K
       XY  = coordinates of triangle K vertices
       NUM = triangle K vertices numbers
   (b) Compute element matrix AK(3,3) and right-hand side bK(3)
   (c) Build global matrix A(Nbpt,Nbpt)
       for i=1:3
       for j=1:3
          A(Num(i),Num(j))=A(Num(i),Num(j))+AK(i,j)
       end
       end
   (d) Build global right-hand side  b(Nbpt)
       for i=1:3
          b(Num(i))=b(Num(i))+bK(i)
       end
end
```

A solution of this exercise is proposed in Sect. 14.8 at page 323. A computed temperature field is displayed in Fig. 14.3(left), representing the variations of temperature within the domain Ω. It corresponds to the following data: $T_D = 50°$ Celsius on the upper part of the oven $(y = 1)$, $T_D = 100°$ on the lower part $(y = -1)$, and no heating sources $(F = 0,\ f = 0)$. Another temperature field is displayed in Fig. 14.3 (right): plotting the temperature variations according to the previous boundary conditions but with four heating resistances (common value is 25 000). It is obvious in Fig. 14.3 that the temperature value in the bumper is far from $250°$, which is supposed to be the ideal one in our study. To fix this problem we have to increase the resistance values. Yes, but by how much? In the previous computation we have used the resistance values as data and obtained the temperature field inside the oven as result; this is referred to as the *direct* formulation of the problem. But what we are interested in is the resistance values that produce the optimal temperature inside the bumper; this is called the *inverse* problem. We shall address the inverse problem in the following section.

14.6 Inverse Problem Formulation

We emphasize now an important property of (14.1)–(14.2): the problem is *linear*. This means that if T' is the unique solution of problem (14.1)–(14.2) corresponding to data $\{F', T'_D, f'\}$, and T'' is the unique solution corresponding to data $\{F'', T''_D, f''\}$, then $\alpha T' + \beta T''$ is the unique solution corresponding to data $\{\alpha F' + \beta F'', \alpha T'_D + \beta T''_D, \alpha f' + \beta f''\}$ for any real numbers α and β. In order to determine the values of the resistances that lead to a correct heating of the object, this propriety is of great interest. Assume that there

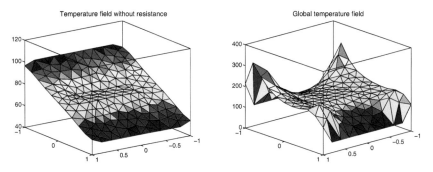

Fig. 14.3 Temperature fields. Left: without resistance; right: with four resistances.

are nwr heating resistances and that the boundary conditions are tempera-
ture value T_D imposed on $\partial\Omega_D$ and heat flux f imposed on $\partial\Omega_N$. Then the
corresponding temperature field T is written as

$$T = T_0 + \sum_{k=1}^{nwr} \alpha_k T_k,$$

where α_k is the kth resistance value and T_k represents the temperature field
when resistance k is the unique resistance heating the oven. These coefficients
α_k are the unknowns of the inverse problem, and we are going to compute
the values corresponding to the desired temperature T_{opt} by minimizing the
quantity

$$J(\alpha) = \int_S \left[T_{opt}(x) - T_0(x) - \sum_{k=1}^{nwr} \alpha_k \, T_k(x) \right]^2 dx.$$

Here $\alpha = (\alpha_1, \ldots, \alpha_{nwr})^t$ and S stands for the bumper. The quadratic
functional J is a strictly convex function of the variable α, and its unique
minimum is reached when $\nabla J(\alpha) = 0$. For $k = 1, 2, \ldots, nwr$, the gradient
kth component is

$$\frac{\partial J}{\partial \alpha_k} = 2 \int_S \left(T_{opt}(x) - T_0(x) - \sum_{k'=1}^{nwr} \alpha_{k'} \, T_{k'}(x) \right) T_k(x) dx,$$

and the minimum is reached when

$$\sum_{k'=1}^{nwr} \alpha_{k'} \int_S T_{k'}(x) \, T_k(x) dx = \int_S (T_{opt}(x) - T_0(x)) \, T_k(x) dx,$$

for $k = 1, 2, \ldots, nwr$. We introduce now the matrix $\tilde{A} \in \mathbb{R}^{nwr \times nwr}$ and the
array $\tilde{b} \in \mathbb{R}^{nwr}$ by

$$\tilde{A}_{k,k'} = \int_S T_k(x) \, T_{k'}(x)dx \text{ and } \tilde{b}_k = \int_S (T_{opt}(x) - T_0(x)) \; T_k(x)dx.$$

The optimal $\tilde{\alpha} = (\tilde{\alpha}_1, \ldots, \tilde{\alpha}_{nwr})^t$ is the unique solution of the linear system

$$\tilde{A}\tilde{\alpha} = \tilde{b}. \tag{14.13}$$

14.7 Implementation of the Inverse Problem

Exercise 14.2 1. Compute and solve the linear system (14.13) arising from the optimization problem.
2. Compute and plot the temperature field corresponding to the optimal value. Comment on the results.

A solution of this exercise is proposed in Sect. 14.8 at page 324. We have first to solve the $nwr+1$ direct problems in order to compute the temperature fields $T_0, T_1, \ldots, T_{nwr}$. They are obtained by the use of $nwr + 1$ calls of the program written to solve the direct problem. Each computation corresponds to a distinct value of the data T_D, f, and F (note that the localization of the resistances in the oven is a geometrical datum of great importance). The corresponding temperature fields are then stored in $nwr + 1$ distinct arrays.

Now we solve problem (14.1)–(14.2) without any heating term $(F = 0)$, but with a temperature $T_D \neq 0$ and thermal flux $f = 0$ given on the boundary. The solution of this problem is denoted by T_0, and displayed in Fig. 14.3 (left). We can see that the boundary conditions are well respected: temperature value is $T = 100$ when $y = -1$ and $T = 50$ when $y = 1$. The heat conductivity coefficient is set to the value $c = 1$ within the air and $c = 10$ within the object (air is a good insulation medium). The vanishing thermal flux on the other parts of the boundary corresponds to isotherm lines parallel to the normal vector when $x = -1$ and $x = 1$.

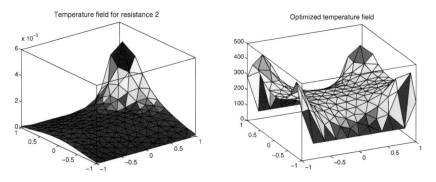

Fig. 14.4 Temperature fields. Left: T_2 field; right: optimized field.

Then we solve *nwr* successive problems (14.1)–(14.2) (one problem by resistance). Each case consists in computing the temperature field when only one resistance is heating the oven. The boundary conditions are temperature $T_D = 0$ on $\partial\Omega_D$ and thermal flux $f = 0$ on $\partial\Omega_N$ for all cases. The *nv* components of array T_k represent the temperature field related to the kth resistance. Figure 14.4 (left) displays the temperature field associated with a single heating resistance. Note the tiny values of the temperature.

We notice again that the boundary conditions are well satisfied: the temperature vanishes when $y = -1$ and $y = 1$ (Dirichlet condition), and the Neumann condition (null flux condition) leads to isotherm lines perpendicular to lines $x = -1$ and $x = 1$.

To solve the inverse problem, we have now to construct the linear system (14.13) and then compute the terms

$$\int_S T_k(x)\, T_{k'}(x)dx$$

from the temperature fields T_k and $T_{k'}$ computed in the previous step. This computation is performed in the same way as before: we first write the integral term on the complete object as summation of integral terms on all triangles of the object:

$$\int_S T_k(x)\, T_{k'}(x)dx = \sum_{K \subset S} \int_K T_k(x)\, T_{k'}(x)dx.$$

Then any integral on K is evaluated using the expression for $T_k(x)$ and $T_{k'}(x)$ in triangle K

$$T_k(x) = T_k(A_3) + [T_k(A_1) - T_k(A_3)]\lambda_1 + [T_k(A_2) - T_k(A_3)]\lambda_2.$$

In this formula λ_i is the ith barycentric coordinate of the point $M(x,y)$ in K, and A_i is one of the vertices of triangle K. So we may write

$$\int_K T_k(x)\, T_{k'}(x)dx = [dl\, T_{k,K}]^t\, [M_K]\, [dl\, T_{k',K}].$$

To avoid any confusion with the notations: T_k represents here the temperature field associated to the resistance k, as computed in the direct problem; $[dl\, T_{k,K}]$ represents an array of the three values of T_k at the vertices of the triangle K.

This leads us to introduce the matrix $[M_K]$, the so-called element mass matrix. The element mass matrix associated with the Lagrange finite triangular element of degree 1 is

$$[M_K] = \frac{\Delta_K}{24} \begin{bmatrix} 2 & 1 & 1 \\ 1 & 2 & 1 \\ 1 & 1 & 2 \end{bmatrix}.$$

So in computing the linear system (14.13), the matrix coefficient $\tilde{A}_{k,k'}$ is obtained by summation over all triangles laying inside the objects of the terms $[dl\ T_{k,K}]^t\ [M_K][dl\ T_{k',K}]$. The right-hand side $\tilde{b}$ is computed in the same way. An example of calculation, corresponding to the case of four heating resistances, is displayed in Fig. 14.5 (left). We may see there the optimized temperature field obtained after computation of coefficients α_k. Figure 14.5 (right) displays the solution corresponding to six heating resistances. The value of the temperature within the rectangle $[-0.5, 0.5] \times [-0.2, 0.2]$ is very near to the target value ($250°$).

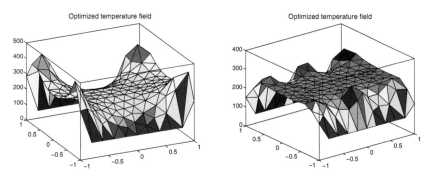

Fig. 14.5 Optimized temperature fields. Left: four resistances; right: six resistances.

It is very important to notice that the optimal value of coefficients α_k depends on the number of resistances but also on the position of these resistances in relation to the object. An interesting development of this study is to try to optimize the position of the resistances inside the oven. The practical aim of such an additional study should be the optimization of the thermal power dissipated by the resistances. In this particular case, we want to optimize the temperature value in the object and the thermal (or electric) energy used to warm the oven. This is modeled by adding a special term to the functional

$$J(\alpha) = \int_S \left(T_{opt}(x) - T_0(x) - \sum_{k=1}^{nwr} \alpha_k\ T_k(x) \right)^2 dx + C \sum_{k=1}^{nwr} \alpha_k^2.$$

Remark 14.3 You may get very small values (even negative) for the coefficients α_k. This means that the corresponding resistances are located too close to the object and then have to cool it instead of heating it.

It may be seen in Fig. 14.6 (left) that a device with six heating resistances produces a larger "well heated" area than with four resistances. Note also that for this simulation we used a rather coarse mesh (173 vertices and 304 triangles). The results are satisfying; they prove the efficiency of the finite element method to solve this problem, and provide a validation of all the procedures, and of the whole process. Nevertheless, in order to get more realistic and more accurate results, it is necessary to solve the problem on a "finer" mesh. We have proceeded to a second computation on a mesh with 732 vertices and 1392 triangles (and still six resistances). The final result is plotted in Fig. 14.6 (right), showing a true improvement, especially around the object *and* the resistances. This improvement, predicted by the finite element method (the smaller the element size, the better is the result) leads to an increase of the computational time.

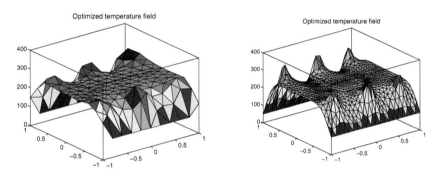

Fig. 14.6 Optimized temperature fields. Left: coarse mesh; right: fine mesh.

14.8 Solutions and Programs

There are two sub-directories **coarse_mesh** and **fine_mesh** allowing to solve each exercise on a coarse (faster) or fine mesh (slower).

14.8.1 Solution of Exercise 14.1

The script **THER_M_oven_ex1** implements the numerical experiment by defining the physical parameters of the problem (localization of the resistances, heat conductivity coefficients, boundary temperature values). It calls the function **THER_F_oven**, which computes the corresponding temperature field.

The function **THER_F_matrix_dir** builds the linear system arising from the heat equation problem. We notice that the right-hand side vanishes outside elements that contain a resistance. The function **THER_F_local** builds the right-hand side for given resistances coordinates. The function **THER_F_elim** takes the boundary conditions into account.

14.8.2 Solution of Exercise 14.2

The function THER_F_oven_ex2 computes the resistances' values corresponding to the optimal temperature field. It calls the function THER_F_matrix_inv computing the matrix and right-hand side of the optimization problem.

Remark 14.4 We also provide an interactive version of the solution of this project, allowing one to realize numerous numerical experiments by changing data through a graphical user interface. To launch the interface, just run the script THER_M_Main from the subdirectory interactive.

14.8.3 Further Comments

In this last section we address the important point of the structure of matrix A. The matrix is sparse because this property is related only to the approximation scheme and the differential operator, which is here similar to the Laplacian[3] operator. The shape of A (see Fig. 14.7 (left)) is not as regular as the one displayed Fig. 9.3 in Chap. 9. This difference results from the use of a finite element mesh with triangles, instead of a rectangular grid. Furthermore, the structure of A is strongly depending on the nodes ordering as can be seen by comparing the matrices obtained on the coarse mesh (Fig. 14.7 (left)) and on a finer mesh (Fig. 14.7 (right)). The obvious difference is due to a reordering of the nodes for the coarse-mesh calculation.

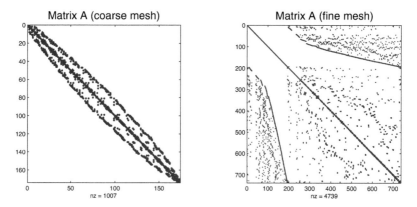

Fig. 14.7 Associated matrix. Left: coarse mesh; right: fine mesh.

---

[3] The value of the heat conductivity coefficient c is not relevant for the matrix structure.

Chapter References

P.G. Ciarlet, *The Finite Element Method for Elliptic Problems* (North Holland, Amsterdam, 1978)

D.N. Norrie, G. de Vries, *The Finite Element Method* (Academic Press, New York, 1973)

J.C. Strikwerda, *Finite Difference Schemes and Partial Differential Equations.* (Wadsworth and Brooks/Cole, 1989)

O.C. Zienkiewicz, *The Finite Element Method in Engineering Science* (McGraw-Hill, London, 1971)

Chapter 15
Fluid Dynamics: Solving the 2D Navier–Stokes Equations

Project Summary

Level of difficulty:	3
Keywords:	Navier–Stokes equations, Helmholtz equation, Poisson equation, projection method, ADI factorization, Fast Fourier transform (FFT)
Application fields:	Incompressible flows, jet flow, Kelvin–Helmholtz instability, vortex dipole

15.1 Introduction

The Navier–Stokes system of partial differential equations (PDEs) contains the main conservation laws that universally describe the evolution of a fluid (i.e. liquid or gas). Even though these laws are known since the nineteenth century, the complete description of their intrinsic properties remains one of the challenging topics of modern physics and mathematics.

In this chapter, we consider some simplifying hypotheses that make the Navier–Stokes equations tractable with relatively simple numerical methods:

- the fluid is *incompressible*, with constant density ($\rho = \rho_0$) and viscosity;
- the flow is *two-dimensional* (2D), with all properties depending on time t and two space variables (x and y),
- we impose *periodic boundary conditions* in space, meaning that all variables are periodic functions of both x and y variables.

This model allows one to study simple, but fascinating, phenomena that turn out to give us an understanding of more complicated flows too. In this project, we shall numerically simulate:

© The Author(s), under exclusive license to Springer Nature Switzerland AG 2023   327
I. Danaila et al., *An Introduction to Scientific Computing*,
https://doi.org/10.1007/978-3-031-35032-0_15

- the Kelvin–Helmholtz instability of a mixing layer, and
- the evolution of a particular vortex structure, the vortex dipole.

From a numerical point of view, this computational project introduces the following algorithms or numerical schemes of more general interest:

- the space discretization using a *staggered grid* and *2D finite difference* schemes;
- the combined *Adams–Bashforth* and *Crank–Nicolson* schemes for the time integration;
- an *alternating direction implicit*, or ADI, method for solving the Helmholtz equation[1];
- a solver for the periodic *Poisson* equation based on the *Fast Fourier Transform* (FFT);
- the *Thomas algorithm* for solving a tridiagonal linear system.

Numerical methods presented in this chapter are mainly focused on finite difference techniques based on elementary schemes introduced in Chaps. 1 and 7 for the time or space discretization of derivatives. Periodic boundary conditions are chosen for their conceptual simplicity: no special treatment of the boundaries is required and using some programming tricks, the vector programming of numerical schemes becomes simple and concise. On the other side, such boundary conditions restrain the field of applications to academic flows (periodic jets or shear layers, arrays of vortices, turbulence in a periodic box etc.). The algorithms of this project offer the advantage to be easily adapted (with minor modifications) to deal with more realistic boundary conditions (inflow-outflow, walls, slipping walls, etc.).

Specific algorithms fully exploiting the periodicity and belonging to Fourier spectral methods (see Chap. 6) are only briefly mentioned. Spectral methods bring the advantage of using FFTs to rapidly compute derivatives in the spectral space. In particular, the Laplace operator is diagonalized and thus the numerical solver for the Poisson equation is greatly simplified. This particular approach is presented in the project as an alternative to the finite difference solver for the Poisson equation. If the spectral approach is very effective for linear PDEs, for the nonlinear PDEs it is necessary to combine direct and inverse FFTs to deal with nonlinear terms. This back and forth between physical and spectral space is the characteristic of pseudo-spectral methods. A large amount of literature exists on pseudo-spectral methods specifically designed for the Navier-Stokes system of equations, starting from the pioneering work by Orszag (1969) (see, for example, the comprehensive book by Canuto et al. (2007)). The present project could also be recast into a spectral oriented approach (instead of the finite difference one), using the same algorithmic projection scheme for the integration of the Navier-Stokes equations (some hints are given in specific parts of the algorithm).

---

[1] We consider here the Helmholtz equation as the general eigenvalue problem for the Laplacian: $-\Delta u = \lambda u$. When $\lambda = k^2 > 0$ we recover the classical Helmholtz equation, while for $\lambda = k^2 < 0$ (i.e. for imaginary k) we obtain a diffusion equation.

15.2 The Incompressible Navier–Stokes Equations

The 2D flow field of an incompressible fluid is completely described by the velocity vector $\boldsymbol{q} = (u(x,y), v(x,y)) \in \mathbb{R}^2$ and the pressure $p(x,y) \in \mathbb{R}$. These functions are a solution of the following conservation laws (see, for instance Batchelor 1988, Hirsch 1988):

- mass conservation:

$$\operatorname{div}(\boldsymbol{q}) = 0, \tag{15.1}$$

  or, written using the explicit form of the *divergence*[2] operator,

$$\frac{\partial u}{\partial x} + \frac{\partial v}{\partial y} = 0. \tag{15.2}$$

- the momentum conservation equations in the compact form[3]

$$\frac{\partial \boldsymbol{q}}{\partial t} + \operatorname{div}(\boldsymbol{q} \otimes \boldsymbol{q}) = -\nabla p + \frac{1}{Re}\Delta \boldsymbol{q}, \tag{15.3}$$

  or, in explicit form,

$$\begin{cases} \dfrac{\partial u}{\partial t} + \dfrac{\partial u^2}{\partial x} + \dfrac{\partial uv}{\partial y} = -\dfrac{\partial p}{\partial x} + \dfrac{1}{Re}\left(\dfrac{\partial^2 u}{\partial x^2} + \dfrac{\partial^2 u}{\partial y^2}\right), \\[2mm] \dfrac{\partial v}{\partial t} + \dfrac{\partial uv}{\partial x} + \dfrac{\partial v^2}{\partial y} = -\dfrac{\partial p}{\partial y} + \dfrac{1}{Re}\left(\dfrac{\partial^2 v}{\partial x^2} + \dfrac{\partial^2 v}{\partial y^2}\right). \end{cases} \tag{15.4}$$

Previous equations are written in the *dimensionless* form, using the following scaled variables:

$$x = \frac{x^*}{L}, \quad y = \frac{y^*}{L}, \quad u = \frac{u^*}{V_0}, \quad v = \frac{v^*}{V_0}, \quad t = \frac{t^*}{L/V_0}, \quad p = \frac{p^*}{\rho_0 V_0^2}, \tag{15.5}$$

where the superscript $(^*)$ denotes variables measured in physical units. The constants L, V_0 are, respectively, the reference length and velocity that characterize the simulated flow.

The dimensionless number Re is called the *Reynolds number* and quantifies the relative importance of inertial (or convective) terms and viscous (or

---

[2] We recall the definitions of the differential operators *gradient*, *curl*, *divergence*, and *Laplacian* for a 2D field: if $\boldsymbol{v} : \mathbb{R}^2 \mapsto \mathbb{R}^2, \boldsymbol{v} = (v_x, v_y)$ and $\varphi : \mathbb{R}^2 \mapsto \mathbb{R}$, then

$$\nabla \varphi = \left(\frac{\partial \varphi}{\partial x}, \frac{\partial \varphi}{\partial y}\right), \quad \operatorname{curl}(\boldsymbol{v}) = \nabla \times \boldsymbol{v} = \left(\frac{\partial v_y}{\partial x} - \frac{\partial v_x}{\partial y}\right)\boldsymbol{k}, \quad \operatorname{div}(\boldsymbol{v}) = \frac{\partial v_x}{\partial x} + \frac{\partial v_y}{\partial y},$$

$$\Delta \varphi = \operatorname{div}(\nabla \varphi) = \frac{\partial^2 \varphi}{\partial x^2} + \frac{\partial^2 \varphi}{\partial y^2}.$$

Recall also that $\operatorname{curl}(\nabla \varphi) = 0$ and $\Delta \boldsymbol{v} = (\Delta v_x, \Delta v_y)$.

[3] We denote by $\otimes$ the tensor product.

diffusive) terms in the flow[4]:

$$Re = \frac{V_0 L}{\nu}, \tag{15.6}$$

where ν is the kinematic viscosity of the flow.

To summarize, the Navier–Stokes system of PDEs that will be numerically solved in this project is defined by (15.2) and (15.4); the initial condition (at $t = 0$) and the boundary conditions will be discussed in the following sections.

15.3 Numerical Algorithm

We start by presenting the *fractional-step method* (Chorin, 1968, Ferziger and Perić, 2002, Kim and Moin, 1985, Orlandi, 1999) as a general algorithm to solve the Navier–Stokes equations. This algorithm belongs to the class of so-called projection methods and has become rather popular in computational fluid dynamics. An extensive presentation of this method in a more general framework can be found in the book by Orlandi (1999), which also provides Fortran programs for the associated algorithms.

We use a fractional-step method that consists of two steps:

1. The predictor step: we solve the momentum equations (15.3) written in the compact form

$$\frac{\partial q}{\partial t} = -\nabla p + \mathcal{H} + \frac{1}{Re} \Delta q, \quad \text{for} \quad q = (u, v) \in \mathbb{R}^2, \tag{15.7}$$

where $\mathcal{H}$ is the vector containing the convective terms

$$-\mathcal{H} = \left(\frac{\partial u^2}{\partial x} + \frac{\partial uv}{\partial y}, \frac{\partial uv}{\partial x} + \frac{\partial v^2}{\partial y} \right), \tag{15.8}$$

and ∇p the pressure gradient vector. Time discretization of (15.7) combines the explicit Adams–Bashforth scheme (for the convective terms $\mathcal{H}$) with the semi-implicit Crank–Nicolson scheme (for the diffusion terms Δq). If δt denotes the time step (supposed constant), the time advancement of the solution from $t_n = n\delta t$ to $t_{n+1} = (n+1)\delta t$ follows the scheme

$$\frac{q^* - q^n}{\delta t} = -\nabla p^n + \underbrace{\frac{3}{2}\mathcal{H}^n - \frac{1}{2}\mathcal{H}^{n-1}}_{\text{Adams–Bashforth}} + \underbrace{\frac{1}{Re} \Delta \left(\frac{q^* + q^n}{2} \right)}_{\text{Crank–Nicolson}}. \tag{15.9}$$

---

[4] The model scalar equations describing convection and diffusion phenomena are discussed in Chap. 1.

In the previous equation, the pressure is treated as an explicit term (computed at time t_n). As a consequence, the velocity vector $\boldsymbol{q}^*$ does not satisfy the mass conservation equation (15.1).

2. The corrector (or projection) step: the velocity $\boldsymbol{q}^*$ is corrected such that the velocity field $\boldsymbol{q}^{n+1}$ is divergence-free (or *solenoidal*). We use the following correction equation:

$$\boldsymbol{q}^{n+1} - \boldsymbol{q}^* = -\delta t \, \nabla \phi. \tag{15.10}$$

The idea of this projection goes back to Chorin (1968) and is based on the so-called Helmholtz decomposition: the velocity $\boldsymbol{q}^*$ is written as the sum of a divergence-free vector field (here $\boldsymbol{q}^{n+1}$) and a curl-free field (here $\nabla \phi$, since $\text{curl}(\nabla \phi) = 0$). Mathematically, this is equivalent to the orthogonal splitting of the functional space $(L^2(\Omega)^3)$ of square integrable functions (see, for instance, Girault and Raviart (2002)). Hence this method is sometimes referred to as the L^2-*projection method*.

We infer from (15.10) that $\text{curl}(\boldsymbol{q}^*) = \text{curl}(\boldsymbol{q}^{n+1})$, meaning that the two fields have the same vorticity $\boldsymbol{\omega} = \text{curl}(\boldsymbol{q})$ (see also Sect. 15.6). Note that the equation for the time evolution of the vorticity, obtained by taking the curl of (15.7), does not depend on the pressure field. This explains why in some versions of the projection method the pressure is used in (15.10) as a correction variable instead of ϕ.

The variable ϕ (related to the pressure, but without any physical meaning) is calculated by taking the divergence of (15.10); using that $\text{div}(\boldsymbol{q}^{n+1}) = 0$ and $\text{div}(\nabla \phi) = \Delta \phi$, we obtain a Poisson equation

$$\Delta \phi = \frac{1}{\delta t} \, \text{div}(\boldsymbol{q}^*). \tag{15.11}$$

To close the algorithm, the pressure for the next time step is updated using:

$$p^{n+1} = p^n + \phi - \frac{\delta t}{2Re} \Delta \phi. \tag{15.12}$$

Equation (15.12) can be obtained as follows: we write (15.9) with an implicit discretization of the pressure term

$$\frac{\boldsymbol{q}^{n+1} - \boldsymbol{q}^n}{\delta t} = -\nabla p^{n+1} + \frac{3}{2}\boldsymbol{\mathcal{H}}^n - \frac{1}{2}\boldsymbol{\mathcal{H}}^{n-1} + \frac{1}{Re}\Delta\left(\frac{\boldsymbol{q}^{n+1} + \boldsymbol{q}^n}{2}\right)$$

and subtract this equation from (15.9). After replacing $\boldsymbol{q}^*$ from (15.10), we get (15.12), up to an additive constant. Note that this constant is discarded by taking the gradient of the pressure in the momentum equations.

To summarize, the numerical algorithm consists of the following steps, rearranged below in the form that will be used in computer programs:

Algorithm 15.1 *to solve the Navier–Stokes equations (15.2)–(15.4). Given the field (u^n, v^n, p^n), compute:*

(A)   *the explicit terms* $\mathcal{H}^n = (\mathcal{H}_u^n, \mathcal{H}_v^n)$:

$$\mathcal{H}_u^n = -\left(\frac{\partial(u^n u^n)}{\partial x} + \frac{\partial(u^n v^n)}{\partial y}\right), \tag{15.13}$$

$$\mathcal{H}_v^n = -\left(\frac{\partial(u^n v^n)}{\partial x} + \frac{\partial(v^n v^n)}{\partial y}\right); \tag{15.14}$$

(B)   *the nonsolenoidal field* $q^* = (u^*, v^*)$ *by solving the Helmholtz equations*

$$\left(I - \frac{\delta t}{2Re}\Delta\right)u^* = u^n + \delta t\left(-\frac{\partial p^n}{\partial x} + \frac{3}{2}\mathcal{H}_u^n - \frac{1}{2}\mathcal{H}_u^{n-1} + \frac{1}{2Re}\Delta u^n\right), \tag{15.15}$$

$$\left(I - \frac{\delta t}{2Re}\Delta\right)v^* = v^n + \delta t\left(-\frac{\partial p^n}{\partial y} + \frac{3}{2}\mathcal{H}_v^n - \frac{1}{2}\mathcal{H}_v^{n-1} + \frac{1}{2Re}\Delta v^n\right); \tag{15.16}$$

(C)   *the variable* ϕ *by solving the Poisson equation*

$$\Delta\phi = \frac{1}{\delta t}\left(\frac{\partial u^*}{\partial x} + \frac{\partial v^*}{\partial y}\right); \tag{15.17}$$

(D)   *the solenoidal field* $q^{n+1} = (u^{n+1}, v^{n+1})$, *with*

$$u^{n+1} = u^* - \delta t\,\frac{\partial\phi}{\partial x}, \tag{15.18}$$

$$v^{n+1} = v^* - \delta t\,\frac{\partial\phi}{\partial y}; \tag{15.19}$$

(E)   *the new pressure:*

$$p^{n+1} = p^n + \phi - \frac{\delta t}{2Re}\Delta\phi. \tag{15.20}$$

Steps (A)–(E) are repeated for each time step.

15.4 Computational Domain, Staggered Grids, and Boundary Conditions

Numerically solving the Navier–Stokes equations is considerably simplified by considering a rectangular domain $L_x \times L_y$ (see Fig. 15.1) with periodic boundary conditions everywhere. The periodicity of the velocity $q(x, y)$ and pressure $p(x, y)$ fields is mathematically written as

$$q(0, y) = q(L_x, y), \quad p(0, y) = p(L_x, y), \quad \forall y \in [0, L_y], \tag{15.21}$$
$$q(x, 0) = q(x, L_y), \quad p(x, 0) = p(x, L_y), \quad \forall x \in [0, L_x]. \tag{15.22}$$

The points at which the solution will be computed are distributed in the domain following a rectangular and uniform 2D grid. Since not all the variables share the same grid in our approach, we first define a *primary* grid (see Fig. 15.1) generated by taking n_x computational points along x and, respectively, n_y points along y:

$$x_c(i) = (i-1)\delta x, \quad \delta x = \frac{L_x}{n_x - 1}, \quad i = 1, \dots, n_x, \tag{15.23}$$

$$y_c(j) = (j-1)\delta y, \quad \delta y = \frac{L_y}{n_y - 1}, \quad j = 1, \dots, n_y. \tag{15.24}$$

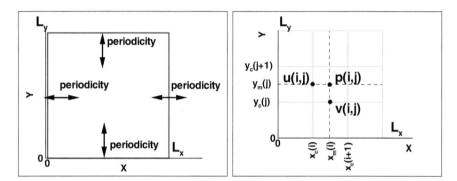

Fig. 15.1 Computational domain, staggered grids, and boundary conditions.

A *secondary* grid is defined by the centers of the primary grid cells:

$$x_m(i) = (i - 1/2)\delta x, \quad i = 1, \dots, n_{xm}, \tag{15.25}$$

$$y_m(j) = (j - 1/2)\delta y, \quad j = 1, \dots, n_{ym}, \tag{15.26}$$

where we have used the shorthand notation $n_{xm} = n_x - 1$, $n_{ym} = n_y - 1$. Inside a computational cell defined as the rectangle $[x_c(i), x_c(i+1)] \times [y_c(j), y_c(j+1)]$, the unknown variables u, v, p will be computed as approximations of the solution at different space locations:

- $u(i, j) \approx u(x_c(i), y_m(j))$ (west face of the cell),
- $v(i, j) \approx v(x_m(i), y_c(j))$ (south face of the cell),
- $p(i, j) \approx p(x_m(i), y_m(j))$ (center of the cell).

This *staggered* arrangement of the variables has the advantage of a strong coupling between pressure and velocity. It also helps (see the references at the end of the chapter) to avoid some problems of stability and convergence experienced with *collocated* arrangements (where all the variables are computed at the same grid points).

15.5 Finite Difference Discretization

In this section, Algorithm 15.1 will be written in a discrete form that will be used in the computer programs of this project. We start by noticing that the periodic boundary conditions take the following discrete form:

$$u(1, j) = u(n_x, j), \ \forall j = 1, \ldots, n_y,$$
$$u(i, 1) = u(i, n_y), \ \forall i = 1, \ldots, n_x, \tag{15.27}$$

and similar relations for v and p. As a consequence, the unknowns of the problem are only the $n_{xm} \times n_{ym}$ values for each variable

$$u(i, j), \quad v(i, j), \quad p(i, j), \quad \text{for} \quad i = 1, \ldots, n_{xm}, \quad j = 1, \ldots, n_{ym}.$$

A very useful programming trick (see also Chaps. 1 and 7) in implementing discrete periodic boundary conditions (15.27) consists in defining the supplementary arrays

$$\begin{cases} ip(i) = i + 1, \ i = 1, \ldots, (n_{xm} - 1), \\ ip(n_{xm}) = 1, \end{cases} \quad \begin{cases} jp(j) = j + 1, \ j = 1, \ldots, (n_{ym} - 1), \\ jp(n_{ym}) = 1, \end{cases}$$
$$\tag{15.28}$$

$$\begin{cases} im(i) = i - 1, \ i = 2, \ldots, n_{xm}, \\ im(1) = n_{xm}, \end{cases} \quad \begin{cases} jm(j) = j - 1, \ j = 2, \ldots, n_{ym}, \\ jm(1) = n_{ym}, \end{cases}$$
$$\tag{15.29}$$

and using the vector capabilities of MATLAB to write the finite difference discretization of the differential operators in a very compact form. For example, to compute $(\partial\psi/\partial x)(i, j)$ for a fixed j and all $i = 1, \ldots, n_{xm}$, the second-order centered finite difference scheme is explicitly written as

$$\frac{\partial\psi}{\partial x}(i, j) \approx \frac{\psi(i + 1, j) - \psi(i - 1, j)}{2\delta x}, \quad i = 2, \ldots, (n_{xm} - 1),$$

with a particular treatment of indices $i = 1$ and $i = n_{xm}$:

$$\frac{\partial\psi}{\partial x}(1, j) \approx \frac{\psi(2, j) - \psi(n_{xm}, j)}{2\delta x}, \quad \frac{\partial\psi}{\partial x}(n_{xm}, j) \approx \frac{\psi(1, j) - \psi(n_{xm} - 1, j)}{2\delta x}.$$

Using the vectors im and ip from (15.28) and (15.29) we can compress the previous relations into a single one:

$$\frac{\partial\psi}{\partial x}(i, j) \approx \frac{\psi(ip(i), j) - \psi(im(i), j)}{2\delta x}, \quad i = 1, \ldots, n_{xm}. \tag{15.30}$$

Remark 15.1 As a general programming rule, in a finite difference scheme with periodic boundary conditions, we shall replace indices $(i + 1)$ by $ip(i)$ and $(i - 1)$ by $im(i)$ (and similarly for j indices).

We are now equipped to present in detail the full discrete form of each step of Algorithm 15.1.

(A) Computation of Explicit Terms

The two components $\mathcal{H}_u^n$ (15.13) and $\mathcal{H}_v^n$ (15.14) of the explicit term $\mathcal{H}^n$ are computed at the same points of the grid as the corresponding velocities. To follow the logic of the discretization below, the reader is invited to add to Fig. 15.1 the adjacent cells $(i, j \pm 1)$, $(i \pm 1, j)$. Using the centered finite difference scheme (15.30) we obtain:

• for the computation of the velocity u (located at $(x_c(i), y_m(j))$):
for $i = 1, \ldots, n_{xm}$, $j = 1, \ldots, n_{ym}$,

$$\frac{\partial u^2}{\partial x}(i,j) \approx \frac{1}{\delta x}\left[\left(\frac{u(i,j) + u(ip(i),j)}{2}\right)^2 - \left(\frac{u(i,j) + u(im(i),j)}{2}\right)^2\right],$$

$$\frac{\partial uv}{\partial y}(i,j) \approx \frac{1}{\delta y}\left[\left(\frac{u(i,j) + u(i,jp(j))}{2}\right)\left(\frac{v(i,jp(j)) + v(im(i),jp(j))}{2}\right)\right.$$
$$\left. - \left(\frac{u(i,j) + u(i,jm(j))}{2}\right)\left(\frac{v(i,j) + v(im(i),j)}{2}\right)\right],$$

$$\mathcal{H}_u^n(i,j) = -\frac{\partial u^2}{\partial x}(i,j) - \frac{\partial uv}{\partial y}(i,j); \tag{15.31}$$

• and similarly for the velocity v (located at $(x_m(i), y_c(j))$):
for $i = 1, \ldots, n_{xm}$, $j = 1, \ldots, n_{ym}$,

$$\frac{\partial v^2}{\partial y}(i,j) \approx \frac{1}{\delta y}\left[\left(\frac{v(i,j) + v(i,jp(j))}{2}\right)^2 - \left(\frac{v(i,j) + v(i,jm(j))}{2}\right)^2\right],$$

$$\frac{\partial uv}{\partial x}(i,j) \approx \frac{1}{\delta x}\left[\left(\frac{u(ip(i),j) + u(ip(i),jm(j))}{2}\right)\left(\frac{v(i,j) + v(ip(i),j)}{2}\right)\right.$$
$$\left. - \left(\frac{u(i,j) + u(i,jm(j))}{2}\right)\left(\frac{v(i,j) + v(im(i),j)}{2}\right)\right],$$

$$\mathcal{H}_v^n(i,j) = -\frac{\partial uv}{\partial x}(i,j) - \frac{\partial v^2}{\partial y}(i,j). \tag{15.32}$$

(B) Computation of the Nonsolenoidal Velocity Field

We first notice that (15.15) and (15.16) can be written in the compact form of a Helmholtz equation:

$$\underbrace{\left(I - \frac{\delta t}{2Re}\Delta\right)\delta q^*}_{\text{Helmholtz operator}} = \underbrace{\delta t\left[-\nabla p^n + \frac{3}{2}\mathcal{H}^n - \frac{1}{2}\mathcal{H}^{n-1} + \frac{1}{Re}\Delta q^n\right]}_{\text{RHS}^n}, \quad (15.33)$$

where we have introduced the notation $\delta q^* = q^* - q^n$. When periodic boundary conditions are imposed, this equation is usually solved using *fast Fourier transforms* (or FFT).[5] We present in the following a different method, the *alternating direction implicit* (or ADI) method, which is easier to implement and has the advantage that it can easily take into account other types of boundary conditions. The Helmholtz operator is approximated as (terms of order $\mathcal{O}(\delta t^2)$ are neglected):

$$\left(I - \frac{\delta t}{2Re}\Delta\right)\delta q^* \approx \left(I - \frac{\delta t}{2Re}\frac{\partial^2}{\partial x^2}\right)\left(I - \frac{\delta t}{2Re}\frac{\partial^2}{\partial y^2}\right)\delta q^*. \quad (15.34)$$

This second-order accurate factorization[6] is used to solve (15.33) in two steps:

$$\left(I - \frac{\delta t}{2Re}\frac{\partial^2}{\partial x^2}\right)\overline{\delta q^*} = \mathbf{RHS}^n \quad (+\text{periodicity along } x), \quad (15.35)$$

$$\left(I - \frac{\delta t}{2Re}\frac{\partial^2}{\partial y^2}\right)\delta q^* = \overline{\delta q^*} \quad (+\text{periodicity along } y). \quad (15.36)$$

It is important to note that the intermediate field $\overline{\delta q^*}$ is physically meaningless, but it requires boundary conditions. We impose periodic boundary conditions. This choice, which seems to be natural for our periodic problem, becomes more difficult for other types of boundary conditions (Dirichlet type, for example) and needs careful analysis.

For the discretization of second derivatives in (15.35) and (15.36) we use a second-order centered finite difference scheme, written here in the general form (see also Chap. 1)

$$\frac{\partial^2 \psi}{\partial x^2}(i,j) \approx \frac{\psi(i+1,j) - 2\psi(i,j) + \psi(i-1,j)}{\delta x^2}, \quad (15.37)$$

$$\frac{\partial^2 \psi}{\partial y^2}(i,j) \approx \frac{\psi(i,j+1) - 2\psi(i,j) + \psi(i,j-1)}{\delta y^2}.$$

Finally, the algorithm used in the programs of this chapter is the following:

---

[5] This is the subject of an exercise of this chapter.
[6] The methods based on this idea are also known as *approximate factorization* or *splitting* methods.

Algorithm 15.2 *(computes u^* using an ADI method):*

- *First step of ADI: for all $j = 1, \ldots, n_{ym}$ solve the linear system*

$$-\beta_x \overline{(\delta u^*)}(i-1, j) + (1 + 2\beta_x)\overline{(\delta u^*)}(i, j) - \beta_x \overline{(\delta u^*)}(i+1, j) = \mathrm{RHS}_u^n(i, j),$$
$$(15.38)$$

where $i = 1, \ldots, n_{xm}$ and $\beta_x = \frac{\delta t}{2Re} \frac{1}{\delta x^2}$. In the previous relation, we take into account the periodicity by imposing

$$\overline{(\delta u^*)}(0, j) = \overline{(\delta u^*)}(n_{xm}, j), \qquad \overline{(\delta u^*)}(n_{xm} + 1, j) = \overline{(\delta u^*)}(1, j).$$

More precisely, we have to solve n_{ym} linear systems with the following matrix of size $n_{xm} \times n_{xm}$:

$$M_x = \begin{pmatrix} 1 + 2\beta_x & -\beta_x & 0 & .. & 0 & 0 & -\beta_x \\ -\beta_x & 1 + 2\beta_x & -\beta_x & .. & 0 & 0 & 0 \\ . & . & . & .. & . & & . \\ 0 & 0 & 0 & .. & -\beta_x & 1 + 2\beta_x & -\beta_x \\ -\beta_x & 0 & 0 & .. & 0 & -\beta_x & 1 + 2\beta_x \end{pmatrix}. \qquad (15.39)$$

This particular matrix pattern will be referred to in the following as a tridiagonal periodic matrix.
An efficient method to solve such systems will be derived later, based on the well-known Thomas algorithm (Algorithm 15.5).
At the end of this step we obtain $\overline{(\delta u^*)}(i, j)$.

- *Second step of ADI: for all $i = 1, \ldots, n_{xm}$ solve the linear system*

$$-\beta_y (\delta u^*)(i, j-1) + (1 + 2\beta_y)(\delta u^*)(i, j) - \beta_y (\delta u^*)(i, j+1) = \overline{(\delta u^*)}(i, j),$$
$$(15.40)$$

where $j = 1, \ldots, n_{ym}$ and $\beta_y = \frac{\delta t}{2Re} \frac{1}{\delta y^2}$. The periodicity requires that

$$(\delta u^*)(i, 0) = (\delta u^*)(i, n_{ym}), \qquad (\delta u^*)(i, n_{ym} + 1) = (\delta u^*)(i, 1).$$

We obtain this time n_{xm} linear systems with tridiagonal and periodic matrices of size $n_{ym} \times n_{ym}$:

$$M_y = \begin{pmatrix} 1 + 2\beta_y & -\beta_y & 0 & .. & 0 & 0 & -\beta_y \\ -\beta_y & 1 + 2\beta_y & -\beta_y & .. & 0 & 0 & 0 \\ . & . & . & .. & . & & . \\ 0 & 0 & 0 & .. & -\beta_y & 1 + 2\beta_y & -\beta_y \\ -\beta_y & 0 & 0 & .. & 0 & -\beta_y & 1 + 2\beta_y \end{pmatrix}. \qquad (15.41)$$

At the end of this step we get $(\delta u^*)(i, j)$ and immediately

$$u^*(i, j) = u(i, j) + (\delta u^*)(i, j).$$

The computation procedure is similar for the other component of the velocity. Considering that it could be helpful to correctly program the algorithm, we present here the details of the computation:

Algorithm 15.3 *(computes v^* using an ADI method):*

- *First step of ADI: for all $j = 1, \ldots, n_{ym}$ solve the linear system*

$$M_x \, \overline{(\delta v^*)}(i, j) = \mathrm{RHS}_v^n(i, j), \tag{15.42}$$

  where $i = 1, \ldots, n_{xm}$ and the matrix M_x is given by (15.39).
  At the end of this step we obtain $\overline{(\delta v^*)}(i, j)$.
- *Second step of ADI: for all $i = 1, \ldots, n_{xm}$, solve the linear system*

$$M_y \, (\delta v^*)(i, j) = \overline{(\delta v^*)}(i, j), \tag{15.43}$$

  where $j = 1, \ldots, n_{ym}$ and the matrix M_y is given by (15.41).
  We obtain $(\delta v^*)(i, j)$ and immediately

$$v^*(i, j) = u(i, j) + (\delta v^*)(i, j).$$

(C) Solving the Poisson Equation

The Poisson equation (15.11) is discretized as

$$\Delta\phi(i, j) = \left(\frac{\partial^2 \phi}{\partial x^2} + \frac{\partial^2 \phi}{\partial y^2} \right)(i, j) = Q(i, j), \tag{15.44}$$

where $i = 1, \ldots, n_{xm}$, $j = 1, \ldots, n_{ym}$, and

$$Q(i, j) = \frac{1}{\delta t} \operatorname{div}(\boldsymbol{q}^*)(i, j) = \frac{1}{\delta t} \left(\frac{\partial u^*}{\partial x} + \frac{\partial v^*}{\partial y} \right)(i, j).$$

To solve this equation, we first use the periodicity along the x direction and expand the variable ϕ in a discrete Fourier series. We recall from Chap. 3 that the decomposition of a periodic function $\varphi : [0, L] \to \mathbb{R}$ using N Fourier modes is theoretically:

$$\varphi(x) = \sum_{k=0}^{N-1} \widehat{\varphi}_k e^{ik\frac{2\pi}{L}x}, \tag{15.45}$$

where i is the imaginary unit ($i^2 = -1$). Note that we used a change of variables to have $2\pi x/L \in [0, 2\pi]$ and thus be able to apply the theory presented in Chap. 3 for 2π-periodic functions. The discrete version of (15.45), using equidistant grid points $x_m = mh$, $m = 0, \ldots, N-1$, with $h = L/N$ and N even, is then:

$$\varphi_m = \sum_{k=0}^{N-1} \widehat{\varphi}_k e^{ik\frac{2\pi}{N}m} = \sum_{k=-N/2}^{N/2-1} \widehat{\varphi}_k e^{ik\frac{2\pi}{N}m} = \sum_{k=-N/2+1}^{N/2} \widehat{\varphi}_k e^{ik\frac{2\pi}{N}m}. \quad (15.46)$$

Equalities in (15.46) are obtained by noticing that $e^{ik\frac{2\pi}{N}m} = e^{ik\frac{2\pi}{N}(m+N)}$ and using the periodicity of the transformed function ($\widehat{\varphi}_k = \widehat{\varphi}_{k+N}$). For example, we can write: $\sum_{k=-N/2}^{N/2-1} = \sum_{k=-N/2}^{-1} + \sum_{k=0}^{N/2-1} = \sum_{k=N/2}^{N-1} + \sum_{k=0}^{N/2-1}$. The last sums in (15.46) are physically relevant, if Fourier modes are regarded in (15.45) as waves $e^{i2\pi x/\lambda}$ with wavelength $\lambda = L/k$. On the discrete grid with space step h, the minimum wavelength that could be captured is $\lambda_{min} = 2h$, corresponding to the maximum wave number $k_{max} = L/(2h) = N/2$. Note that k_{max} is proportional to the inverse of the Nyquist critical frequency $f_c = 1/(2h)$, well known in signal processing (f_c is the maximum frequency that could be captured by a evenly spaced sampling of a signal). The correspondence between the position x_m of the physical points and their wave numbers k_m is displayed in Fig. 15.2. Note the two equivalent definitions of wave numbers, corresponding to sums in (15.46).

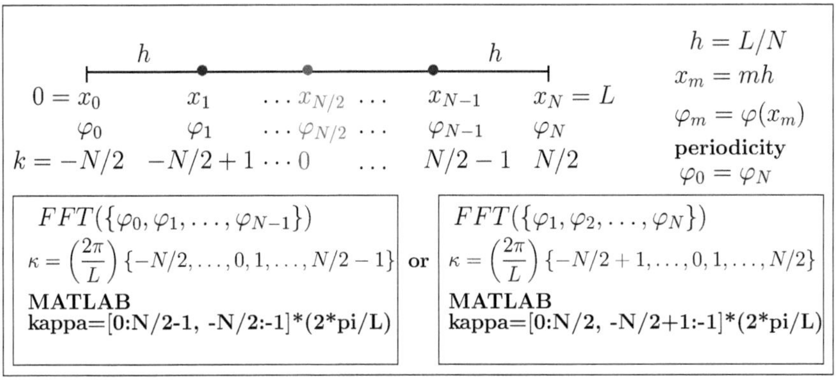

Fig. 15.2 Correspondence between physical points x and wave numbers κ in a Fast Fourier Transform (FFT) of a periodic function $\varphi : [0, L] \to \mathbb{R}$ using N Fourier modes (with N even). Equivalent definitions of wave numbers. Note that if $L = 2\pi$, then $\kappa = k$.

We switch now from the mathematical theory to the practical application of the Fourier decomposition. Since in MATLAB the first index of arrays is one and not zero, we shift indices by replacing $k = l - 1$ and $m = i - 1$. The Fourier decomposition (15.46) becomes:

$$\varphi_i = \sum_{l=1}^{N} \widehat{\varphi}_l e^{i(l-1)\frac{2\pi}{N}(i-1)}, \, i = 1, \dots, N, \quad (15.47)$$

for equidistant grid points $x_i = (i - 1)h$, $i = 1, \dots, N$, $h = L/N$.

Spectral coefficients $\widehat{\varphi}_l$ are computed using optimized Fast Fourier Transforms (FFTs) (built-in function `fft` in MATLAB). The advantage of using a Fourier series expansion is to *diagonalize* the Laplace operator and thus reduce the initial 2D problem (15.44) to a 1D problem. Indeed, a direct derivation of (15.45) leads to

$$\left(\widehat{\frac{d^2\varphi}{dx^2}}\right)_l = -\kappa_l^2 \widehat{\varphi}_l, \quad \kappa_l = \frac{2\pi}{L}(l-1) = \frac{1}{\delta x}\frac{2\pi}{N}(l-1), \, l = 1,\ldots,N, \quad (15.48)$$

meaning that the Fourier coefficients of the second derivative are calculated directly from the Fourier coefficients of the function itself. Using an inverse FFT (`ifft` in MATLAB) on the vector $(-\kappa_l^2 \widehat{\varphi}_l)$ provides the discrete values of the spectral approximation of the second derivative. Coefficients κ_l are the *wave numbers* for our function of period L. When using this spectral approach to approximate derivatives, we must be careful on how the wave numbers are arranged (this depends on the FFT package). In MATLAB, the vector of wave numbers corresponding to (15.48) contains first the positive values (see Fig. 15.2). We use the following expression for the wave-number vector:

$$\text{kappa = [0:N/2-1, -N/2:-1]*(2*pi/L)}. \quad (15.49)$$

In the following, we use a different approach than the spectral one. We start from the Fourier spectral representation of the unknowns and approximate the second derivative using finite difference schemes (see also, Orlandi 1999). The fully spectral approach to solve the Poisson equation will be discussed in Exercise 15.5.

We apply (15.47) to our field ϕ and obtain the following decomposition:

$$\phi(i,j) = \sum_{l=1}^{n_{xm}} \widehat{\phi}_l(j) e^{i\frac{2\pi}{n_{xm}}(i-1)(l-1)}, \quad \forall i = 1,\ldots,n_{xm}. \quad (15.50)$$

Considering an approximation of $\partial^2\phi/\partial x^2$ by second-order centered differences, we obtain

$$\frac{\partial^2\phi}{\partial x^2}(i,j) \approx \frac{\phi(i+1,j) - 2\phi(i,j) + \phi(i-1,j)}{\delta x^2}$$

$$= \frac{1}{\delta x^2} \sum_{l=1}^{n_{xm}} \widehat{\phi}_l(j) e^{i\frac{2\pi}{n_{xm}}(i-1)(l-1)} \left(e^{i\frac{2\pi}{n_{xm}}(l-1)} - 2 + e^{-i\frac{2\pi}{n_{xm}}(l-1)} \right)$$

$$= \sum_{l=1}^{n_{xm}} \widehat{\phi}_l(j) e^{i\frac{2\pi}{n_{xm}}(i-1)(l-1)} \underbrace{\frac{2}{\delta x^2} \left[\cos\left(\frac{2\pi}{n_{xm}}(l-1)\right) - 1 \right]}_{K_l}. \quad (15.51)$$

Coefficients K_l are the *modified wave numbers*:

$$K_l = \frac{2}{\delta x^2} \left[\cos\left(\frac{2\pi}{n_{xm}}(l-1) \right) - 1 \right] = -\frac{4}{\delta x^2} \sin^2\left(\frac{\pi}{n_{xm}}(l-1) \right). \quad (15.52)$$

Note that the presence of periodic trigonometric functions in the expression of K_l makes equivalent the calculation of the vector of modified wave numbers by varying l=[1:N] or l=[0:N/2-1, -N/2:-1]. Note also that, as expected, expression (15.52) of K_l is an approximation of $(-\kappa_l^2)$, with κ_l from (15.48). Since $\sin x \sim x$ for x close to zero, we infer that for low wave numbers this approximation is close to the spectral representation of the second derivative.

Using a similar Fourier series expansion to (15.50) for the right-hand-side function Q (which is also periodic), we find that solving the initial problem (15.44) is equivalent to solving n_{xm} 1D equations

$$\frac{\partial^2}{\partial y^2} \widehat{\phi}_l(j) + K_l \widehat{\phi}_l(j) = \widehat{Q}_l(j), \quad l = 1, \ldots, n_{xm}. \quad (15.53)$$

At this point of our numerical algorithm, equations (15.53) can be solved using either a similar Fourier expansion along y or a finite difference scheme to discretize the second derivative. Since the first method can be applied only for periodic boundary conditions, we choose the second one, which can be easily adapted to more general cases that can be simulated using the programs of this project (a *wall* boundary condition in y, for example). With this choice, (15.53) becomes, for $j = 1, \ldots, n_{ym}$,

$$\frac{1}{\delta y^2} \widehat{\phi}_l(j-1) + \left(-\frac{2}{\delta y^2} + K_l \right) \widehat{\phi}_l(j) + \frac{1}{\delta y^2} \widehat{\phi}_l(j+1) = \widehat{Q}_l(j). \quad (15.54)$$

This equation must be supplemented with discrete boundary conditions for $j = 1$ and $j = n_{ym}$. In our case, we naturally use the periodicity of the function $\widehat{\phi}$.

It is important to note that special treatment is required for the wave number $l = 1$ for which $K_l = 0$. For this value, it is easy to see that the matrix of the system (15.54) is singular and our formulation is not well-posed! Indeed, the solution of the Poisson equation with periodic boundary conditions is determined up to an additive constant. This constant is exactly the first term (or the average value) of the discrete Fourier expansion (15.50). Since the absolute value of the pressure is of no significance for an incompressible flow (we saw that only the gradient of the pressure appears in the equations), we shall not worry about this constant, which will be freely fixed! Nevertheless, a reasonable choice would be to impose $\widehat{\phi}_0(j) = 0$, for $j = 1, \ldots, n_{ym}$, which yields zero average solutions $\widehat{\phi}_l$. The final algorithm to solve the Poisson equation is presented in great detail in the following.

Algorithm 15.4 *To compute the pressure correction ϕ at points of coordinates $(x_m(i), y_m(j))$:*

- *Compute the array $Q(i, j)$, for $i = 1, \ldots, n_{xm}$ and $j = 1, \ldots, n_{ym}$, by*

$$Q(i,j) = \frac{1}{\delta t} \left(\frac{u^*(ip(i),j) - u^*(i,j)}{\delta x} + \frac{v^*(i,jp(i)) - v^*(i,j)}{\delta y} \right). \quad (15.55)$$

- *Apply a fast Fourier transform FFT to each column of the array Q:*

$$\widehat{Q}(l,j) = FFT(Q(i,j)), \quad l = 1, \ldots, n_{xm}, \quad j = 1, \ldots, n_{ym}. \quad (15.56)$$

   The reader is of course aware that the values $\widehat{Q}$ are complex!
- *For $l = 1$ impose $\widehat{\phi}_1(j) = 0$, for $j = 1, \ldots, n_{ym}$.*
- *For each $l = 2, \ldots, n_{xm}$, solve the linear system $M_l \widehat{\phi}_l = \widehat{Q}(l,j)^T$ of a tridiagonal matrix of size $(n_{ym} \times n_{ym})$:*

$$M_l = \frac{1}{\delta y^2} \begin{pmatrix} -2 + \delta y^2 K_l & 1 & 0 \ldots 0 & 0 & 1 \\ 1 & -2 + \delta y^2 K_l & 1 \ldots 0 & 0 & \\ . & . & \ldots & . & . \\ 0 & 0 & 0 \ldots 1 & -2 + \delta y^2 K_l & 1 \\ 1 & 0 & 0 \ldots 0 & 1 & -2 + \delta y^2 K_l \end{pmatrix}, \quad (15.57)$$

   where

$$K_l = \frac{2}{\delta x^2} \left[\cos \left(\frac{2\pi}{n_{xm}} (l-1) \right) - 1 \right].$$

- *Build the array $\widehat{\Phi}(l,j)$, with rows that are the already computed vectors $\widehat{\phi}_l$.*
- *Apply an inverse Fourier transform (IFFT) to obtain the final solution*

$$\phi(i,j) = IFFT(\widehat{\Phi}(l,j)), \quad i = 1, \ldots, n_{xm}, \quad j = 1, \ldots, n_{ym}. \quad (15.58)$$

(D) Computation of the Solenoidal Field

After solving the Poisson equation for the pressure correction, it is easy to correct the velocity field:

- for $i = 1, \ldots, n_{xm}, \quad j = 1, \ldots, n_{ym}$,

$$u^{n+1}(i,j) = u^*(i,j) - \delta t \frac{\phi(i,j) - \phi(im(i),j)}{\delta x}, \quad (15.59)$$

- for $i = 1, \ldots, n_{xm}, \quad j = 1, \ldots, n_{ym}$,

$$v^{n+1}(i,j) = v^*(i,j) - \delta t \frac{\phi(i,j) - \phi(i,jm(j))}{\delta y}. \quad (15.60)$$

(E) Computation of the Pressure Field

Using (15.20), the new pressure field is computed as

- for $i = 1, \ldots, n_{xm}, \quad j = 1, \ldots, n_{ym}$,

$$p^{n+1}(i,j) = p^n(i,j) + \phi(i,j) - \frac{\delta t}{2Re} \left[\frac{\phi(ip(i),j) - 2\phi(i,j) + \phi(im(i),j)}{\delta x^2} \right.$$
$$\left. + \frac{\phi(i,jp(j)) - 2\phi(i,j) + \phi(i,jm(j))}{\delta y^2} \right]. \tag{15.61}$$

Finally, the pressure gradient is updated for the next time step:

- for $i = 1, \ldots, n_{xm}, \quad j = 1, \ldots, n_{ym}$,

$$\frac{\partial p^{n+1}}{\partial x}(i,j) = \frac{p^{n+1}(i,j) - p^{n+1}(im(i),j)}{\delta x}, \tag{15.62}$$

- for $i = 1, \ldots, n_{xm}, \quad j = 1, \ldots, n_{ym}$,

$$\frac{\partial p^{n+1}}{\partial y}(i,j) = \frac{p^{n+1}(i,j) - p^{n+1}(i,jm(j))}{\delta y}. \tag{15.63}$$

Calculation of the Time Step

The last point to discuss for our numerical algorithm is how to compute the value of the time step δt. Since we use a semi-implicit scheme, the time step value will be bounded through an inequality called the *CFL condition*[7]. This condition comes from a stability analysis of the scheme, which is far from a trivial matter when one is dealing with Navier–Stokes equations.[8] For the applications considered in this project, a fair CFL condition would be

$$dt = \frac{\text{cfl}}{\max\left(\left| \frac{u}{\delta x} \right| + \left| \frac{v}{\delta y} \right| \right)}, \tag{15.64}$$

where cfl < 1 is a constant that controls the time step value. In practice, we shall use, when possible, a constant time step, computed from the condition (15.64) applied to the initial flow field.

---

[7] Courant–Friedrichs–Lewy

[8] Exact CFL conditions are derived for scalar convection and wave equations in Chap. 1. The CFL condition is also discussed for the Navier–Stokes equations with zero viscosity (i.e. the Euler equations) in Chap. 12.

15.6 Flow Visualization

An important point for numerically solving the Navier–Stokes equations is the postprocessing of the obtained data. Various interesting physical information can be extracted from a numerical field. For the unsteady flows considered in this project, we use visualization techniques offering an intuitive picture (even for a nonspecialist user) of the flow evolution.

A simple way to visualize the simulated flow is to calculate the *vorticity vector field* $\boldsymbol{\omega}$ by taking the *curl* of the velocity. As we shall see, this is an effective visualization mean for flows dominated by large vortices (the reader can find nice illustrations of vortex flows in the remarkable *Album of Fluid Motion* by Van Dyke (1982)). For 2D flows, the vorticity vector has a single nonzero component, perpendicular to the flow evolution plane:

$$\omega = \frac{\partial v}{\partial x} - \frac{\partial u}{\partial y}. \tag{15.65}$$

The discrete values of ω are computed at points $(x_c(i), y_c(j))$ by

$$\omega(i,j) = \frac{v(i,j) - v(im(i),j)}{\delta x} - \frac{u(i,j) - u(i,jm(j))}{\delta y}. \tag{15.66}$$

The isocontours of vorticity (i.e. the lines of points at which the variable takes the same given value)[9] allow one to identify vortical structures in the flow.

A second visualization method consists in following the evolution of a *passive tracer (or scalar)* in the flow. This numerical technique is equivalent to experimental visualizations using smoke (for gases) or dye (for liquids). As suggested by its name, the passive scalar does not affect the flow field evolution; it is just transported by the velocity field, following a convection-diffusion equation

$$\frac{\partial \chi}{\partial t} + \frac{\partial \chi u}{\partial x} + \frac{\partial \chi v}{\partial y} = \frac{1}{Pe} \Delta \chi, \tag{15.67}$$

where the dimensionless number Pe (Peclet number) quantifies the diffusion properties of the passive tracer χ. The values $\chi(i,j)$ are computed at the cell centers $(x_m(i), y_m(j))$ following the same numerical scheme as for momentum equations; this calculation is done at the end of each time step, allowing one to use the velocity values of the updated (solenoidal) field.

---

[9] MATLAB built-in functions `contour` or `pcolor` draw isocontours for a given 2D solution field.

15.7 Initial Condition

At this point, we are able to advance the numerical solution in time, but we still have to make precise the starting point of the computation, or the initial condition. We shall see that the initial field will be constructed so as to trigger the unsteady flow that we wish to simulate. In principle, the initial condition must be compatible with the Navier–Stokes equations; in practice, we prescribe only the initial velocity field and set the pressure to zero values everywhere. The correct pressure field will be established by the calculation after the first time step.

In this project, we shall simulate two classes of relatively simple flows that illustrate basic mechanisms found in more general and complex real flows.

Dynamics of a 2D Jet: The Kelvin–Helmholtz Instability

The Kelvin–Helmholtz instability generally occurs in flows where shear is present. The basic example for this instability is the flow of two parallel streams of different velocities and, eventually, densities. This flow can be obtained in a simple experiment: put in a long rectangular transparent box two immiscible liquids with large density difference and start to slowly incline the box. The denser liquid will start to flow in the lower part of the box, pushing the lightest liquid into the upper part. Very nice patterns, called Kelvin *cat eyes*, form at the interface between the two liquids (see Fig. 15.3 for a sketch and Fig. 15.6 for a numerical simulation).

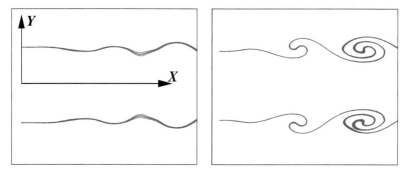

Fig. 15.3 Evolution of the Kelvin–Helmholtz instability in a 2D jet: perturbation of the shear layer forming the contour of the jet (left) and roll-up of Kelvin (*cat eyes*) vortices (right).

This phenomenon also occurs in jet flows that are generated by the injection of fluid into a quiescent environment. The instability develops in the shear layer between the injected fluid and the fluid at rest. Our numerical simulation will start from an initial condition setting the velocity profile corresponding to the shear layers forming the contour of the jet:

$$v(x, y) = 0, \qquad u(x, y) = u_1(y)(1 + u_2(x)), \qquad (15.68)$$

where u_1 is the mean velocity profile

$$u_1(y) = \frac{U_0}{2} \left(1 + \tanh \left(\frac{1}{2} P_j \left(1 - \frac{|L_y/2 - y|}{R_j} \right) \right) \right), \qquad (15.69)$$

and u_2 the perturbation that triggers the Kelvin–Helmholtz instability

$$u_2(x) = A_x \sin \left(2\pi \frac{x}{\lambda_x} \right). \qquad (15.70)$$

Note that both velocity profiles are compatible with the periodicity condition at the boundaries. Physically, the periodicity in the y-direction means that we simulate an infinite array of parallel 2D jets. The parameters $U_0, P_j, R_j, A_x, \lambda_x$ will be specified later for numerical applications.

Evolution of a Vortex Dipole

Vortices generated by the Kelvin–Helmholtz instability all rotate in the same sense, i.e. they have vorticity of the same sign. A vortex dipole is a pair of vortices of opposite signs. This configuration is encountered in many areas of practical interest (meteorological and coastal flows, trailing vortices from aircraft, 2D turbulence, swirled injection in stratified charge engines). We consider here symmetric dipoles for which the two vortices have the same vorticity magnitude. This is a stable structure that propagates along its axis of symmetry with a quasi-constant translation velocity generated by a self-induction mechanism. The reader interested in studying vortex motion is referred to Batchelor (1988) and Saffman (1992).

We shall numerically construct a vortex dipole by superimposing the velocity fields of two individual vortices (see Fig. 15.4). Each vortex, defined by its center (x_v, y_v), size l_v, and intensity ψ_0, is analytically described by the following *stream-function*:

$$\psi(x, y) = \psi_0 \, exp \left(-\frac{(x - x_v)^2 + (y - y_v)^2}{l_v^2} \right). \qquad (15.71)$$

The stream-function is used to derive the velocity components as

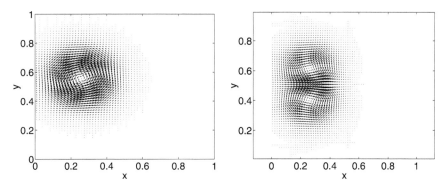

Fig. 15.4 Velocity field of a single vortex (left) and of a vortex dipole (right).

$$\begin{cases} u = \dfrac{\partial\psi}{\partial y} = -2\dfrac{(y - y_v)}{l_v^2}\,\psi(x, y), \\[2ex] v = -\dfrac{\partial\psi}{\partial x} = 2\dfrac{(x - x_v)}{l_v^2}\,\psi(x, y). \end{cases} \tag{15.72}$$

The dipole is now *assembled* by taking two vortices of the same size l_v but opposite intensities $\pm\psi_0$ and placing them symmetrically about a chosen line, which will be the propagation direction. For example, a dipole propagating to the right along the x axis will be defined by (see Fig. 15.4)

vortex 1 : $+\psi_0, l_v, x_v, y_v = L_y + a$,

vortex 2 : $-\psi_0, l_v, x_v, y_v = L_y - a$,

where a is the distance separating the vortex centers.

It is important to note that this is not a rigorous method to construct a vortex dipole, since the initial condition is not compatible with the Navier–Stokes equations. However, the velocity and pressure fields will be automatically adjusted to satisfy the equations after the first time step of the numerical simulation. The reader can test more rigorous analytical models for the vortex dipole as, for example, the Lamb–Chaplygin dipole, which corresponds to a steady solution of the 2D Euler equations (see Batchelor 1988, Saffman 1992). Besides, our initial condition does not satisfy the assumed periodicity of the problem: a more rigorous method to simulate vortex configurations in a periodic box consists in summing the velocity fields of the vortex with those of the four *image vortices* (placed symmetrically with respect to the four boundaries). Supplementary programs with more involved vortex initial conditions are provided in the folder NSE_SUPP.

15.8 Step-by-Step Implementation

The survivors of previous lengthy theoretical developments may now start to implement the numerical algorithm to simulate some physical flows. Since this is a delicate process, we shall proceed step by step to construct our *Navier–Stokes code.* We adopt the programming strategy of building special-ized program modules that will be first validated on simpler problems for which an exact solution is known. We start with some preliminary questions.

15.8.1 Solving a Linear System with Tridiagonal, Periodic Matrix

Since there exist different MATLAB built-in functions to solve linear systems, this part is not compulsory for the following numerical developments. Never-theless, we consider that the reader should be aware of the structure of the involved systems and efficient algorithms to solve them.[10] Moreover, the algo-rithms in this section can be used in applications using other (less-friendly) programming languages.

We shall use the particular pattern (tridiagonal, periodic) of matrices (15.39), (15.41), (15.57) to build an efficient numerical algorithm to solve the corresponding linear systems. We start by presenting the well-known Thomas algorithm for solving tridiagonal systems.

Algorithm 15.5 *Thomas algorithm for tridiagonal systems.[11] The tridiago-nal system*

$$
\begin{pmatrix}
b_1 & c_1 & 0 & . & . & 0 & 0 \\
a_2 & b_2 & c_2 & 0 & . & 0 & 0 \\
. & . & . & . & & . & . \\
. & . & . & . & . & . & . \\
0 & 0 & 0 & 0 & a_{n-1} & b_{n-1} & c_{n-1} \\
0 & 0 & 0 & 0 & . & a_n & b_n
\end{pmatrix}
\begin{pmatrix}
X_1 \\
X_2 \\
. \\
. \\
X_{n-1} \\
X_n
\end{pmatrix}
=
\begin{pmatrix}
f_1 \\
f_2 \\
. \\
. \\
f_{n-1} \\
f_n
\end{pmatrix}
$$

is solved by introducing the following recurrence relation:

$$
\begin{cases}
X_k = \gamma_k - \dfrac{c_k}{\beta_k} X_{k+1}, & k = 1, \ldots, (n-1), \\[2mm]
X_n = \gamma_n.
\end{cases}
\tag{15.73}
$$

---

[10] It is always interesting to know what happens behind the *magical* MATLAB command $x = A\backslash b$ that solves the system $Ax = b$.

[11] We can easily show that this algorithm is a particular form of the Gauss elimination method.

Inserting these relations in the initial form of the system, we can calculate the coefficients γ_k and β_k:

$$
\begin{cases}
\beta_1 = b_1, \\[2mm]
\beta_k = b_k - \dfrac{c_{k-1}}{\beta_{k-1}} a_k, \quad k = 2, \ldots, n,
\end{cases}
$$

$$
\begin{cases}
\gamma_1 = \dfrac{f_1}{\beta_1} = \dfrac{f_1}{b_1}, \\[2mm]
\gamma_k = \dfrac{f_k - a_k \gamma_{k-1}}{\beta_k}, \quad k = 2, \ldots, n.
\end{cases}
$$

After computing the coefficients γ_k and β_k, the unknowns X_k are immediately obtained from (15.73) by a backward substitution starting from the known value $X_n = \gamma_n$.

We now note that a periodic tridiagonal matrix has supplementary nonzero coefficients in the upper-right and lower-left corners. The idea of the following algorithm is to eliminate these *intruders* and to work with tridiagonal systems.

Algorithm 15.6 *Thomas algorithm for tridiagonal, periodic systems. The system*

$$
\begin{pmatrix}
b_1 & c_1 & 0 & . & 0 & 0 & | a_1 \\
a_2 & b_2 & c_2 & . & 0 & 0 & | 0 \\
. & . & . & . & . & & | 0 \\
. & . & . & . & & . & | 0 \\
0 & 0 & 0 & . & a_{n-1} & b_{n-1} & | c_{n-1} \\
\hline
c_n & 0 & 0 & . & 0 & a_n & | b_n
\end{pmatrix}
\begin{pmatrix}
X_1 \\
X_2 \\
. \\
. \\
X_{n-1} \\
\hline
X_n
\end{pmatrix}
=
\begin{pmatrix}
f_1 \\
f_2 \\
. \\
. \\
f_{n-1} \\
\hline
f_n
\end{pmatrix}
$$

is rewritten as[12]

$$
\begin{pmatrix}
b_1^* & c_1 & 0 & . & 0 & 0 & 0 & | v_1 \\
a_2 & b_2 & c_2 & . & 0 & 0 & 0 & | 0 \\
. & . & . & . & . & & . & | 0 \\
. & . & . & . & & . & . & | 0 \\
0 & 0 & 0 & . & a_{n-1} & b_{n-1} & c_{n-1} & | 0 \\
0 & 0 & 0 & . & 0 & a_n & b_n^* & | v_n \\
\hline
-1 & 0 & 0 & . & 0 & 0 & -1 & | 1
\end{pmatrix}
\begin{pmatrix}
X_1 \\
X_2 \\
. \\
. \\
X_{n-1} \\
X_n \\
\hline
X^*
\end{pmatrix}
=
\begin{pmatrix}
f_1 \\
f_2 \\
. \\
. \\
f_{n-1} \\
f_n \\
\hline
0
\end{pmatrix}
, \quad (15.74)
$$

where

---

[12] This decomposition is similar to the Shermann–Morrison formula.

$$\begin{cases} v_1 = a_1, \\ v_n = c_n, \\ b_1^* = b_1 - a_1, \\ b_n^* = b_n - c_n, \end{cases} \qquad and \qquad X^* = X_1 + X_n.$$

An equivalent form of (15.74) is

$$\underbrace{\begin{pmatrix} b_1^* & c_1 & 0 & . . & 0 & 0 \\ a_2 & b_2 & c_2 & 0 & . & 0 & 0 \\ . & . & . & . & . & . \\ 0 & 0 & 0 & 0 & . & a_n & b_n^* \end{pmatrix}}_{M^*} \begin{pmatrix} X_1 \\ X_2 \\ \cdot \\ X_n \end{pmatrix} + \begin{pmatrix} v_1 \\ 0 \\ \cdot \\ v_n \end{pmatrix} X^* = \begin{pmatrix} f_1 \\ f_2 \\ \cdot \\ f_n \end{pmatrix}$$

together with

$$X^* = X_1 + X_n.$$

We now seek a solution of the form

$$X_k = X_k^{(1)} - X_k^{(2)} \cdot X^*, \quad k = 1, \ldots, n, \tag{15.75}$$

with the vectors $X^{(1)}$ and $X^{(2)}$ solutions of two tridiagonal systems of size n:

$$\begin{cases} M^* \cdot X^{(1)} = (f_1, f_2, \ldots f_{n-1}, f_n)^T, \\ M^* \cdot X^{(2)} = (v_1, 0, \ldots 0, v_n)^T. \end{cases} \tag{15.76}$$

Finally, the supplementary unknown is calculated as

$$X^* = \frac{X_1^{(1)} + X_n^{(1)}}{1 + X_1^{(2)} + X_n^{(2)}}. \tag{15.77}$$

To summarize, the algorithm consists of the following steps:

- *solve the two tridiagonal systems (15.76) using the Thomas algorithm 15.5; note that the program for this step can be optimized since both systems share the same matrix M^*;*
- *compute X^* from (15.77);*
- *compute the final solution using (15.75).*

Exercise 15.1 Write a MATLAB function

```
function fi=NSE_F_trid_per_c2D(aa,ab,ac,fi)
```

that solves simultaneously m systems with tridiagonal, periodic matrices. Algorithm 15.6 is used to solve each system j (for $1 \le j \le m$) defined as follows:

```
for i=1,...,n
   aa(j,i)*X(j,i-1)+ab(j,i)*X(j,i)+ac(j,i)*X(j,i+1)=fi(j,i),
with periodicity condition X(j,1)=X(j,n).
```

Hint: use vector programming to apply the relations of the algorithm simultaneously to all m systems; for example, the program lines computing the coefficients b_1^*, b_n^* of the matrix M^* from (15.74) are written as

```
ab(:,1)=ab(:,1)-aa(:,1);
ab(:,n)=ab(:,n)-ac(:,n);
```

which implies that the computation is done for all row indices.
    Test this function using as model the MATLAB script NSE_M_test_trid.m.[13]

15.8.2 Solving the Unsteady Heat Equation

Study of the 2D unsteady heat equation provides the ideal framework for testing the procedures that will constitute the core of this project: the Helmholtz and Poisson solvers. We consider the unsteady heat equation (see Chap. 1 for the 1D heat equation)

$$\frac{\partial u}{\partial t} - \Delta u(t, x, y) = f(x, y), \quad \text{for} \quad (x, y) \in \Omega = [0, L_x] \times [0, L_y], \quad (15.78)$$

with periodic boundary conditions and initial condition $u(0, x, y) = u^0(x, y)$. This equation will be numerically integrated in time until a steady (equilibrium) solution is reached. This steady solution satisfies

$$-\Delta u_s(x, y) = f(x, y), \quad \text{for} \quad (x, y) \in [0, L_x] \times [0, L_y], \quad (15.79)$$

with the same periodic boundary conditions. The steady solution $u_s(x, y)$ may be interpreted as the limit for $t \to \infty$ of the unsteady solution $u(t, x, y)$.
    We adopt in this section the method of *manufactured solutions* (see also Chap. 7) to test the programs. We set the right-hand-side function

$$f(x, y) = (a^2 + b^2) \sin(ax) \cos(by), \quad \text{where} \quad a = \frac{2\pi}{L_x}, b = \frac{2\pi}{L_y}, \quad (15.80)$$

which satisfies the periodicity along x and y. For this choice, the exact solution of (15.79) is

$$u_{ex}(x, y) = \sin(ax) \cos(by). \quad (15.81)$$

---

[13] Although this script is intended to be straightforward, some comments may be helpful: the (diagonal) vectors aa,ab,ac are filled with random values; for each j, the matrix A of the system is reconstructed and transformed into a *diagonal dominant* matrix that is known to be invertible; the right-hand side of the system is computed as $f = A * \tilde{X}$, where $\tilde{X}$ is arbitrarily set; every system is first solved using the MATLAB syntax $X = A \backslash f$ to obtain a reference solution; the function NSE_F_trid_per_c2D is finally validated if the returned solution is exactly $\tilde{X}$. This is a commonly used technique to test programs that solve linear systems.

Indeed, it can be easily checked that f was chosen such that $f(x, y) = -\Delta u_{ex}$. Using f as input data in programs, we get a numerical solution that has to fit the exact (analytical) solution. If this is not the case, debugging is necessary!

Explicit Solver

The simplest method to solve (15.78) is based on the explicit Euler scheme (see Chap. 1)

$$u^{n+1} = u^n + \delta t \left(f + \Delta u^n\right). \tag{15.82}$$

Assuming that the solution is computed at grid points (x_c, y_m) (see Fig. 15.1), the discrete form of the scheme becomes $(i = 1, \ldots, n_{xm}, j = 1, \ldots, n_{ym})$:

$$
u^{n+1}(i, j) = u^n(i, j) + \delta t \left[f(i, j) + \frac{u(ip(i), j) - 2u(i, j) + u(im(i), j)}{\delta x^2} \right.
$$
$$
\left. + \frac{u(i, jp(j)) - 2u(i, j) + u(i, jm(j))}{\delta y^2} \right]. \tag{15.83}
$$

The time integration starts from the initial condition $u^0 = 0$ and stops when the convergence to a steady solution is reached. We impose the following numerical convergence criterion:

$$\varepsilon = \|u^{n+1} - u^n\|_2 < 10^{-6}, \tag{15.84}$$

where the norm is defined as $\|\varphi\|_2 = \left(\int_\Omega \varphi^2 \, dx \, dy\right)^{1/2}$.

The drawback of this scheme in computing steady solutions is that the time step value is limited by a stability (CFL) condition. For the 2D heat equation, the CFL condition is written as (see, for instance, Hirsch (1988)):

$$\delta t \left(\frac{1}{\delta x^2} + \frac{1}{\delta y^2}\right) = \frac{\mathrm{cfl}}{2}, \quad \mathrm{cfl} \le 1. \tag{15.85}$$

Exercise 15.2 Compute the solution of the unsteady heat equation (15.78) using the explicit scheme (15.83). Compare the obtained steady solution to the exact solution (15.81). Use the following input parameters:
$L_x = 1, L_y = 2, n_x = 21, n_y = 51, \mathrm{cfl} = 1$.
Hints:
• the grid parameters may be defined as global variables;
• write modular programs with specialized functions that can be reused for subsequent applications; for example, write separate functions to compute f, Δu^n, to plot the solution, etc.;
• use a `while` loop for the time advancement, which allows one to easily implement the convergence criterion;
• avoid `for` loops and use vector programming, more compact and easier to compare with mathematical relations; for example, the following function

computes the discrete values of the Laplacian Δu^n using directly the array u (of size $n_{xm} \times n_{ym}$) and the vectors ip, jp, im, jm defined by (15.28)-(15.29):

```
function hc=NSE_F_calc_lap(u)
global dx dy
global im ip jp jm ic jc
hc =...
(u(ip,jc)-2*u+u(im,jc))/(dx*dx)+(u(ic,jp)-2*u+u(ic,jm))/(dy*dy);
```

• plot in the same figure the isocontours of the numerical steady solution and the exact solution (15.81).

A solution of this exercise is proposed in Sect. 15.9 at page 357.

Figure 15.5 displays a typical result for the isocontours of the steady solution. Note that the numerical and exact solutions are difficult to distinguish in the plot (a more quantitative comparison can be made by computing $\|u - u_{ex}\|_2$). The slow convergence to the steady solution is also illustrated in the same figure. We conclude that the explicit solver is easy to implement but requires small time steps and, consequently, large computational times. This suggests that an implicit solver able to take larger time steps is more appropriate for this problem.

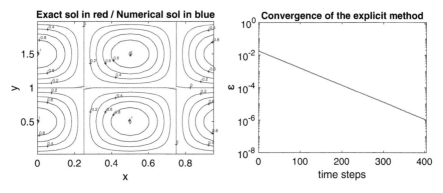

Fig. 15.5 Test of the explicit solver for the unsteady heat equation. Superposition of isocontours of the steady numerical and exact solutions (left) and history of the convergence to the steady solution (right), with $\varepsilon = \|u^{n+1} - u^n\|_2$.

Implicit Solver

We use the combined Adams–Bashforth and Crank–Nicolson schemes described previously to discretize (15.78)

$$\frac{u^{n+1} - u^n}{\delta t} = \underbrace{\frac{3}{2}\mathcal{H}^n - \frac{1}{2}\mathcal{H}^{n-1}}_{\text{Adams–Bashforth}} + \underbrace{\frac{1}{2}\Delta\left(u^{n+1} + u^n\right)}_{\text{Crank–Nicolson}}, \tag{15.86}$$

where in this case, the term $\mathcal{H}^n = \mathcal{H}^{n-1} = f(x, y)$ does not depend on time. We finally get the Helmholtz equation

$$\left(I - \frac{\delta t}{2}\Delta\right)\delta u = \delta t\left(f + \Delta u^n\right), \quad \text{with} \quad \delta u = u^{n+1} - u^n, \tag{15.87}$$

which is solved using the ADI method with the following steps:

$$\begin{cases} \left(I - \dfrac{\delta t}{2}\dfrac{\partial^2}{\partial x^2}\right)\overline{\delta u} = \delta t\left(f + \Delta u^n\right) + \text{periodicity along } x, \\[2mm] \left(I - \dfrac{\delta t}{2}\dfrac{\partial^2}{\partial y^2}\right)\delta u = \overline{\delta u} \qquad\qquad + \text{periodicity along } y. \end{cases} \tag{15.88}$$

Using second-order centered finite differences to discretize second derivatives, we obtain two linear systems with tridiagonal, periodic matrices. These systems are solved using the function NSE_F_trid_per_c2D written for the previous exercise.

Remark 15.2 The semi-implicit solver defined in this section is unconditionally stable, allowing arbitrarily large time steps δt. Compared to the explicit solver, it requires much less computational time to reach the steady solution (more work per time step but very few time steps to converge).

Exercise 15.3 Resume Exercise 15.2 and implement the implicit solver. The time step will be computed using (15.85) taking cfl $= 100$. Evaluate the necessary computing time to reach the steady solution and compare the explicit solver. Hint: noticing that the coefficients of the matrices involved in the ADI steps are constant in time, optimize the function NSE_F_trid_per_c2D by:
• storing the coefficients of the matrices in vectors and not in 2D arrays;
• computing all the quantities not depending on the right-hand side only once, before the while loop.
A solution of this exercise is proposed in Sect. 15.9 at page 358.

Exercise 15.4 Consider now the following nonlinear convection–diffusion equation:

$$\frac{\partial u}{\partial t} + \frac{\partial u^2}{\partial x} - \Delta u = f(x, y), \quad \text{for} \quad (x, y) \in [0, L_x] \times [0, L_y], \tag{15.89}$$

with periodic boundary conditions and initial condition $u^0(x, y) = 0$.

1. Choose the analytical form of the right-hand-side function $f(x, y)$ such that (15.81) is the steady solution of (15.89).

2. Use the implicit scheme (15.86) to solve this equation. Compare the results to the exact solution. Hint: use the previous program and modify only the function computing $\mathcal{H}$ (be careful, for this equation $\mathcal{H}$ varies in time, and consequently, $\mathcal{H}^n \neq \mathcal{H}^{n-1}$).
3. The time step is constant and computed from (15.85). Find the stability limit (cfl_{max}) for a given space discretization.

A solution of this exercise is proposed in Sect. 15.9 at page 358.

15.8.3 Solving the Steady Heat Equation Using FFTs

We now write and test the necessary functions for solving the Poisson equation (Algorithm 15.4). We use as a test case the heat equation (15.79).

Exercise 15.5 1. Solve the steady heat equation (15.79) with the right-hand side (15.80) using the Poisson solver described in Sect. 15.5 (i.e. FFT along x and finite differences along y). Compare to the exact solution.[14] Input parameters: $L_x = 1, L_y = 2, n_x = 65, n_y = 129$.
2. Optimize (see Exercise 15.3) the function solving the tridiagonal system.
3. Solve numerically the same equation using the same finite difference approach and two FFTs (i.e. FFT along x and FFT along y).
4. Write a spectral solver for the stationary heat equation using the Fourier decomposition (15.47) and its property (15.48). Compare with the solver using the finite difference approach by computing $\|u_{num} - u_{ex}\|_2$.
A solution of this exercise is proposed in Sect. 15.9 at page 358.

15.8.4 Solving the 2D Navier–Stokes Equations

We are now ready to assemble all the modules previously developed to solve the Navier–Stokes equations and simulate the flows described in Sect. 15.7.

Exercise 15.6 Write a Navier–Stokes solver for 2D periodic flows. Hints for the structure of the program:

• define the global variables,
• set the input parameters,
• build the 2D grid and related arrays,
• define the arrays to store the flow variables and initialize them to zero,

---

[14] As we have already seen, the numerical solution u_{num} is computed up to an additive constant. To compare to the exact solution u_{ex}, we have to calculate this constant by imposing that the two solutions be identical at a chosen point ($i = j = 1$, for example). We then compare u_{ex} to $u_{num} + (u_{ex}(1,1) - u_{num}(1,1))$.

- set the initial condition corresponding to the simulated flow (see below the parameters for the suggested run cases),
- visualize the initial field,
- compute the time step,
- compute the variables for the optimization of the ADI method and Poisson solver (i.e. all variables or coefficients not depending on time),
- start the time loop:

  – solve the momentum equation for u,
  – solve the momentum equation for v,
  – compute the divergence of the nonsolenoidal field,
  – solve the Poisson equation,
  – correct the velocity field,
  – compute the pressure,
  – update the pressure gradient for the next step,
  – solve the equation for the passive scalar,
  – check that the divergence of the velocity field is zero,
  – visualize the flow field by plotting the isocontours of vorticity and passive scalar.

- end of the time loop.

A solution of this exercise is proposed in Sect. 15.9 at page 359.

Run cases. The expected results are illustrated in the figures at the end of the chapter.

1. 2D jet: Kelvin–Helmholtz instability; input parameters

$$L_z = 2, L_y = 1, n_x = 65, n_y = 65, \text{cfl} = 0.2,$$

$$Re = 1000, Pe = 1000, U_0 = 1, P_j = 20, R_j = L_y/4, A_x = 0.5, \lambda_x = 0.5L_x.$$

The initial field for the passive scalar is identical to the field of u.
2. Same configuration, but changing cfl $= 0.1, \lambda_x = 0.25L_x$.
3. Vortex dipole:

$$L_z = 1, L_y = 1, n_x = 65, n_y = 65, \text{cfl} = 0.4, Re = 1000, Pe = 1000;$$

vortex 1:
$$\psi_0 = +0.01, x_v = L_x/4, y_v = L_y/2 + 0.05,$$
$$l_v = 0.4\sqrt{2} \min\{x_v, y_v, L_x - x_v, L_y - y_v\};$$

vortex 2:
$$\psi_0 = -0.01, x_v = L_x/4, y_v = L_y/2 - 0.05,$$
$$l_v = 0.4\sqrt{2} \min\{x_v, y_v, L_x - x_v, L_y - y_v\}.$$

The initial field for the passive scalar is set to a large stripe, placed in the middle of the computational domain. For example, take

$$\begin{cases} \chi(i,j) = 1, & \text{if } n_{xm}/2 - 10 \leq i \leq n_{xm}/2 + 10, \\ \chi(i,j) = 0, & \text{otherwise.} \end{cases}$$

4. Add to the previous configuration a second dipole propagating in the opposite direction.
5. Imagine other flow configurations with several dipoles in the computational domain.

15.9 Solutions and Programs

The MATLAB scripts for this project are organized in two directories:

- **NSE_QP** containing the solution scripts for all preliminary questions (Exercises 15.1 to 15.5),
- **NSE_QNS** containing the Navier–Stokes solver (Exercise 15.6).

There is also a third directory named **NSE_INTERFACE** in which a different programming philosophy is illustrated. All the solution scripts of the project are called from a graphical user interface (GUI) and the results are displayed interactively. A supplementary Navier–Stokes run case is computed in this version. To launch the interface, just run the script Main from the subdirectory **Tutorial**.

Solution of Exercise 15.1 (Solving a Tridiagonal, Periodic System)

The MATLAB script NSE_F_trid_per_c2D.m contains the function that solves simultaneously m linear systems with tridiagonal, periodic matrices of size n. Numerous comments in the script are intended to guide the reader through the steps of Algorithm 15.6. Memory storage was optimized using a minimum number of arrays for computing intermediate coefficients. Note also the vector programming of the algorithm.
This function is called (and tested) by the script NSE_M_test_trid.m.

Solution of Exercise 15.2 (Explicit Solver for the Unsteady Heat Equation)

The script NSE_M_test_Heat_exp.m is straightforward to read and execute. Nevertheless, it is useful to indicate the specialized functions called by this program:

- NSE_F_calc_lap: computes the Laplacian Δu;
- NSE_F_Heat_fsource: computes the right-hand term (or source term) f;
- NSE_F_Heat_exact: computes the exact solution;

- NSE_F_norm_L2: computes the norm $\|u\|_2$;
- NSE_F_visu_isos: plots in the same figure the isocontours of the numerical and exact solutions.

Solution of Exercise 15.3 (Implicit Solver for the Unsteady Heat Equation)

The script NSE_M_test_Heat_imp.m inherits the structure of the program implementing the explicit solver. Since the implicit method requires to solve tridiagonal, periodic systems, two functions optimizing this part were added: NSE_F_ADI_init and NSE_F_ADI_step. The optimization starts from the observation that in the general solver NSE_F_trid_per_c2D all the computations not depending on the right-hand side fi can be done only once, outside the time loop. This is the role of the function NSE_F_ADI_init, returning the vectors ami, api, alph, xs2, which do not change during the time integration; these vectors are used in NSE_F_ADI_step (called inside the time loop) to compute the solution of the linear system for a given (time-dependent) vector fi.

Solution of Exercise 15.4 (Implicit Solver for the Nonlinear Convection–Diffusion Equation)

The script solution NSE_M_test_ConvDiff_imp.m for the nonlinear problem inherits the structure of the previous program (NSE_M_test_Heat_imp.m). The only difference is the computation of the term $\mathcal{H}$ for every time step by calling inside the time loop the function NSE_F_ConvDiff_calc_hc.m. The exact solution (15.81) (the same as for the linear heat equation) is implemented in NSE_F_ConvDiff_exact.m. The expression of the source term f in NSE_F_ConvDiff_fsource.m takes into account the new nonlinear term.

Solution of Exercise 15.5 (Solving the Steady Heat Equation Using FFTs)

Algorithm 15.4 is implemented in the script NSE_M_test_HeatS_FD_1fft.m. The script solving a tridiagonal system is optimized by splitting the algorithm into two parts, corresponding to the functions NSE_F_Phi_init and NSE_F_Phi_step. This is similar to the optimization of the ADI method, with the difference that this time the coefficients of the matrix change from one system to another (because of the dependence on the wave number), and consequently, they are stored in 2D arrays. The script NSE_M_test_HeatS_FD_2fft.m solves the same problem using two FFTs (along the x and y directions). The Laplace operator in (15.79) is diagonalized using 2 FFTs, resulting in the explicit relation:

$$(K_x(i,j) + K_y(i,j))\,\widehat{u}_s(i,j) = -\widehat{f}(i,j),\ i = 1,\dots,nxm,\ j = 1,\dots,nym,$$
$$(15.90)$$

with modified wave numbers K_x and K_y given by (15.52) for each direction. In matrix form, Eq. (15.90) becomes: `Kxy * Uhat=Fhat`. Since in MATLAB it is possible to perform the FFT on a matrix (the FFT is applied to each column), the matrix `Fhat` is directly computed as `Fhat=fft(fft(-F)')`. Note the transpose after the first FFT. The matrix `Kxy` has to be carefully assembled taking into account the arrangement of the 2D arrays:

```
Kx=(cos(2*pi/nxm*(ic-1))-1)*2/(dx*dx);
Ky=(cos(2*pi/nym*(jc-1))-1)*2/(dy*dy);
Kxy=Kx'*ones(1,nym)+ones(nxm,1)*Ky;
```

After the correction of the elements corresponding to the singular wave number ($i = j = 1$), we simply compute `Uhat=Uhat./Kxy'`; and finally apply the inverse FFTs to obtain the solution: `u=ifft(ifft(Uhat)')`;.

The spectral solver `NSE_M_test_HeatS_SP_2fft.m` is identical to the previous program, excepting for the expression of the matrix `Kxy`. Following the theoretical part, in the spectral approach the modified wave numbers K_x and K_y in (15.90) have to be replaced with $(-\kappa_x^2)$ and $(-\kappa_y^2)$, respectively. Expressions for κ_x and κ_y are obtained from (15.49), applied to each direction. Note the *spectral precision* obtained with this program ($\varepsilon = \|u_{\text{num}} - u_{ex}\|_2 = 3.14e-16$), compared to the second-order precision obtained using the finite difference approach ($\varepsilon = 4.8e-4$).

Solution of Exercise 15.6 (Solving the 2D Navier–Stokes Equations)

The main program `NSE_M_QNS.m` allows one to choose between the four suggested run cases. Comments in the program body identify each step of the numerical algorithm. The program calls the main functions written for the preliminary questions (`NSE_F_ADI_init`, `NSE_F_ADI_step`, `NSE_F_Phi_init`, `NSE_F_Phi_step`) and the following specific functions:

- `NSE_F_init_KH`: initializes the flow field for the Kelvin–Helmholtz (2D jet) run cases;
- `NSE_F_init_vortex` builds the flow field corresponding to an individual vortex; the vortex dipoles are obtained by superimposing the individual vortex fields;
- `NSE_F_visu_vort` visualizes the vorticity field (color images of isocontours);
- `NSE_F_visu_sca` visualizes the passive tracer leading to images similar to experimental ones;
- `NSE_F_print_div` prints the divergence of the velocity field to verify whether the computation is stable. The divergence must be close to the

machine precision zero value. We set this value to 10^{-15} (calculations are done with double-precision accuracy) and check that for each time step the divergence of the velocity stays around this value; if this is not the case, run the same computation with smaller time steps.

It is beyond the scope of this project, which focuses essentially on numerics, to get into a detailed physical description of the simulated flows. We discuss, however, some interesting physical features illustrated in the following figures.

The evolution of the Kelvin–Helmholtz instability is shown in Figs. 15.6 and 15.7. The Kelvin *cat eyes* vortices form progressively in the two shear regions of the jet; their spatial distribution is dictated by the wavelength (λ_x) of the initial perturbation. At this point, one might question whether a periodic simulation could be realistic. Since in real jet flows the instability progressively grows downstream of the injection point, our periodic computational box may be regarded as a fixed frame that zooms in the shear layer region while traveling downstream with the mean velocity of the flow. Periodic simulations offer useful information on the evolution of vortical structures in jet or shear-layer flows that fit very well to experimental results. The reader who wishes to pursue this study further could attempt to simulate the next stage of the Kelvin–Helmholtz instability, which consists in the pairing of neighboring vortices.

The first vortex dipole run case is illustrated in Fig. 15.8. The dipole effectively propagates along the horizontal axis toward the right boundary; it could be interesting to continue the simulation and see how the periodicity makes the dipole reenter the computational box from the left. The velocity induced by the dipole triggers the movement of the passive scalar (initially at rest), with a nice mushroom pattern forming. This kind of structure has been reported in studies of flow dynamics in oceanography, meteorology, and combustion.

The last run case (see Fig. 15.9) shows the head-on interaction between two dipoles of the same intensity. The result is the *partner interchange* with the formation of two new dipoles propagating in the perpendicular direction. The simulation may be performed for larger values of the final integration time to see a second collision (due to the periodicity, a dipole leaving the domain reenters through the opposite boundary). The reader may wonder whether this is a never-ending evolution!

Other interesting run cases could be imagined and simulated with this Navier–Stokes solver (see the cover of the book illustrating a case for which the initial condition is an array of individual vortices). We refer to many existing fluid mechanics books as an obvious source of inspiration.

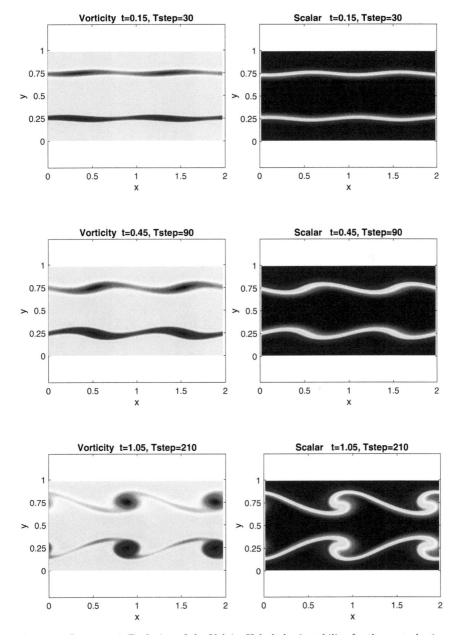

Fig. 15.6 Run case 1. Evolution of the Kelvin–Helmholtz instability for the perturbation wavelength $\lambda_x/L_x = 0.5$.

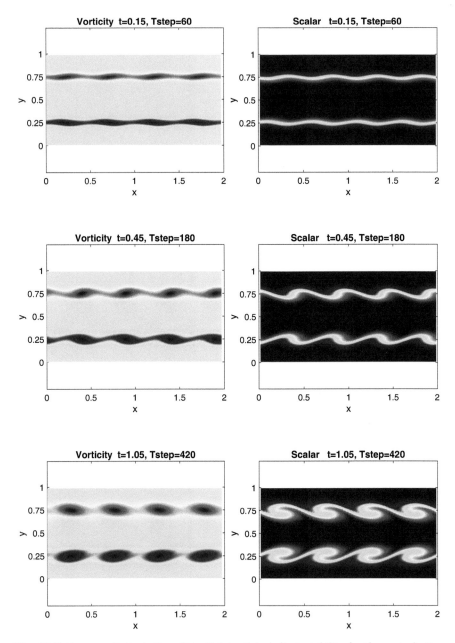

Fig. 15.7 Run case 2. Evolution of the Kelvin–Helmholtz instability for the perturbation wavelength $\lambda_x/L_x = 0.25$.

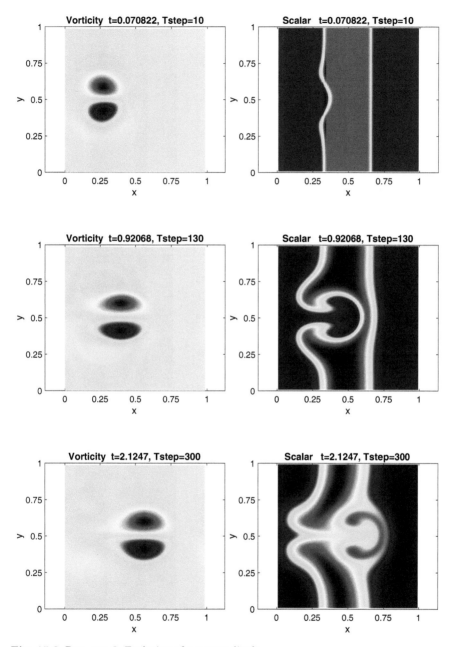

Fig. 15.8 Run case 3. Evolution of a vortex dipole.

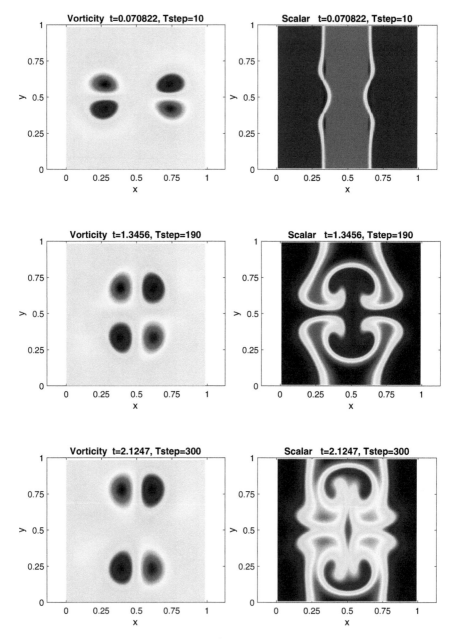

Fig. 15.9 Run case 4. Head-on collision of two identical vortex dipoles.

Chapter References

G.K. Batchelor, *An Introduction to Fluid Dynamics*, 7th edn. (Cambridge University Press, 1988)

C. Canuto, M.Y. Hussaini, A. Quarteroni, T.A. Zang, *Spectral Methods. Evolution to Complex Geometries and Applications to Fluid Dynamics* (Springer, Berlin, Heidelberg, 2007)

A.J. Chorin, Numerical solution of the Navier-Stokes equations. Math. Comput. **23**, 341–354 (1968)

J.H. Ferziger, M. Perić, *Computational Methods for Fluid Dynamics* (Springer, 2002)

V. Girault, P.-A. Raviart, *Finite Element Methods for Navier-Stokes Equations. Theory and Algorithms* (Springer, 1986)

C. Hirsch, *Numerical Computation of Internal and External Flows* (Wiley, 1988)

J. Kim, P. Moin, Application of a fractional step method to incompressible Navier-Stokes equations. J. Comput. Phys. **59**, 308 (1985)

P. Orlandi, *Fluid Flow Phenomena* (Kluwer Academic Publishers, 1999)

S.A. Orszag, Numerical methods for the simulation of turbulence. Phys. Fluids **12**(12), II-250–II-257 (1969)

P.G. Saffman, *Vortex Dynamics* (Cambridge University Press, 1992)

M. Van Dyke, *An Album of Fluid Motion* (The Parabolic Press, 1982)

Index

Index of Programs

© The Editor(s) (if applicable) and The Author(s), under exclusive license
to Springer Nature Switzerland AG 2023
I. Danaila et al., *An Introduction to Scientific Computing*,
https://doi.org/10.1007/978-3-031-35032-0